Lecture Notes in Mathematics

Edited by A. Dold and B. Eckmann

Subseries: Fondazione C.I.M.E., Firenze
Adviser: Roberto Conti

947

Algebraic Threefolds

Proceedings of the 2nd 1981 Session
of the Centro Internazionale Matematico Estivo (C.I.M.E.),
Held at Varenna, Italy, June 15–23, 1981

Edited by Alberto Conte

Springer-Verlag
Berlin Heidelberg New York 1982

Editor
Alberto Conte
Istituto di Geometria "C.Segre"
Università di Torino, Via Principe Amedeo 8
10123 Torino, Italy

AMS Subject Classifications (1980): 14-06, 14J10, 14J30

ISBN 3-540-11587-0 Springer-Verlag Berlin Heidelberg New York
ISBN 0-387-11587-0 Springer-Verlag New York Heidelberg Berlin

Printing and binding: Beltz Offsetdruck, Hemsbach/Bergstr.
2146/3140-543210

INTRODUCTION

This volume contains the Proceedings of the CIME conference on Algebraic Threefolds which was held at the "Villa Monastero" in Varenna (Como) during the period 14-23 June 1981.

The only exceptions are Collino's two papers which were not presented at the session, but which were included instead of his original paper because of their obvious interest.

I hope that the aim to give a complete survey of the known results in the theory of three-dimensional algebraic varieties has been attained.

I wish to thank all the participants, who were able to create a warm and stimulating atmosphere, and the CIME staff for the excellent administrative and secretarial job done.

Alberto Conte

C.I.M.E. Session on Algebraic Threefolds

List of participants

A. Albano, Ist. di Geom. Univ., via Principe Amedeo 8, 10123 Torino

E. Ambrogio, Ist. di Geom. Univ., via Principe Amedeo, 8, 10123 Torino

E. Arbarello, via di Porta Pinciana 4, 00187 Roma

F. Bardelli, Ist. Mat. U. Dini, V.le Morgagni 67/A, 50134 Firenze

A. Beauville, Ecole Polytechnique,Centre de Math., 91128 Palaiseau Cedex,France

M. Beltrametti, Ist. Mat., via L. B. Alberti 4, 16132 Genova

M.T. Bonardi, Ist. Mat., via L. B. Alberti 4, 16132 Genova

G. Canuto, Ist. Mat. Univ., via Mezzocannone 8, 80134 Napoli

F. Catanese, Ist. Mat., via Buonarroti 2, 56100 Pisa

G. Ceresa, Ist. di Geom. Univ., via Principe Amedeo 8, 10123 Torino

C.H. Clemens, Dept. of Math., The Univ. of Utah, Salt Lake City, Utah 84112,USA

C. Ciliberto, Piazzetta Arenella 7/L, 80128 Napoli

A. Collino, Ist. di Geom. Univ.,via Principe Amedeo 8, 10123 Torino

A. Conte, Ist. di Geom. Univ., via Principe Amedeo 8, 10123 Torino

A. Del Centina, Ist. Mat. U. Dini, V.le Morgagni 67/A, 50134 Firenze

G. Elencwajg, Départ. de Math., IMSP, Parc Valrose, 06034 Nice Cedex, France

D. Epema, Math. Instituut, Wassenaarseweg 80, 2333 AN Leiden

H. Esnault, 9, Av. de la Soeur Rosalie, 75013 Paris

M.L. Fania, Univ. of Notre Dame, Dept. of Math., PO Box 398, Notre Dame, Ind. 46556, USA

G. Ferrarese, via dei Faggi 9, 10156 Torino

P. Francia, Ist. Mat., via L. B. Alberti 4, 16132 Genova

A. Gimigliano, V.le Repubblica 85, 50019 Sesto F. (Firenze)

D. Harbater, Dept.of Math., Univ.of Pennsylvania, Philadelphia, PA 19104,USA

A. Hefez, Av. N.S. de Copacabana 959/703, 22060 Rio de Janeiro, Brasil

A. Howard, Univ.of Notre Dame, Dept. of Math., South Bend, Ind. 46556, USA

T. Kalker, Math. Instituut, Wassenaarseweg 80, 2333 AN, Leiden

S. Kilambi, Départ. de Math., Univ. de Montréal, C.P. 6128, Montréal, Canada

S.L. Kleiman, 2-278, M.I.T., Cambridge, MA 02139, USA

A. Lanteri, Ist. Mat. Univ., via C. Saldini 50, 20133 Milano

M. Levine, Dept. of Math., Univ. of Pennsylvania, Philadelphia, PA 19104, USA

M.G. Marinari, via Colombara 4/18, 16016 Cogoleto (GE), Italy

J.Y. Merindol, Fac. des Sciences, Blvd Lavoisier, 49045 Augers Cedex, France

S. Mori, Dept.of Math.,Fac.of Sci., Nagoya Univ., Chikusa-ku, Nagoya 464, Japan

J.P. Murre, Math. Instituut, Wassenaarseweg 80, 2300 RA Leiden

A. Nowakowski, Inst. of Math., Univ. of Lodz, ul. Banacha 22, Lodz, Poland

E. Oeljeklaus, Frauzinsstrasse 6, D-2800 Bremen 1, BRD

P. Oliverio, Scuola Normale Superiore, P.za dei Cavalieri 7, 56100 Pisa

F. Oort, Math. Inst., Budapestlaan 6, 3508 TA - Utrecht, De Uithof (Holland)

M. Palleschi, via Bergognone 27, 20144 Milano

S. Pantazis, Dept. of Math., Harvard Univ., Cambridge, MA 02138, USA

U. Persson, Inst. Mittag-Leffler, Auravagen 17, S-182 62 Djursholm, Sweden

L. Picco Botta, Ist. di Geom. Univ., via Principe Amedeo 8, 10123 Torino

R. Piene , Inst. of Math., PB 1053, Blindern, Oslo 3

E. Previato, Sem. Mat. Univ., Via Belzoni 7, 35100 Padova

P. Puts, Philipslaan 21, Rÿswÿk (ZH), Holland

M. Raimondo, Ist. Mat., Via L. B. Alberti 4, 16132 Genova

D. Romagnoli, Ist. di Geom. Univ., via Principe Amedeo 8, 10123 Torino

D. Schaub, Fac.des Sciences, Blvd Larovisier, 49045 Augers Cedex, France

E. Sernesi, Ist. Mat., Città Universitaria, 00100 Roma

M.E. Serpico, Ist. Mat., via L. B. Alberti 4, 16132 Genova

R. Smith, Math. Dept., Boyd Graduate Students Bldg, Univ. of Georgia,
Athens, GA 30602, USA

A. Sommese, Dept. of Math., Univ. of Notre Dame, Notre Dame, Ind. 46656, USA

M. Teicher, IAS, Princeton, N.J., USA

A. Toet, Flanorpad 2^A, 2333 AN Leiden

R. Treger, Str. Weizmann 12/28, Nathanya, Israel

C. Turrini, Ist. Mat. Univ., via C. Saldini 50, 20133 Milano

K. Ueno, Dept. of Math., Fac. of Sciences, Kyoto Univ., Kyoto, Japan

I. Vainsencher, R. Navegantes 1017/201, B. Viagem, Recife, Brasil

L. Verdi, Ist. Mat. U. Dini, V.le Morgagni 67/A, 50134 Firenze

A. Verra, Ist.di Geom. Univ., via Principe Amedeo 8, 10123 Torino

E. Viehweg, 9, Av. de la Soeur Rosalie, 75013 Paris

A. Vitter, Math. Dept., Tulane Univ., New Orleans, LA 70118, USA

G.G. Weill, 300 E 33rd St., New York, NY 10016, USA

G. Welters, D.to de Geometria y Topologia, Universidad de Barcelona,
P.za Universidad, Barcelona

P. Wilson, Dept. of Pure Math., Univ. of Cambridge, 16 Mill Lane,
Cambridge, England CB2 1SB

TABLE OF CONTENTS

Bimeromorphic Geometry of Algebraic and Analytic Threefolds.

Kenji UENO

The birational classification of algebraic threefolds has been developped recently and some part of the classification of algebraic surfaces due to Enriques has been generalized to that of algebraic threefolds ([19],[24]). Moreover, some part of the classification of analytic surface due to Kodaira has been generalized to that of analytic threefolds and very strange phenomena of analytic threefolds were found ([22]).

The main purpose of the present notes is to give an introduction to the classification theory of analytic threefolds. The reason why I do not restrict the subject to algebraic threefolds is the following. Firstly, the proof of the inequality of Kodaira dimensions for a fibre space is based on the Hodge theory.(See Theorem 3.2 and Theorem 3.5 below). This inequality is the most important in the classification theory and we cannot avoid the analytic theory. Secondly, it is interesting to know difference between algebraic and analytic threefolds. Thirdly, non-Kähler manifold give some hints on algebraic manifolds in positive characteristics.

In the present notes we shall use the following convention.

By a complex variety we mean a reduced compact complex space.

By an algebraic variety we mean a complete algebraic variety defined over the complex number field $\mathbb{C}$.

We often omit the word "compact" if it is clear from the context.

A morphism $f : V \longrightarrow W$ of complex varieties is called a fibre space, if f is surjective and has connected fibres. A fibre space $f : V \longrightarrow W$ is said to be bimeromorphic to a fibre space $g : X \longrightarrow Y$, if there exist bimeromorphic mappings $h_1 : V \rightarrow X$ and $h_2 : W \rightarrow Y$

with $g \circ h_1 = h_2 \circ f$. We shall also use the following notation.

Ω_M^ν : the sheaf of germs of holomorphic ν-forms on a complex manifold M.

K_M : the canonical bundle of a complex manifold M.

$h^0(M, F) = \dim_{\mathbb{C}} H^0(M, F)$

$g_\nu(M) = h^0(M, \Omega_M^\nu)$

§1 Certain bimeromorphic invariants.

§2 Fibrations attached to bimeromorphic invariants.

§3 Kodaira dimensions for fibre spaces.

§4 Classification of algebraic threefolds.

§5 Classification of analytic threefolds.

1. Certain bimeromorphic invariants.

Let $\mathbb{C}(M)$ be the meromorphic function field of a complex variety M. It is well-known that $\mathbb{C}(M)$ is an algebraic function field and tr.deg $\mathbb{C}(M) \leq \dim M$. The number tr. deg $\mathbb{C}(M)$ is denoted by $a(M)$ and we call it the algebraic dimension of M. If $a(M) \geq 1$, then there is a non-singular projective manifold W and a surjective meromorphic mapping $\varphi : M \longrightarrow W$ such that φ induces an isomorphism of the function fields $\varphi^* : \mathbb{C}(W) \xrightarrow{\sim} \mathbb{C}(M)$. We let $\hat{M}$ be a bimeromorphically equivalent model of M such that $\hat{\varphi} : \hat{M} \longrightarrow W$ be a surjective morphism. The morphism $\hat{\varphi} : \hat{M} \longrightarrow W$ is called an algebraic reduction of M. This is unique up to bimeromorphic equivalence. Note that every fibre of $\hat{\varphi}$ is connected.

Let D be a Cartier divisor over a normal complex variety V. Put $\mathbb{N}(D, V) = \{ m \geqslant 1 \mid h^0(V, \mathcal{O}(mD)) \geqslant 1 \}$. We define the D-dimension $\kappa(D, V)$ of D by

$$\kappa(D, V) = \begin{cases} \max\limits_{m \in \mathbb{N}(D,V)} \dim \Phi_{|mD|}(V), & \text{if } \mathbb{N}(D, V) \neq \emptyset \\ -\infty, & \text{if } \mathbb{N}(D, V) = \emptyset, \end{cases}$$

where $\Phi_{|mD|}$ is a meromorphic mapping associated with the complete linear system $|mD|$. If D is a Cartier divisor on a non-normal variety V, we define $\kappa(D, V)$ as $\kappa(\iota^* D, V^*)$ where $\iota : V^* \longrightarrow V$ is the normalization of V. From the definition it follows the following facts.

D - 0) If D and E are linearly equivalent, we have $\kappa(D, V) = \kappa(E, V)$. Moreover, we have $\kappa(mD, V) = \kappa(D, V)$.

D - 1) $\kappa(D, V) \leq a(V)$.

D - 2) If $D \geq D'$ for effective divisors D, D', then $\kappa(D, V) \geq \kappa(D', V)$.

D - 3) D_1 (resp. D_2) is a Cartier divisor on a complex variety V_1

(resp. V_2). Then $\kappa(p_1^*D_1 + p_2^*D_2, V_1 \times V_2) = \kappa(D_1, V_1) + \kappa(D_2, V_2)$, where p_i is the projection of $V_1 \times V_2$ to the i-th factor.

We can also prove the following fact.

D - 4) Put $R[D, V] = \bigoplus_{m=0}^{\infty} H^0(V, O(mD))$ and consider $R[D,V]$ as a commutative algebra over $\mathbb{C} = H^0(V, O_V)$. If $\kappa(D, V) \geq 0$, then we have

$$\kappa(D, V) = \text{tr. deg}_{\mathbb{C}} R[D,V] - 1.$$

Moreover $\kappa(D, V) = -\infty$ if and only if $\text{tr. deg}_{\mathbb{C}} R[D,V] = 0$.

For a complex line bundle L over a complex variety V, we can define the L-dimension $\kappa(L, V)$ similary. For a Cartier divisor D by $[D]$ we mean the corresponding line bundle. We have $\kappa(D, V) = \kappa([D], V)$. Hence, in the following, often we do not distinguish Cartier divisors from line bundles.

The following results are important. Their proofs can be found in Ueno [18].

Lemma 1.1. Let $f : V \longrightarrow W$ be a fibre space over a normal variety W and D a Cartier divisor on W. Then there is an isomorphism

$$f^* : H^0(W, O_W(D)) \xrightarrow{\sim} H^0(V, O_V(f^*D)).$$

Especially, we have

$$\kappa(D, W) = \kappa(f^*D, V).$$

Corollary 1.2. Let D be a Cartier divisor on a normal variety W. If $f : W' \longrightarrow W$ is a modification, we have an isomorphism

$$f^* : H^0(W, O_W(D)) \xrightarrow{\sim} H^0(W', O_{W'}(f^*D)),$$

and

$$\kappa(D, W) = \kappa(f^*D, W').$$

Actually we have a stronger result.

Theorem 1.3. If $f : V \longrightarrow W$ be a surjective morphism (not necessarily a fibre space), then for any Cartier divisor

$$\kappa(D, W) = \kappa(f^*D, V).$$

Proposition 1.4. For a Cartier divisor D on a normal variety, put

$W_m = \Phi_{|mD|}(V)$ for each $m \in \mathbb{N}(D, V)$. Then there exists a positive integer m_0 such that for each $m \in \mathbb{N}(D, V)$ with $m \geq m_0$, $\mathbb{C}(W_m) = \mathbb{C}(W_{m_0})$ and $\mathbb{C}(W_{m_0})$ is algebraically closed in $\mathbb{C}(V)$.

Corollary 1.5. For every integer $m \in \mathbb{N}(D, V)$ with $m \geq m_0$, we let $\pi : V^* \longrightarrow V$ be a modification such that $\varphi_m = \Phi_{|mD|} \circ \pi : V^* \longrightarrow W_m$ is a morphism. Then general fibres of φ_m are connected.

We have much stronger results on φ_m. (See Theorem 2.4 below.)

Theorem 1.6. For a fibre space $f : V \longrightarrow W$ of non-singular varieties and a Cartier divisor D on V, there exists a Zariski open subset U of W such that we have

$$\kappa(D, V) \leq \kappa(D_w, V_w) + \dim W$$

for each $w \in U$, where $V_w = f^{-1}(w)$ and $D_w = D|_{V_w}$.

Now let us define the Kodaira dimension of a variety. For a compact complex manifold M, by K_M we denote the canonical line bundle (or a canonical divisor) of M. We use the notation $\kappa(M)$ instead of $\kappa(K_M, M)$ and call it the Kodaira dimension of M. We also use the notation $\mathbb{N}(M)$ instead of $\mathbb{N}(K_V, M)$. For a singular variety M we define $\kappa(M)$ by $\kappa(M) = \kappa(M^*)$, where M^* is a non-singular model of M. This is well-defined by virtue of the following lemma.

Lemma 1.7. If two complex manifolds M_1 and M_2 are bimeromorphically equivalent, there is a natural isomorphism

$$H^0(M_1, \mathcal{O}(mK_{M_1})) \xrightarrow{\sim} H^0(M_2, \mathcal{O}(mK_{M_2})),$$

for each positive integer m.

Put $P_m(M) = h^0(M, \mathcal{O}(mK_M))$ for a positive integer m and a complex manifold M and call it the m-genus of M. We often use the notation $p_g(M)$ instead of $P_1(M)$ and call it the geometric genus of M. For a singular variety M, we define $P_m(M)$ (resp. $p_g(M)$) by $P_m(M) = P_m(M^*)$(resp. $p_g(M) = p_g(M^*)$) where M^* is a non-singular model of M. These are well-defined by virtue of Lemma 1.7. From the definition and the above properties D - 3), D - 4) we obtain the following results.

K - 1) $\kappa(M) \leq a(M)$

K - 2) $\kappa(M) = -\infty$ if and only if $P_m(M) = 0$, $m = 1,2,\ldots$.
$\kappa(M) = 0$ if and only if $P_m(M) \leq 1$, $m = 1,2,\ldots$, and $P_n(M) = 1$ for a positive integer n.

K - 3) $\kappa(M_1 \times M_2) = \kappa(M_1) + \kappa(M_2)$

K - 4) Let $R[M] = \bigoplus_{m=0}^{\infty} H^0(V, O(mK_V))$ be the canonical ring of M.

If $\kappa(M) \geq 0$, we have

$$\kappa(M) = \text{tr. deg}_{\mathbb{C}}\, R[M] - 1 .$$

Moreover $\kappa(M) = -\infty$ if and only if $\text{tr. deg}_{\mathbb{C}}\, R[M] = 0$.

Note that by Zariski [27] , R [D, V] may not be finitely generate and by Wilson [26] , R [M] is not finitely generated for a certain compact complex manifold which is non-algebraic and non-Kähler.

Now we state the important properties of the Kodaira dimensions. Some of them are direct consequences from the above properties of D-dimensions.

<u>Theorem 1.8</u>. Let $f : V \to W$ be a surjective morphism between complex manifolds with same dimensions. Then we have

$$P_m(V) \geq P_m(W), \quad m = 1,2,\ldots\ldots\ldots, \quad \kappa(V) \geq \kappa(W).$$

Moreover, if f is finite unramified, we have

$$\kappa(W) = \kappa(V) .$$

The last statement of the theorem is a special case of Theorem 1.3, since in our case $K_V = f^*K_W$. Theorem 1.6 is restated in the following form.

<u>Theorem 1.9</u>. For a fibre space $f : V \to W$ of non-singular varietie there is an open subset U of W such that for each point $w \in U$, $V_w = f^{-1}(w)$ is non-singular and we have

$$\kappa(V) \leq \kappa(V_w) + \dim W .$$

This theorem says that if V has a structure of a fibre space over W with $\kappa(V_w) = 0$ (resp. $\kappa(V_w) = -\infty$) for general fibres, then we have $\kappa(V) \leq \dim W$ (resp. $\kappa(V) = -\infty$).

Finally, we introduce the Albanese torus of a complex manifold M. A complex torus $A(M)$ and a morphism $\alpha : M \longrightarrow A(M)$ are called the <u>Albanese torus</u> of M and an <u>Albanese mapping</u>, respectively, if they enjoy the following universal property.

(A) : For any complex torus T and a morphism $f : M \longrightarrow T$, there is a unique Lie group homomorphism $h : A(M) \longrightarrow T$ such that $f(z) = h\cdot\alpha(z) + \underline{a}$ for any point $z \in M$ where $\underline{a}$ is an element of T.

By definition, an Albanese mapping is unique up to translations of $A(M)$, if it exists. By Blanchard [3], every compact complex manifold has the Albanese torus. The number $t(M) = \dim A(M)$ is called the <u>Albanese dimension</u> of M. From the existence theorem of the Albanese torus due to Blanchard, we have the following

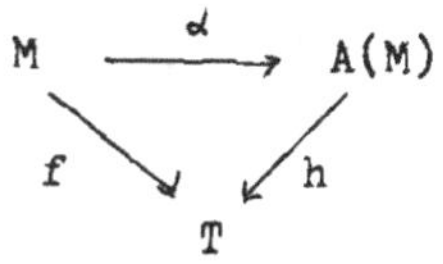

<u>Proposition 1.10</u>. We have

$$t(M) \leq h^0(M, dO_M) \leq h^0(M, \Omega^1_M) .$$

Moreover, if M is a bimeromorphic image of a Kähler manifold (e.g. M is an algebraic manifold or $a(V) = \dim M$), then the equalities hold.

The last statement is due to the fact that under the hypothesis the cohomology groups $H^*(M, \mathbb{C})$ carry the Hodge structure. (See Ueno [18] Corollary 9.3.).

Since any meromorphic mapping from a complex manifold to a complex torus is holomorphic, we have the following lemma.

<u>Lemma 1.11</u>. If two complex manifolds M_1 and M_2 are bimeromorphic, then their Albanese tori are isomorphic and their Albanese mappings $\alpha_1 : M_1 \longrightarrow A(M_1)$ and $\alpha_2 : M_2 \longrightarrow A(M_2)$ are bimeromorphically equivalent. Hence, a fortiori $t(M_1) = t(M_2)$.

2. Fibrations attached to bimeromorphic invariants.

In the above section we attached several meromorphic (or holomorphic) mappings to a complex variety. In this section we shall study properties of the fibre spaces associated with these meromorphic (or holomorphic) mappings.

We begin with the algebraic reduction of M.

Theorem 2.1. Let $\varphi : M^* \longrightarrow W$ be the algebraic reduction of a comp variety M where M^* is a non-singular model of M. Then there exis a dense subset U of W such that for each point $w \in U$, the fibre $M^*_w = \varphi^{-1}(w)$ is non-singular and we have

$$\kappa(M^*_w) \leq 0 .$$

Moreover if $a(M) = \dim M - 1$, the equality holds, that is, M^*_w is an elliptic curve.

More generally we have the follcwing

Theorem 2.2. Let $\varphi : M^* \longrightarrow W$ be the same as above. For each comple line bundle L on M^* there exists a dense subset U of W such that for each point $w \in U$, we have

$$\kappa(L_w, M^*_w) \leq 0 ,$$

where $M^*_w = \varphi^{-1}(w)$, $L_w = L|_{M^*_w}$.

A proof of Theorem 2.2 is very simple. Assume the contrary. Then the sheaf $\varphi_* O_{M^*}(L)$ is non-zero and locally free of rank $r \geqslant 2$ on a certain Zariski open set of W. Then we have a meromorphic mapping $\pi : M^* \longrightarrow \mathbb{P}(\varphi_* O_{M^*}(L))$ and a commutative diagramm

$$\begin{array}{ccc} \pi : M^* & \longrightarrow & \mathbb{P}(\varphi_* O_{M^*}(L)) \\ & \searrow \quad \circlearrowleft \quad \swarrow & \\ & W & \end{array} .$$

From our assumption it follows that $\dim \varphi(M^*) > \dim W$. On the other-hand, since W is algebraic and $\varphi_* \mathcal{O}_{M^*}(L)$ is coherent, $\mathbb{P}(\varphi_* \mathcal{O}_{M^*}(L))$ is algebraic. Hence $\varphi(M^*)$ is algebraic and we have $a(M) = a(M^*) \geq a(\varphi(M^*)) > \dim W = a(M)$. This is a contradiction. This proves the theorem.

Theorem 2.2 gives many informations about the structure of fibres of an algebraic reduction. For example $L = K_{M^*}$ gives the first part of Theorem 2.1 and $L = K^{-1}_{M^*}$ gives the last part of Theorem 2.1.

<u>Theorem 2.3</u>. Let $\varphi : M^* \longrightarrow W$ be the algebraic reduction of a complex variety M. Assume $a(M) = \dim M - 2$. Then a general fibres of φ is bimeromorphically equivalent to one of the following surfaces 1) rational surface, 2) ruled surface of genus 1, 3) surface of class VII, 4) elliptic surface with a trivial cononical bundle, 5) Enriques surface, 6) hyperelliptic surface, 7) complex torus, 8) K 3 surface.

This is almost a corollary to Theorem 2.1 and the classification of surfaces. We only need to show that a ruled surface of genus 2 does not appear. This is due to Kuhlmann [15]. Assume that a general fibre of φ is a ruled surface of genus $g \geq 2$. Then all regular fibres of φ are ruled surfaces of genus g and $\varphi_* \Omega^1_{M^*}$ is a locally free sheaf of rank g on a Zariski open set of W. We have a meromorphic mapping $\pi : M^* \longrightarrow \mathbb{P}(\varphi_* \Omega^1_{M^*})$ and a commutative diagram

$$\begin{array}{ccc} \pi : M^* & \longrightarrow & \mathbb{P}(\varphi_* \Omega^1_{M^*}) \\ & \varphi \searrow \quad \swarrow & \\ & W & \end{array} .$$

If we restrict π on a general fibre $M^*_x = \varphi^{-1}(x)$ of φ, $\pi|_{M^*_x}$ is a composition of the Albanese mapping $\alpha : M^*_x \longrightarrow A(M^*_x)$ and the canonical mapping of the image curve $\alpha(M^*_x) = C_x$ which is a non-singular curve of genus g. Hence $\dim \pi(M^*) > \dim W$. As we have $a(M) = a(M^*) \geq$

$\geq \dim \pi(M^*) > a(M)$, this is a contradiction. This porves the theorem.

Next we shall consider the fibration induced by $\Phi_{|mD|}$ for a Cartier divisor D on a normal variety W. The following theorem is the most important.

Theorem 2.4.(Iitaka [10].) Let D be a Cartier divisor on a normal variety V. Assume $\kappa(D, V) > 0$. Then there are a compact complex manifold V^*, a modification $\pi : V^* \longrightarrow V$, a projective manifold W^* and a surjective morphism $f : V^* \longrightarrow W^*$ which enjoy the following properties

1) $\dim W^* = \kappa(D, V)$,

2) there exists a dense subset U which is a complement of at most countable union of algebraic subsets in W^* such that for each $u \in U$, $V_u^* = f^{-1}(u)$ is irreducible and non-singular,

3) $\kappa(\pi^* D_u, V_u^*) = 0$ for each $u \in U$, where $\pi^* D_u = \pi^* D|_{V_u^*}$,

4) if there exist $\pi^{\#} : V^{\#} \longrightarrow V$ and $f^{\#} : V^{\#} \longrightarrow W^{\#}$ which satisfy the above properties 1), 2), 3) then $f^{\#} : V^{\#} \longrightarrow W^{\#}$ is bimeromorphically equivalent to $f : V^* \longrightarrow W^*$,

5) moreover, $f : V^* \longrightarrow W^*$ is bimeromorphically equivalent to $\Phi_{|mD|} : V \longrightarrow W_m$ for a sufficiently large $m \in N(D, V)$.

Proof. Since we can replace D by a member in $|nD|$ for a positive integer n, we can assume that D is effective. By Proposition 1.4 we take a positive integer m_0 such that for each $m \geq m_0$, $\mathbb{C}(W_m) = \mathbb{C}(W_{m_0})$ and $\mathbb{C}(W_{m_0})$ is algebraically closed in $\mathbb{C}(V)$. Fix an integer $m \geq m_0$ and consider a meromorphic mapping

$$\Phi_{|mD|} : V \longrightarrow W_m \subset \mathbb{P}^{\dim |mD|} .$$

We let $\pi : V^* \longrightarrow V$(resp. $\tau : W^* \longrightarrow W_m$) be a modification of V (resp. W_m) such that V^*(resp. W^*) is non-singular, there exists a morphism $f : V^* \longrightarrow W^*$ which is bimeromorphically equivalent to $\Phi_{|mD|}$ and we have a commutative diagram

$$\begin{array}{ccc} V & \xrightarrow{\varphi = \Phi_{|mD|}} & W_m = W \\ \pi \uparrow & & \uparrow \tau \\ V^* & \xrightarrow[f]{} & W^* \end{array} .$$

Replacing D by π^*D, we may assume that $V = V^*$ and $\tau \circ f = \Phi_{|mD|}$. Put $\varphi = \Phi_{|mD|}$, $W = W_m$, $L = [mD]$. To each hyperplane H_λ of $\mathbb{P}^{\dim|mD|}$ there corresponds a divisor $E_\lambda \in |mD|$ and this correspondence is given by

$$E_\lambda = E_\lambda^* - F_\lambda$$

where $E_\lambda^* = \varphi^* H_\lambda$ and F is the fixed component of $|mD|$. Our assumption implies that $|E_\lambda^*|$ is free from base points and fixed components.

Now we need to distinguish the algebraic structure of W and the analytic structure of W. For that purpose, we use the notation W^{alg} and W^{an}. $\varphi_* \mathcal{O}_V(L^{\otimes n})$ is an analytic coherent sheaf on W^{an}. By GAGA there is an algebraic coherent sheaf F_n on W^{alg} such that $F_n^{an} \simeq \varphi_* \mathcal{O}_V(L^{\otimes n})$. Since $W_\lambda^{alg} = W^{alg} - H_\lambda^{alg}$ is an affine variety, F_n is spanned by global sections $\psi_0, \psi_1, \ldots, \psi_M \in H^0(W^{alg}, F_n)$ as $\mathcal{O}_{W^{alg}}$-module. Since there is a natural inclusion

$$H^0(W_\lambda^{alg}, F_n) \subset \bigoplus_{e=1}^{\infty} H^0(W^{alg}, F_n(eH_\lambda^{alg})) ,$$

we can assume that

$$\psi_i \in H^0(W^{alg}, F_n(eH_\lambda^{alg})), \ i = 0, 1, \ldots, M ,$$

for a sufficient large e. Then we have

$$H^0(W^{alg}, F_n(eH_\lambda^{alg})) \simeq H^0(W^{an}, f_*\mathcal{O}_V(L^{\otimes n})(enH_\lambda))$$

$$\simeq H^0(V, L^{\otimes n}(enE_\lambda^*)) \subset H^0(V, L^{\otimes n}(enE_\lambda)) .$$

Hence we consider ψ_i as an element of $H^0(V, L^{\otimes n}(enE_\lambda))$.

Take $\eta \in H^0(V, \mathcal{O}_V([enE_\lambda]))$ with $(\eta) = enE_\lambda$. Then $\eta\psi_i$

is an element of $H^0(V, L^{\otimes n} \otimes [enE]\ ) \hookrightarrow H^0(V, L^{\otimes(e+1)n})$ by the definition of E_λ. Let $\{\varphi_0, \varphi_1, \ldots, \varphi_N\}$ be a basis of $H^0(V, L)$ and we let $\Psi_{M+1}, \ldots, \Psi_{M'}$ be all monomials of degree $(e+1)n$ in $\varphi_0, \varphi_1, \ldots, \varphi_N$. Hence $\Psi_j \in H^0(V, L^{\otimes(e+1)n})$. Consider a meromorphic mapping $h^{(n)}$ defined by

$$\begin{array}{cccl} h^{(n)} : V & \longrightarrow & \mathbb{P}^{M'} & \\ \cup\!\!\!| & & \cup\!\!\!| & \\ z & \longmapsto & (\eta(z)\Psi_0(z):\cdots:\eta(z)\Psi_M(z):\Psi_{M+1}(z):\cdots:\Psi_{M'} \end{array}$$

Put $X_n = h^{(n)}(V)$. Note that by the choice of $\Psi_0, \ldots, \Psi_M$ we have

$\mathbb{C}(W) = \mathbb{C}(\frac{\Psi_1}{\Psi_0}, \ldots, \frac{\Psi_M}{\Psi_0})$. Then there are inclusions of fields

$\mathbb{C}(W) = \mathbb{C}(W_m) \subset \mathbb{C}(X_n) \subset \mathbb{C}(W_{(e+1)nm})$. As $m \geq m_0$, $\mathbb{C}(W_m) = \mathbb{C}(W_{(e+1)nm})$ Hence $\mathbb{C}(W) = \mathbb{C}(X_n)$. Let $g_n : X_n \longrightarrow W$ be a meromorphic mapping induced from the projection $\mathbb{P}^{M'} \longrightarrow \mathbb{P}^{M''}$ to the last $M' - M$ coordinates. Since the inclusion $\mathbb{C}(W) \subset \mathbb{C}(X_n)$ is induced from g_n, g_n is a <u>birational</u> mapping.

$$\begin{array}{ccccl} V & \xrightarrow{\ h^{(n)}\ } & X_n \subset \mathbb{P}^{M'} & \ni & (x_0 : x_1 : \cdots : x_{M'}) \\ & \searrow^{\varphi} & \downarrow g_n & & \downarrow \\ & W \underset{\text{Segre imbedding}}{\xrightarrow{\ \simeq\ }} & W \subset \mathbb{P}^{M''} & \ni & (x_{M+1} : x_{M+2} : \cdots : x_{M'}) \end{array}$$

On the other hand, there is an algebraic subset S_n of W such that $f_*\mathcal{O}_V(L^{\otimes n})$ is locally free on $W - S_n$ and that there is an isomorphism

$$f_*\mathcal{O}_V(L^{\otimes n}) \underset{\mathcal{O}_{W,w}}{\otimes} \mathbb{C} \xrightarrow{\sim} H^0(V_w, L_w^{\otimes n})$$

for each point $w \in W - S_n$. From the choice of $\Psi_0, \ldots, \Psi_M$, we infer that $\Psi_i|_{V_w}$, $i = 0, \ldots, M$ span $H^0(V_w, L_w^{\otimes n})$ for every point $w \in W - \{S_n \cup H_\lambda\}$. Let T be an algebraic subset of W such that $\varphi|_{\varphi^{-1}(W-T)} : \varphi^{-1}(W - T) \to W - T$ is smooth. Then for each point $w \in W - \{S_n \cup H_\lambda \cup T\}$, the restriction $h_w^{(n)} = h^{(n)}|_{V_w} : V_w \to \mathbb{P}^{M'}$ is bimeromorphically equivalent to $\Phi_{|L_w^{\otimes n}|}$. But, since g_n is birational, there exists an algebraic subset A_n of W such that g_n induces an isomorphism between a Zariski open set of X_n and $W - A_n$. Then for each $w \in W - \{S_n \cup H_\lambda \cup T \cup A_n\}$, $h_w^{(n)}(V_w)$ is the point $g_n^{-1}(w)$. Hence $h^0(V_w, \mathcal{O}_{V_w}(nmD_w)) = h^0(V_w, \mathcal{O}_{V_w}(L_w^{\otimes n})) = 1$. Now put $Y = H_\lambda \cup T \cup \bigcup_{n=1}^{\infty}\{S_n \cup A_n\}$. Since Y is a countable union of nowhere dense subset of W, $W - Y$ is dense in W and foe each $w \in W - Y$, we have

$$h^0(V_w, \mathcal{O}_{V_w}(nmD_w)) = 1 , \quad n = 1,2,3,\ldots \quad .$$

Finally, there exists a Zariski open subset W_0^* of W^* such that τ induces an isomorphism between W_0^* and $\tau(W_0^*)$. Then for each point $u \in W_0^*$, we have $f^{-1}(u) = \varphi^{-1}(w)$, $w = \tau(u)$. Since W_0^* is dense in W^*, $U = W_0^* \cap \tau^{-1}(W - Y)$ is dense in W^*. From the above results, we infer easily that the properties 1), 2), 3), 5) holds. The proof of the property 4) is left to the reader. QED.

Remark 2.5. The proof shows that there is a Zariski open subset $\mathcal{O}$ of W^* which contains U such that for each point $w \in \mathcal{O}$, we have $\kappa(D_w, V_w^*) \geq 0$.

The fibre space $f : V^* \longrightarrow W^*$ is called the Iitaka fibration associated with a Cartier divisor D. When V is non-singular and $D = K_V$, the Iitaka fibration is called the canonical fibration. As for the canonical fibration Theorem 2.4 has the following form.

Theorem 2.4a.(Iitaka[10]). For a non-singular manifold V, assume $\kappa(V) \geq 1$. Then the properties 1),2),4),5) in Theorem 2.4 hold and moreover we have

3) $\kappa(V_u^*) = 0$ for each $u \in U$.

Next we shall consider the Albanese mapping. Let $\alpha : M \longrightarrow A(M)$ be the Albanese mapping of a complex manifold M. We consider the Stein factorization of $\alpha : M \longrightarrow \alpha(M)$

$$\begin{array}{ccc} M & \xrightarrow{\alpha} & (M) \\ & \searrow_{\beta} \quad {}^{\gamma}\nearrow & \\ & W & \end{array} \quad .$$

The fibre space $\beta : M \longrightarrow W$ is called the <u>Albanese fibration</u>. Since $\gamma : W \longrightarrow \alpha(M)$ is finite, by Theorem 1.8 we have

$$\kappa(W) \geq \kappa(\alpha(M)) \ .$$

The following theorem is important fcr the classification theory.

<u>Theorem 2.6</u>. For the Albanese mapping $\alpha : M \longrightarrow A(M)$, we have always $\kappa(\alpha(M)) \geq 0$. Moreover α is surjective if and only if $\kappa(\alpha(M)) = 0$.

To prove the theorem we need the following

<u>Proposition 2.7</u>. Let B be a subvariety of a complex torus A. Then there exists a complex subtorus A_1 of A and a projective variety W which is a subvariety of an abelian variety such that

1) B is an analytic fibre bundle over W whose fibre is A_1,

2) $\kappa(W) = \dim W = \kappa(B)$.

Furthermore, if B is an algebraic variety, then there exist finite unramified coverings $\widetilde{B}$ and $\widetilde{W}$ of B and W, respectively, such that $\widetilde{B} = A_1 \times \widetilde{W}$.

For the proof see Ueno [18], Theorem 10.9.

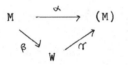

3. Kodaira dimensions for fibre spaces.

The following conjecture was raised by Iitaka.

Conjecture $C_{m,n}$. Let $\varphi : V \longrightarrow W$ be a fibre space with $\dim V = m$ and $\dim W = n$. Then we have

$$\kappa(V) \geq \kappa(W) - \kappa(V_w),$$

where $V_w = \varphi^{-1}(w)$ is a general fibre of φ.

It turned out that the conjecture is not correct in the category of complex varieties. (See below.) But if V is algebraic or Kähler, in many cases the conjecture holds. The conjecture has deep connection with the theory of variations of Hodge structures.

Definition 3.1. A complex variety V is called a varietyiin the class C if V is a meromorphic image of a compact Kähler manifold M.

In the definition we do not assume $\dim M = \dim V$. A compact complex manifold in C enjoys several nice properties which has a compact Kähler manifold. Later we shall consider such properties to study algebraic reductions and Albanese fibrations. Here we only use the fact that the cohomology group $H^*(V\ \mathbb{C})$ of a compact complex manifold in C satisfies the Hodge decomposition theorem and also we can define the primitive cohomology which has the same properties as a Kähler manifold.

The following theorem is very helpful to prove the conjecture $C_{n,1}$.

Theorem 3.2.(Fujita [6]). Let $\varphi : V \longrightarrow W$ be a fibre space where V and W are non-singular and W is a curve. Assume that V is in C and for a general fibre V_w of φ, $P_g(V_w) \geq 1$. Then $\varphi_*\omega_{V/W}$ is pseudo semi-positive, that is any quotient invertible sheaf L of $\varphi_*\omega_{V/W}$ we have $\deg L \geq 0$, where $\omega_{V/W} = O_V(K_V \otimes \varphi^* K_W^{-1})$.

Proof. Since $\varphi_*\omega_{V/W}$ is torsion free and W is a curve, $F = \varphi_*\omega_{V/W}$ is locally free. Let W^0 be the maximal Zariski open subset of W

such that $f|_{\varphi^{-1}(W^0)} : \varphi^{-1}(W^0) \longrightarrow W^0$ is smooth. Put $V^0 = \varphi^{-1}(W^0)$, $W - W^0 = \{\beta_1, \ldots, \beta_s\}$. For any open set U of W we also put $U^0 = U \cap W^0$.

Now we introduce a natural inner product $(\ ,\ )_0$ of $F|_{W^0}$. For every open set $U \subset W^0$, a section $\tau \in H^0(U, F)$ defines uniquely a holomorphic n-form τ_x on the fibre $V_x = \varphi^{-1}(x)$ for each point $x \in U$. For two sections $\tau, \tau' \in H^0(U, F)$, put

$$(\tau, \tau')_{0,x} = (\sqrt{-1})^{n^2} \int_{V_x} \tau_x \wedge \overline{\tau'_x} .$$

This defines a natural inner product on $F|_{W^0}$.

Now let L be a quotient line bundle of F and $\alpha : F \longrightarrow L$ the quotient homomorphism. Put $M = \mathrm{Ker}\,\alpha$.

(3.1). $$0 \longrightarrow M \longrightarrow F \longrightarrow L \longrightarrow 0 .$$

Choose a sufficiently fine open covering $\{U_\lambda\}_{\lambda \in \Lambda}$ of W. Then there are sections $\tau_\lambda, \tau_{\lambda,1}, \ldots, \tau_{\lambda,\ell} \in H^0(U, F)$ such that $\alpha(\tau_\lambda)$ is a base of $L|U_\lambda$ and $\{\tau_{\lambda,1}, \ldots, \tau_{\lambda,\ell}\}$ is a bases of $M|U_\lambda$. Then a matrix $(h_{\lambda i\bar{j}}(x)) = ((\tau_{\lambda,i}, \tau_{\lambda,j})_x)$ is positive definite on U_λ^0. Put $(h^{\bar{i}j}(x)) = (h_{i\bar{j}}(x))^{-1}$ and $\omega_\lambda = \tau_\lambda - \sum_{i=1}^{\ell} (\tau_\lambda, \tau_{\lambda,j}) h^{\bar{j}i} \tau_{\lambda,i}$. Then $\omega_\lambda \in H^0(U_\lambda^0, C^\infty(F))$ and $(\omega_\lambda, \tau_{\lambda,j}) = 0$ for $j = 1, 2, \ldots, \ell$. Hence ω_λ gives a C^∞-splitting of the exact sequence (3.1) on U^0. Now transition functions $\{\ell_{\lambda\mu}\}$ of L with respect to the open covering $\{U_\lambda\}_{\lambda \in \Lambda}$ is given by

$$\alpha(\tau_\lambda) = \ell_{\lambda\mu}\, \alpha(\tau_\mu).$$

As $\alpha(\omega_\lambda) = \alpha(\tau_\lambda)$, we have $\omega_\lambda - \ell_{\lambda\mu}\omega_\mu \in H^0(U_\lambda^0 \cap U_\mu^0, C^\infty(M))$. Since $(\omega_\lambda, t) = 0$ for any $t \in H^0(U_\lambda^0, M)$, we have

$$\omega_\lambda = \ell_{\lambda\mu}\,\omega_\mu .$$

Put $g_\lambda(x) = (\omega_\lambda, \omega_\lambda)_x$ for each $x \in U_\lambda^0$. Then $g_\lambda(x) = |\ell_{\lambda\mu}(x)|^2 g$ for $x \in U_\lambda^0 \cap U_\mu^0$ and $\{g_\lambda\}$ defines a hermitian inner product $(\ ,\ )^0$

on $L|_{W^0}$. Hence $\Theta_0 = \frac{1}{2\pi\sqrt{-1}}\partial\bar{\partial} \log g_\lambda$ defines a C -form on W^0.

Claim.1. Θ_0 is a positive semi-definite type (1,1)-form. Hence $\int_U \Theta_0 \geq 0$ for any open set $U \subset W^0$.

This is a direct consequence of a theorem of Griffiths .
Unfortunately the hermitian metric defined by $\{g_\lambda\}$ may not be extendable to that of L on W. Hence we need to consider a behaviour of g near the points $\{\wp_1, \ldots, \wp_s\} = W - W^0$. Let $\wp$ be one of the points $\{\wp_1, \ldots, \wp_s\}$ and U a coordinate neighbourhood of W with center $\wp$..

Claim.2. Let $\{\tau_0, \tau_1, \ldots, \tau_\ell\}$ be a basis of $\varphi_*\omega_{V/D}$. Then there exist a neighbourhood U' of $\wp$ in U and a positive number N such that

$$h_\mu(x) \geq N$$

for all $\mu = (\mu_0, \mu_1, \ldots, \mu_\ell) \in \mathbb{C}^{\ell+1}$ with $\sum_{i=0}^{\ell} |\mu_i|^2 = 1$ where

$$h_\mu(x) = ((\Sigma\mu_i\tau_i)_x, (\Sigma\mu_i\tau_i)_x)^0 .$$

Since the proof of the claim given in Fujita [6] is difficult to follow, we shall give a proof. Note that if the claim is true for a basis of $\varphi_*\omega_{V/D}$, it is also true for other bases. Therefore it is enough to prove the claim for a suitable basis. Since the problem is local we assume that $U = D = \{t \in \mathbb{C} \mid |t| < \varepsilon\}$ and $\wp = 0$.

First we assume that the singular fibre $V_0 = \varphi^{-1}(0)$ is a reduced divisor with normal crossings. Every element $\omega \in H^0(U, F)$ induces a holomorphic n-form on the non-singular part V_0^0 of V_0. Put $S = \{\mu \in \mathbb{C}^{\ell+1} \mid \sum_{i=0}^{\ell} |\mu_i|^2 = 1\}$. Then for each $\mu \in S$, $\omega(\mu) = \sum_{i=0}^{\ell} \mu_i\tau_i$ does not vanish identically on an irreducible component of V_0^0, since $\{\tau_0, \tau_1, \ldots, \tau_\ell\}$ is a basis of $F|_D$. (Otherwise we can write $\tau = t\tau'$, $\tau' \in H^0(U, F)$, where $\tau' = \sum a_i(t)\tau_i$ and $a_i(\tau)$'s are holomorphic on D. This is a contradiction.) Let p be a point of V_0^0 at which $\omega_0 = \omega|_{V_0^0}$ does not vanish identically. Take a small coordinate neighbourhood $\tilde{U}$ of p in V with local coordinates $(t, z_1, \ldots, z_n)$, $|z_i| < \varepsilon'$, $i = 1, \ldots, n$. Then we can write

$$\omega(\mu)_t = \omega(\mu)|_{V_t} = A(t, z, \mu)\, dz_1 \wedge \cdots \wedge dz_n,$$

where $A(t,z,\mu)$ is holomorphic in (t, z) and linear in μ.

Now we have

$$(\omega(\mu)_t, \omega(\mu)_t)^0 = (\sqrt{-1})^{n^2} \int_{V_t} \omega(\mu)_t \wedge \overline{\omega(\mu)}_t \geq (\sqrt{-1})^{n^2} \int_{V_t \cap U} \omega(\mu)_t \wedge \overline{\omega(\mu)}_t$$

$$= (\sqrt{-1})^{n^2} \int_{|z_i| < \varepsilon'} |A(t, z, \mu)|^2\, dz_1 \wedge \cdots \wedge dz_n \quad .$$

The last term is a continuous function in t and μ. Since $A(0,z,\mu) \neq 0$ there are a positive number N and an open neighbourhood Δ of μ such that

$$(\omega_t(\nu), \omega_t(\nu))^0 \geqslant N$$

for all $\nu \in \Delta$ and a sufficiently small ε'. Since S is compact we obtain the desired result.

Next we consider a general case. By the stable reduction theorem ([13], Chap II, III), we can find a finite ramified covering

$$\begin{array}{ccc} \pi : \hat{D} = \{ s \mid |s| < \varepsilon^{1/m} \} & \longrightarrow & D = \{ t \mid |t| < \varepsilon \} \\ \overset{\psi}{s} & \longmapsto & s^m = \overset{\psi}{t} \end{array}$$

and a non-singular model $\hat{\varphi} : \hat{V} \to \hat{D}$ of $V \times_D \hat{D} \longrightarrow \hat{D}$ such that the fibre of $\hat{\varphi}$ over the origin is reduced. Moreover we can assume that the cyclic group G of order n operates on $\hat{V}$ and $\hat{V}/G \longrightarrow \hat{D}/G = D$ is bimeromorphically equivalent to $\varphi : V \longrightarrow D$. There are natural isomorphisms

$$H^0(V, K_V) \xrightarrow{\sim} H^0(\hat{V}, K_{\hat{V}})^G \quad ,$$

$$\varphi_* \omega_{V/D} \xrightarrow{\sim} (\hat{\varphi}_* \omega_{\hat{V}/\hat{D}})^G,$$

where the isomorphisms are induced from a meromorphic mapping $\psi : \hat{V} \longrightarrow V$ over $\pi : \hat{D} \longrightarrow D$. Let $\{ \hat{\varphi}_0, \ldots, \hat{\varphi}_\ell \}$ be a basis of $\hat{\varphi}_* \omega_{\hat{V}/\hat{D}}$.

Then $\{\hat{\varphi}_0 \otimes ds, \ldots, \hat{\varphi}_\ell \otimes ds\}$ can be considered as a basis of $\hat{\varphi}_* \omega_{\hat{V}}$. Let us consider the action of G on $\hat{\varphi}_* \omega_{\hat{V}}$. For a generator g of G, we have

$$g^*(\hat{\varphi}_i \otimes ds) = \sum_{j=0}^{\ell} a_{ij}(s)\, \hat{\varphi}_j \otimes ds, \quad i = 0, \ldots, \ell$$

where the matrix $A(s) = (a_{ij}(s))$ is invertible on $\hat{D}$. The matrix $A(s)$ statisfies the equation

$$A(e_m^{m-1}s)A(d_m^{m-2}s) \cdots A(e_m s)A(s) = I_{\ell+1}, \quad e_m = \exp(2\pi\sqrt{-1}/m) .$$

By a suitable basis change, we can assume that $A(s)$ is a constant diagonal matrix

$$(3.2) \qquad M = \begin{pmatrix} e_m^{a_0} & & 0 \\ & \ddots & \\ 0 & & e_m^{a_\ell} \end{pmatrix}, \quad 0 < a_i \le m, \quad i = 0, 1, \ldots, \ell .$$

(See, for example, [21], p.15). Then $\{ s^{m-a_0} \hat{\varphi}_0 \otimes ds, \ldots, s^{m-a} \hat{\varphi} \otimes ds$ is a basis of $(\hat{\varphi}_* \omega_{\hat{V}})^G$ as O_D-module. By the above isomorphism, there is a basis $\{\varphi_0, \ldots, \varphi_\ell\}$ of $\varphi_* \omega_V$ such that

$$(3.3) \qquad s^{m-a_i}\, \hat{\varphi}_i \otimes ds = \psi^*(\varphi_i \otimes dt), \quad i = 0, 1, \ldots, \ell .$$

Note that we may assume that ψ is holomorphic on $\psi^{-1}(\hat{D} - \{0\})$ and induces an isomorphism between $\hat{V}_s$ and V_t where $t = s^m$. Therefore, for $\omega(\mu) = \sum_{i=0}^{\ell} \mu_i \varphi_i$, by (3.3) we have

$$(\omega(\mu)_t, \omega(\mu)_t) = (\sqrt{-1})^{n^2} \int_{V_t} \omega(\mu)_t \wedge \overline{\omega(\mu)}_t = (\sqrt{-1})^{n^2} \int_{\hat{V}_s} \psi_s^* \omega(\mu)_t \wedge \overline{\omega(\mu)}_t$$

$$= m^2 (\sqrt{-1})^{n^2} \int_{\hat{V}_s} \Big(\sum \frac{\mu_i}{s^{a_i-1}} \hat{\varphi}_i\Big) \wedge \overline{\Big(\sum \frac{\mu_i}{s^{a_i-1}} \hat{\varphi}_i\Big)} .$$

Put $\rho(s) = \sum_{i=0}^{\ell} \left| \frac{\mu_i}{s^{a_i-1}} \right|^2$. Then the last term is equal to

$$\rho(s)\, m^2 (\sqrt{-1})^{n^2} \int_{\hat{V}_s} \Big(\sum \frac{\mu_i}{\sqrt{\rho(s)}\, s^{a_i-1}} \hat{\varphi}_i\Big) \wedge \overline{\Big(\sum \frac{\mu_i}{\sqrt{\rho(s)}\, s^{a_i-1}} \hat{\varphi}_i\Big)},$$

and from the first part of the proof there are positive number N and a sufficiently small positive number ε'' such that

$$(\sqrt{-1})^{n^2} \int_{\hat{V}_s} \Big(\sum \frac{\mu_i}{\sqrt{\rho(s)}\, s^{a_i-1}} \hat{\varphi}_i\Big) \wedge \overline{\Big(\sum \frac{\mu_i}{\sqrt{\rho(s)}\, s^{a_i-1}} \hat{\varphi}_i\Big)} \geqslant N ,$$

for any s with $0 < |s| < \varepsilon''$. Hence we have

$$(\omega(\mu)_t, \omega(\mu)_t) \geq \rho(s) m^2 N \geq m^2 N \quad ,$$

if $|t| = |s^m| < \varepsilon'^m < 1$. This proves the claim.

Now we choose $\tau, \tau_1, \ldots, \tau_\ell$ as above and put

$$\omega = \tau - \sum_{i=1}^{\ell} (\tau, \tau_j) h^{\bar{j}i} \tau_i \ ,$$

$$g(t) = (\omega, \omega)_t$$

where $(h^{\bar{j}i}) = (h_{i\bar{j}})^{-1}$, $h_{i\bar{j}}(t) = (\tau_i, \tau_j)_t$. Then from the above claim, for any t with $0 < |t| < \varepsilon$ we have

$$g(t) \geq N \ ,$$

for a suitable positive number N and a small number ε. Hence we ha

$$(3.4) \qquad \int_0^{2\pi} \log g(r e^{i\theta}) \, d\theta \geq 2\pi \log N \ .$$

On the other hand if we put

$$I(r) = \frac{1}{2\pi\sqrt{-1}} \int_{|z|=r} \partial \log g, \qquad F(r) = \int_0^{2\pi} \log g(r e^{i\theta}) \, d\theta \ ,$$

we have

$$I(r) = \frac{-r}{4\pi} F'(r) \ .$$

Hence from (3.4) it follows the following

Claim 3. $\lim_{r \to +0} I(r) \geq 0.$

Let $D_{i,\varepsilon}$ be a disk of radius ε with center P_i and we introduce an inner product $(\ ,\ )$ of L on W in such a way that on $W - \bigcup_{i=1}^{s} D_{i,\varepsilon/2}$ it coincides with $(\ ,\ )^0$. We let θ be the Chern form constructed by the same method as above from the inner product $(\ ,\ )$. Then we have

$$\deg L = \int_W \theta = \int_{W - \bigcup_{i=1}^{s} D_{i,\varepsilon}} \theta_0 + \sum_{i=1}^{s} \int_{\partial D_{i,\varepsilon}} \log g_i \ .$$

From Claim 1 and Claim 3 we infer that $\deg L \geq 0$. This proves the theorem.

Remark 3.3. 1) The above proof shows that Claim 3 is independent of Kählerian property of V. On the other hand to prove Claim 1 the theory of Hodge structure is essential. Actually Theorem 3.2 does not necessarily hold for non Kähler fibre spaces.

2) In the above theorem, if $\deg L = 0$ for every quotient invertible sheaf L of $\varphi_* \omega_{V/W}$, then in Claim 1 we have $\Theta_0 \equiv 0$. This means that the period mapping of holomorphic n-forms of our family $\varphi_0 : V_0 \longrightarrow W_0$ is constant.

Corollary 3.4. If $\varphi : V \longrightarrow C$ is a fibre space over a curve of genus ≥ 2 for a general fibre V_x of φ, $p_g(V_x) \geq 1$ and V is in the class C, then we have

$$\kappa(V) = \kappa(V_x) + 1 .$$

That is $C_{m,1}$ holds.

Proof: Since K_C is ample, from Theorem 3.2 it follows that every quotient invertible sheaf of $\varphi_* O_V(K_V)$ is of positive degree. Therefore, by a theorem of Hartshorne [8], $\varphi_* O_V(K_V)$ is an ample vector bundles. This implies $\kappa(V) > 0$. Now we prove the corollary by induction on dim V. Consider the canonical fibration $f : V^* \longrightarrow Y$ of V. By theorem 2.4', a general fibre $f^{-1}(y)$ of f is of Kodaira dimension zero. Put $\hat{\varphi} = \varphi \cdot \pi$ where $\pi : V^* \longrightarrow V$ is a modification. If $\hat{\varphi}(f^{-1}(y)) = C$, then consider the Stein factorization of $\hat{\varphi}|_{f^{-1}(y)}$

$$\begin{array}{ccc} f^{-1}(y) & \xrightarrow{\hat{\varphi}|_{f^{-1}(y)}} & C \\ \psi \downarrow & \nearrow & \\ C' & & \end{array}$$

.

By induction for a general point $z \in C'$ we have

$$\kappa(f^{-1}(y)) = 1 + \kappa(\psi^{-1}(z)) \geq 1 ,$$

since $\kappa(f^{-1}(y)) = 0$ implies $\kappa(\psi^{-1}(z)) \geq 0$ for a general point $z \in C'$. This is a contradiction. Hence $\hat{\varphi}(f^{-1}(y))$ is a point. Therefore, we have a meromorphic mapping $h : Y \longrightarrow C$ where $\hat{\varphi} = h \circ f$. Hence we have a surjective morphism $f_x : \hat{\varphi}^{-1}(x) \longrightarrow h^{-1}(x)$ for a point $x \in C$. By Theorem 1.9, for a general point x of C, we have

$$\kappa(\varphi^{-1}(x)) \leq \dim h^{-1}(x) ,$$

since a general fibre of f_x is of Kodaira dimension zero. Thus we have

$$\kappa(\varphi^{-1}(x)) = \kappa(\hat{\varphi}^{-1}(x)) \leq \dim W - 1 .$$

As $\dim W = \kappa(V)$ we have

$$\kappa(V) \geq \kappa(\varphi^{-1}(x)) + 1 .$$

Again by Theorem 1.9, we have the oppositive inequality. Therefore we prove the corollary.

For analytic threefolds we have the following best possible result

Theorem 3.5. Let $\varphi : V \longrightarrow C$ be a fibre space with $\dim V = 3$ and $\dim C = 1$. Then the conjecture $C_{3,1}$ holds except general fibres of φ are bimeromorphically equivalent to non-algebraic complex tori or non-algebraic K 3 surfaces.

The proof is divided into three parts according to the Kodaira dimensions of general fibres of φ. If $\kappa(V_x) = 2$ for a general fibre the theorem is due to Viehweg [25]. If $\kappa(V_x) = 1$, then the proof is reduced to the study of a fibre space of elliptic curves over an algebraic surfaces.

If $\kappa(V_x) = 0$, then we use the deep results on the classification theory of analytic surfaces. First we can show that it is enough to consider the case where all smooth fibre of φ are minimal. Moreover, if $p_g(V_x) = 1$, then thanks to a result due to Kodaira [14], we can show that $H^2(V_x, \mathbb{C})$ carries a Hodge structure and if $b_1(V_x)$ is odd, it carries a polarized Hodge structure. Therefore we can use Theorem 3.2

and the moduli theory of such surfaces. In case $p_g(V_x) = 0$, we need to consider more. Here we just give an outline of our proof in case V_x is an Enriques surface.

If V_x is Enriques, $p_g(V_x) = 0$ and $K_{V_x}^2$ is trivial. Hence $\varphi_* \omega_{V/C}^{\otimes 2}$ is invertible. First we show $\deg(\varphi_* \omega_{V/C}^{\otimes 2}) \geq 0$. Note that each local section τ of $\varphi_* \omega_{V/C}^{\otimes 2}$ induces an element $\tau_x \in H^0(V_x, K_{V_x}^{\otimes 2})$. Then we introduce a norm on $\varphi_* \omega_{V/C}^{\otimes 2}|C^0$ by

$$\|\tau_x\| = \int_{V_x} (\tau_x \wedge \overline{\tau}_x)^{1/2} \quad .$$

Since $K_{V_x}^{\otimes 2} = \mathbb{1}$ for a smooth fibre V_x, there exist a two-sheeted unramified covering $g : \widetilde{V}_x \longrightarrow V_x$ and a holomorphic 2-form ω_x on $\widetilde{V}_x$ such that $K_{\widetilde{V}_x} = \mathbb{1}$ and $g^* \tau_x = (\omega_x)^{\otimes 2}$. Then we have

$$\|\tau_x\| = \int_{V_x} \omega_x \wedge \overline{\omega_x} \quad .$$

On the other hand, V_x is an algebraic K 3 surface. From this, it can be shown that Claim 1 and Claim 3 do hold in our situation. Hence $\deg(\varphi_* \omega_{V/C}^{\otimes 2}) \geq 0$. Thus, if $g(C) \geq 2$, we obtain the desired result. If $g(C) = 1$, and $\deg(\varphi_* \omega_{V/C}^{\otimes 2}) > 0$, taking a finite unramified covering $h : \widetilde{C} \to C$, and $\widetilde{V} = V \times_C \widetilde{C}$, we have $\widetilde{\varphi}_* \omega_{\widetilde{V}/\widetilde{C}}^{\otimes 2} = h^* \varphi_* \omega_{V/C}^{\otimes 2}$. Hence $p_g(\widetilde{V}) \geq 2$ if $\deg h \geq 2$. Therefore $\kappa(\widetilde{V}) \geq 1$. By Theorem 1.8 we have $\kappa(V) = \kappa(\widetilde{V}) \geq 1$. Therefore assume $g(C) = 1$ and $\deg(\varphi_* \omega_{V/C}^{\otimes 2}) = 0$. In this case the moduli of algebraic K 3 surfaces V_x are constant. From this it follows that $P_2(V) \geq 1$. Hence we have $\kappa(V) \geq 0$. This is the desired result.

In this way we can prove the theorem. The detailed proof can be found in Ueno[23].

4 Classification of algebraic threefolds.

In this section we classify algebraic threefolds using the results in the previous sections. At the moment the structure theorem of algebraic threefolds with $\kappa = 0$, $g_1 = 0$ and $\kappa = -\infty$, $g_1 = 0$ are still missing.

An n-dimensional (algebraic) manifold with $\kappa = n$ is called of general type. By Theorem 2.4a, the canonical fibration is birational for such a manifold, that is, a pluricanonical mapping is birational.

Theorem 4.1. Let V be an n-dimensional (algebraic) manifold and we let $f : V^* \longrightarrow W$ be the canonical fibration.

1) If $\kappa(V) = n-1$, then every smooth fibre of f is a non-singular elliptic curve.

2) If $\kappa(V) = n-2$, then every smooth fibre is a non-singular surface with $\kappa = 0$.

Proof. By Theorem 2.2a, a general fibre of f is a manifold with $\kappa = 0$. For a curve the Kodaira dimension is invariant under smooth deformations. For surfaces, Iitaka [9] shows that this is also the case. Hence we obtain the desired results.

Remark 4.2. As we shall see in the next section, the Kodaira dimensio is not necessarily invariant under smooth deformations in the analytic category.

Next we consider algebraic threefolds with $\kappa = 0$.

Theorem 4.2. Let V be an algebraic threefold with $\kappa = 0$. Then the Albanese mapping $\alpha : V \longrightarrow A(V)$ is surjective. Hence we have $g_1(V) \leq 3$.

1) If $g_1(V) = 3$, then the Albanese mapping α is birational.

2) If $g_1(V) = 2$, then $\alpha : V \longrightarrow A(V)$ is birationally equivalent to an étale fibre bundle over $A(V)$ whose fibre is an elliptic curve. In this case we have $p_g(V) = 0$.

3) If $g_1(V)=1$, then $\alpha : V \to A(V)$ is birationally equivalent to an étale fibre bundle over $A(V)$ whose fibre is an algebraic surface with $\kappa = 0$.

<u>Proof</u>. Let $\beta : V \to W$ be the Albanese fibration. By Theorem 2.6, if the Albanese mapping is not surjective we have $\kappa(W) > 0$. On the other hand, by Theorem 1.9 we have $\kappa(V_w) \geq 0$ for a general fibre of β. Hence from Theorem 3.5 and Theorem 3.6, we obtain

$$\kappa(V) \geq \kappa(V_w) + \kappa(W) > 0.$$

This is a contradiction. Hence the Albanese mapping is surjective. Again applying the same argument, we conclude that $\kappa(W) = 0$ and $\kappa(V_w) = 0$ for a general fibre of β. Moreover, we have

$$g_1(V) \geq g_1(W) \geq g_1(A(V)) = g_1(V).$$

Hence we have $\kappa(W) = 0$, $g_1(W) = \dim W$. If $\dim W \leq 2$, that is $g_1(V) \leq 2$, then by the theory of surfaces and curves, W is birationally equivalent to an Ablelian variety. From the universal mapping property of the Albanese mapping, we infer that W is birationally equivalent, hence isomorphic to $A(V)$. This shows that α has connected fibres if $g_1(V) \leq 2$.

Now let us consider the case $g_1(V) = 3$. As α is surjective, $p_g(V) = 1$. We let $\sum_{i=1}^{m} n_i E_i$ be the effective canonical divisor of V. First assume that $\operatorname{codim}\alpha(E_i) \geqslant 2$, $i = 1,2,\ldots,m$. Put $V' = V - \bigcup_{i=1}^{m} \alpha^{-1}(\alpha(E_i))$, $A(V)' = W - \bigcup_{i=1}^{m} \alpha(E_i)$, $\alpha' = \alpha|_{V'}$. Then $\alpha' : V' \to A(V)'$ is finite and étale. Since $\operatorname{codim}\alpha(E_i) \geqslant 2$, $i = 1,2,\ldots,m$, the fundamental groups of $A(V)$ and $A(V)'$ are isomorphic. Therefore the covering $\alpha' : V' \to A(V)'$ can be extended to a finite étale covering $\gamma : \widetilde{V} \to A(V)$. $\widetilde{V}$ is an abelian variety. On the other hand there is a natural inclusion $\iota : V' \to \widetilde{V}$ and this can be extended to a bimeromorphic mapping. Therefore V is bimeromorphically equivalent to an abelian variety.Hence the Albanese mapping is bimeromorphic.

Next assume that codim $\alpha(E) = 1$, where E is one of E_i's. By a finite succession of blowing ups, we may assume that E is non-singular. Moreover we assume that the Albanese variety $A(V)$ contains ı non-trivial abelian subvariety. Then by Proposition 2.7 we have $\kappa(\alpha(E)) = 2$. Hence $\kappa(E) = 2$. That is E is a surface of genera type. From the classification theory of algebraic surfaces we infer that $p_g(E) - g_1(E) + 1 > 0$. From the exact sequence

$$0 \to O_V(-E) \to O_V \to O_E \to 0,$$

we obtain

$$H^2(V, O_V) \xrightarrow{\rho} H^2(E, O_E) \to H^3(V, O_V(-E)) \to H^3(V, O_V) \to 0 .$$

The Serre duality and our assumption imply

$$h^3(V, O_V(-E)) = h^0(V, O_V(K + E))$$

$$1 = h^0(V, O_V(K_V)) \leq h^0(V, O_V(K + E)) \leq h^0(V, O_V(2K_V)) = 1$$

Hence the mapping ρ is surjective. Therefore we have

$$g_2(V) \geq p_g(E).$$

Moreover it is easy to show

$$p_g(E) \geq p_g(\alpha(E)) \geq g_2(A(V)) = 3$$

$$g_1(E) \geq g_1(\alpha(E)) \geq g_1(A(V)) = 3 .$$

Now we show $g_2(V) = 3$. Since α is surjective, we have $g_2(V) \geq 3$. Put $\tau_i = \alpha^*(dz_i)$, $i = 1,2,3$, where (z_1, z_2, z_3) is a system of glob coordinates of $A(V)$. For a holomorphic 2-form ω on V, we can fi constants a_i, $i = 1,2,3$ such that

$$\tau_i \wedge \omega = a_i \alpha^*(dz_1 \wedge dz_2 \wedge dz_3) ,$$

since $p_g(V) = 1$. Put $\omega' = a_1 \tau_2 \wedge \tau_3 - a_2 \tau_1 \wedge \tau_3 + a_3 \tau_1 \wedge \tau_2$. As we have $(\omega - \omega') \wedge \tau_i = 0$, $i = 1,2,3$, $\omega = \omega'$. Thus we have $g_2(V) = 3$. Fron the above inequalities, we infer $p_g(E) = 3$. As we have $p_g(E) - g_1(E) + 1 > 0$, we also have $g_1(E) = 3$. Thus we have $\chi(E, O_E) = 1$.

The fact $g_1(E) = 3$ implies that the natural inclusion $\iota : E \to$ induces an isomorphism $\iota_*: H_1(E, \mathbb{Q}) \simeq H_1(V, \mathbb{Q})$. We can find a subgroup I of finite index in $H_1(V, \mathbb{Z})$ such that $\iota_*^{-1}(\Gamma) \subsetneqq H_1(E, \mathbb{Z})$. Let us consider a finite étale abelian covering $f : \tilde{V} \rightarrow V$ which corresponds the group Γ. From Theorem 1.8 and the first part of our theorem, we infer that $\tilde{V}$ has numerical invariants $p_g(\tilde{V}) = 1$, $\kappa(\tilde{V}) = 0$ and $g_1(V) = 3$. We let $\tilde{\alpha} : \tilde{V} \longrightarrow A(\tilde{V})$ be the Albanese mapping. It is easy to show that $\tilde{V}$ and an irreducible compaonent $\tilde{E}$ of $f^{-1}(E)$ satisfy our assumption. Hence we have $\chi(\tilde{E}, \mathcal{O}_{\tilde{E}}) = 1$. On the other hand, as we have assumed $\iota_*^{-1}(\Gamma) \subsetneqq H_1(E, \mathbb{Z})$, $f|_{\tilde{E}} : \tilde{E} \longrightarrow E$ is a non-trivial étale covering. Therefore we must have $\chi(\tilde{E}, \mathcal{O}_{\tilde{E}}) = (\deg f|_{\tilde{E}}) \cdot \chi(E, \mathcal{O}_E) \geq 2$. This is a contradiction.

$$\begin{array}{ccc} \tilde{V} & \xrightarrow{\tilde{\alpha}} & A(\tilde{V}) \\ {\scriptstyle f}\downarrow & & \downarrow \\ V & \xrightarrow{\alpha} & A(V) \end{array}$$

It remains to consider the case where $A(V)$ is not simple. In this case we can construct a surjective morphism $h : V \rightarrow B$ from V to a two-dimensional abelian variety with connected fibres. Then from Theorem 3.5 and the following lemma, we obtain the desired result.

<u>Lemma 4.3.</u> Let $f : V \longrightarrow A$ be a surjective morphism from an n-dimensional algebraic variety V to an (n-1)-dimensional Abelian variety. Assume that general fibres of f are non-singular elliptic curves. Then $\kappa(V) = 0$, if and only if $f : V \rightarrow A$ is birationally equivalent to an étale fibre bundle over A whose fibre is an elliptic curve.

This is a special case of Proposition 4.1 of Ueno [21] and the proof can be found there.

Our statement 2) can be reduced from Lemma 4.3. If $p_g(V) = 1$ in this case, then the structure group of the fibre bundle is a subgroup of the translations of the elliptic curve. Then we have $g_1(V) = 3$. This is a contradiction.

The statement 3) can be proved by using Theorem 3.5 and similar lemmas to Lemma 4.3. The proof can be found in Ueno [23]. Q.E.D.

<u>Remark 4.4.</u> For an algebraic manifold V of dimension $n \geq 3$, Kawamata

[11] and Kawamata-Viehweg [12] obtained the similar results to the first part of the theorem and the statement 1).

The structure of algebraic threefold with $\kappa = 0$, $g_1 = 0$ seems very complicated. Several examples of such threefolds can be found in Ueno [20] .

Theorem 4.5. Let V be an algebraic threefold with $\kappa = -\infty$. If $g_1(V) \geq 1$, then V is uniruled.

Proof. Let $\beta : V \longrightarrow W$ be the Albanese fibration. We have $\kappa(W) \geq 0$. From Theorem 3.5 it follows that the Kodaira dimension of general fibres is of $-\infty$. If dim $W = 2$, then general fibres are $\mathbb{P}^1$. Therefore in this case there is a finite ramified covering $f : \hat{W} \longrightarrow W$ such that $V \times_W \hat{W}$ is birationally equivalent to $W \times \mathbb{P}^1$. Hence V is uniruled. If dim W =1, then general fibres of are rational surfaces or ruledsurfaces. In this case, using the theory of Hilbert scheme, we infer that we can find a finite ramified covering $f : \hat{W} \longrightarrow W$ and a surjective morphism $g : S \longrightarrow \hat{W}$ of a surface S to $\hat{W}$ such that $V \times_W \hat{W}$ is birationally equivalent to $S \times \mathbb{P}^1$. (For the proof, see, for example, the appendix to Ashikaga [1].)

For algebraic threefolds with $\kappa = -\infty$, $g_1 = 0$, we do not know the structure. It is conjectured that the above theorem also holds in this case.

5 Classification of analytic threefolds.

In this section we consider the structure of analytic threefolds with $\kappa = 0$ or $k = -\infty$. First we give an example which shows that non-Kähler threefolds may have strange properties.

Example 5.1. Let C be a hyperelliptic curve of genus 2 and $\pi : C \to \mathbb{P}^1$ the double covering. Put $L = \pi^* \mathcal{O}_{\mathbb{P}^1}(1)$, $F = L^{\ell}$. Then we have $K_C = L^{g-1}$. For any point $t \in \mathrm{Pic}^0(C)$, we let $[t]$ be the line bundle of degree 0 on C corresponding to the point t. Put $F_t = F \otimes [t]$. There exists a line bundle $\mathcal{F}$ on $C \times \mathrm{Pic}^0(C)$ such that restriction $\mathcal{F}|_{C \times t}$ is isomorphic to F_t. Assume $m \geq g$. Then F_t is generated by its global sections and we have $h^0(F_t) = 2m - g + 1$, $h^1(F_t) = 0$. Hence $p_* \mathcal{F}$ is locally free where $p : C \times \mathrm{Pic}^0(C) \to \mathrm{Pic}^0(C)$ is the natural projection. We can find an open neighbourhood U of the origin of $\mathrm{Pic}^0(C)$ and two holomorphic sections φ, ψ of $\mathcal{F}$ over $p^{-1}(U)$ such that $\varphi_t = \varphi|_{p^{-1}(t)}$, $\psi_t = \psi|_{p^{-1}(t)}$, considered as elements of $H^0(C, F_t)$ have no common zero on C for each point $t \in U$. Put

$$I_1 = \begin{pmatrix} 1 & 0 \\ 0 & 1 \end{pmatrix}, \quad I_2 = \begin{pmatrix} 0 & 1 \\ -1 & 0 \end{pmatrix}, \quad I_3 = \begin{pmatrix} 0 & \sqrt{-1} \\ \sqrt{-1} & 0 \end{pmatrix}, \quad I_4 = \begin{pmatrix} -1 & 0 \\ 0 & -\sqrt{-1} \end{pmatrix}.$$

It is easy to show that $\Lambda_t = \sum_{i=2}^{4} \mathbb{Z}\, I_i \binom{\varphi_t}{\psi_t}$ is a lattice of each fibre of a rank 2 vector bundle $V_t = F_t \oplus F_t$ over C for each point $t \in U$. The group Λ_t acts on V_t as translations in each fibre. We have a quotient manifold $M_t = V_t / \Lambda_t$. Let $\pi_t : M_t \to C$ be the natural morphism. From our construction we infer that π_t is smooth, each fibre of π_t is a complex 2-torus and we have

$$\omega_{M_t/C} = K_{M_t} \otimes \pi_t^* K_C^{-1} = \pi_t^* F_t^{-2} \quad .$$

The collection $\mathcal{M} = \bigcup_{t \in U} M_t$ is a complex analytic family over U. We let $f : \hat{C} \to C$ be a double covering ramified at the divisor (s)

where s is a generic element of $H^0(C, L^{2k})$. Then we have $K_C = f^*L^{g-1+k}$. Put $\hat{M}_t = M_t \times_C \hat{C}$. Then $\hat{\pi}_t : \hat{M}_t \to \hat{C}$ is smooth and we have

$$\omega_{\hat{M}_t/\hat{C}} = \hat{\pi}_t^* f^* F_t^{-2}, \quad K_{\hat{M}_t} = \hat{\pi}_t^* f^*(L^{g-1+h-2} \otimes [-2t]).$$

Hence if we put $k = 2\ell - g + 1$, then we have

$$K_{\hat{M}_t} = \pi_t^* f^* [-t] .$$

Put $U' = U \cap \mathrm{Pic}^0(C)_{tor}$ where $\mathrm{Pic}^0(C)_{tor}$ is the set of all points of finite order. The set U' is dense in U and so is $U^* = U - U'$. We have

$$\kappa(\hat{M}_t) = \begin{cases} 0, & \text{if } t \in U' \\ -\infty, & \text{if } t \in U^{*\prime} \end{cases} .$$

More precisely we have

$$P_m(\hat{M}_t) = \begin{cases} 1, & \text{if } 2mt = 0 \text{ in } \mathrm{Pic}^0(C) \\ 0, & \text{otherwise} . \end{cases}$$

Hence Kodaira dimension and plurigenera are not necessarily invariant under smooth deformations. (For other examples see Nakamura [17].) Moreover we have : for analytic threefolds with $\kappa = 0$, we cannot find a positive integer m such that $P_m = 1$, even if they are in the same complex analytic family. By a theorem of Griffiths [7], M_t and $\hat{M}_t$ are not Kähler, since $\pi_{t*}\omega_{M_t/C}$ and $\hat{\pi}_{t*}\omega_{\hat{M}_t/\hat{C}}$ are negative. Furthermore, for the fibre space $\hat{\pi}_t : \hat{M}_t \to \hat{C}$, $C_{3,1}$ does not hold, since $\kappa(C) = 1$ and fibres of $\hat{\pi}_t$ are complex tori, hence $\kappa = 0$, but we have $\kappa(M_t) \leq 0$. (See Theorem 3.5 above.)

Thus the above example show strange phenomena of analytic threefolds. But by virtue of Theorem 3.5, we have structure theorems of analytic threefold of $\kappa = 0$ or $-\infty$ with positive Albanese dimensio[n]

<u>Theorem 5.2</u>. Let M be an analytic threefold with $\kappa(M) = 0$, $t(M) \geq$ ·

1) If Albanese mapping is surjective, then the Albanese mapping has connected fibres and M has the following structure.

a) If $t(M) = 3$, then α is bimeromorphic.

b) If $t(M) = 2$, then $\alpha : M \to A(M)$ is bimeromorphically equivalent to an analytic fibre bundle over $A(M)$ whose fibre is an elliptic curve.

c) If $t(M) = 1$, then any smooth fibre of M_x of the Albanese mapping α is a surface with $\kappa = 0$. If, moreover, M is a bimeromorphic image of Kähler manifold, then $\alpha : M \to A(M)$ is bimeromorphically equivalent to an analytic fibre bundle over $A(M)$ whose fibre is a surface with $\kappa = 0$.

2) If the Albanese mapping is not surjective, the image $\alpha(M) = C$ is a non-singular curve of genus $g \geq 2$. The Albanese mapping has connected fibres and a general fibre M_x of α is bimeromorphically equivalent to a non-algebraic complex torus or K 3 surface.

Remark 5.3. 1) If M is a meromorphic image of a Kähler manifold and $\kappa(M) = 0$, $t(M) \geq 1$, then the Albanese mapping is surjective.

2) Example 5.1 shows that there exists an analytic threefold M with $\kappa(M) = 0$, $t(M) = 1$ such that all fibres of the Albanese mapping $\alpha : M \to A(M)$ are complex tori but $\alpha : M \to A(M)$ is not bimeromorphically equivalent to an analytic fibre bundle over $A(M)$.

3) Example 5.1 also shows that an analytic threefold which belongs to the class in Theorem 5.2, 2) may have arbitrary big $t(M)$.

4) There exists an analytic threefold M belonging to the class in Theorem 5.2, 2) such that general fibres of the Albanese mapping are K 3 surfaces. Such threefold is obtained by the threefold in example by taking the quotient by the standard involution of each fibre of $\hat{\pi}_t : \hat{M}_t \to \hat{C}$, $\in U'$ and resolving its singularities. We can also obtain another example by using the gravitation instanton attached to a K 3 surface. ([2])

Finally we consider the case where threefolds have negative Kodaira dimension.

Theorem 5.4. Let M be an analytic threefold with $\kappa(M) = -\infty$ and $t(M) \geq 1$. Then we have $\dim \alpha(M) \leq 2$.

1) If $\dim \alpha(M) = 2$, then a general fibre of the Albanese fibration

$\beta : M \longrightarrow W$ is $\mathbb{P}^1$.

2) If $\dim \alpha(M) = 1$, then $\alpha(M) = C$ is a non-singular curve of genus $t(M)$ and $\alpha : M \longrightarrow C$ has connected fibres. Moreover general fibre of α is a surface with $\kappa = -\infty$ or bimeromorphically equivalent to a complex torus or a K 3 surface. Moreover, if M is a meromorphic image of a Kähler manifold, a general fibre of is a rational or ruled surface.

The proofs of Theorem 5.2 and Theorem 5.4 are based on Theorem 3.5 and the classification of analytic surfaces. Details will be found in Ueno [23].

On the above theorems show that an analytic threefold which is a meromorphic image of a Kähler manifold has similar properties as an algebraic threefold has. Fujiki [5] and Campana [4] studied such manifolds of arbitrary dimension and obtained many interesting facts.

Unfortunately, the structure of threefolds with $\kappa = 0$ and $t = 0$, or $\kappa = -\infty$ and $t = 0$ seems very complicated. There are partial results due to Mabuchi [16].

References

1 Ashikaga, T., Rhe deformation behaviour of the Kodaira dimension of algebraic manifold. Tohoku Math. J. 33 (1981), 193-214.

2 Atiyah, M. F., Hitchin, N. J., and I. M. Singer., Self-duality in four-dimensional Riemannian geometry. Proc. Royal Soc. London A 362 (1978), 425-461.

3 Blanchard, A., Sur les varietes analytiques comples. Ann. Sci. Ecole Norm. Sup. 73 (1956), 157-202.

4 Campana, F., Application de l'espace des cycles a la classification bimeromorphe des espaces analytiques Kahleriens compacts, preprint 1980.

5 Fujiki, A., On the structure of compact complex manifolds in C, to appear.

6 Fujita, T., On Kähler fibre spaces over curves, J. Math. Soc. Japan, 30 (1978), 779-794.

7 Griffiths, P.A., Integrals of algebraic manifolds III. Pub Math. IHES 38 (1970), 126-180.

8 Hartshorne, R., Ample vector bundles on curves. Nagoya Math. J. 43 (1971), 73-89.

9 Iitaka, S., Deformations of compact complex surfaces, II, J. Math. Soc. Japan, 22 (1970), 247-261.

10 Iitaka, S., On D-dimensions of algebraic varieties. J. Math. Soc. Japan, 23 (1971), 356-373.

11 Kawamata, Y., Characterization of abelian varieties, Compositio Math. 43 (1981), 253-276.

12 Kawamata, Y and E. Viehweg., On a characterization of an abelian variety in the classification theory of algebraic varieties. Compositio Math. 41 (1980), 355-360.

13 Kempf, G., Kundsen, F., Mumford, D. and V. Saint-Donat, Toroidal embeddings I, Lecture Notes in Math. 339, Springer, 1973.

14 Kodaira, K., On the structure of compact complex analytic surfaces, I. Amer. J. Math., 87 (1964) 751-798.

15 Kuhlmann, N., Ein Satz uber Regelflache vom Geschlecht 2, Archiv Math. 29 (1977), 619-620.

16 Mabuchi, T., Invariant β and uniruled threefolds, to appear.

17 Nakamura, I., Complex parallelisable manifolds and their small deformations. J. Diff. Geometry 10 (1975), 85-112.

18 Ueno, K., Classification theory of algebraic varieties and com compact complex spaces, Lecture Notes in Math. 439, Springer, 1975

19 Ueno, K., Classification of algebraic varieties, II, Proceedings of the International symposium on Algebraic Geometry, Kyoto, 1977 693-708, Kinokumiya Shoten, 1978.

20 Ueno, K., Birational geometry of algebraic threefolds, Journees de geometric algebrique d'Angers 1979, 311-322, Sijthoff Noordhoff 1980.

21 Ueno, K., On algebraic fibre spaces of abelian varieties., Math Ann. 237 (1978), 1-22.

22 Ueno, K., On three-dimensional compact complex manifolds with non-positive Kodaira dimension, Proc. Japan Acad., 56 (1980), 479-483.

23 Ueno, K., On the structure of three-dimensional compact complex manifolds with positive Albanese dimension, in preparation.

24 Viehweg, E., Klassifikationstheorie algebraischer Varietäten der Dimension drei, Compositio Math. 41 (1980), 361-400.

25 Viehweg, E., Die Additivität der Kodaira Dimension für projektiv Faserräume über Varietäten des allgemeinen Typs, to appear.

26 Wilson, P. M. H., On the canonical ring of algebraic varieties, to appear.

27 Zariski, O., The theorem of Riemann-Roch for high multiples of an effective divisor on an algebraic surface.Ann. of Math. 76 (1962), 560-615.

CLASSIFICATION OF FANO THREEFOLDS ACCORDING TO FANO AND ISKOVSKIH

J.P. Murre

INTRODUCTION

These lectures are based entirely upon the papers of Iskovskih ([I1] and [I2]) and Shokurov ([S1] and [S2]) and contain no innovations from the part of the lecturer.

Our presentation has greatly benefitted by the survey paper [I3]. For the theorem of Shokurov, on the existence of lines, we have oriented ourselves upon the Mittag-Leffler report of Miles Reid ([R]).

We have tried to keep these notes close to the oral exposition and informal. We have not aimed at completeness; the lectures and the notes serve only as an introduction to the original papers of Iskovskih, Shokurov and Reid.

For a short introduction to the subject itself and for some historical remarks, see 1.1.

CONTENTS

§1. GENERAL THEORY OF FANO THREEFOLDS

1.1. INTRODUCTION

Fano varieties are varieties with ample anticanonical class. In dimension one these varieties are simply the rational curves, in dimension two they are the well-known Del Pezzo surfaces. In dimension three (threefolds) the study of this type of varieties was started by Fano in the beginning of this century ([F1]). At least for questions of rationality the theory of varieties of dimension larger than two is considerably more complicated than for surfaces. Fano did have the insight that among this class of varieties there are varieties which are non-rational, in spite of the fact that they have geometric genus, plurigenera and irregularity all zero, and he devoted much effort to prove this for, among others, the quartic threefolds and the cubic threefolds. We know now that it took more than half a century before "rigorous" proofs for the non-rationality of these varieties were given by Iskovskih-Manin ([I-M]), respectively Clemens-Griffiths ([C-G]).

In a long series of papers Fano has studied the threefolds which carry now his name, both from the birational and from the biregular point of view (see for instance [F1,2,...,5]). In these lectures we are only concerned with the biregular part of the theory. Fano has started and carried out, at least under certain restrictions, a biregular classification of these varieties. His work has been continued, among others, by Roth (see [Ro] and also the survey article of Gallarati [G]).

In spite of the beautiful geometry and ingenious ideas, the work of Fano (and Roth) contains serious lacunes and, judged to present standards, is in an extremely unsatisfactory state. Recently however Iskovskih has taken up again the theory of Fano. Using modern tools he has been able to justify and amplify the work of Fano and he has given a complete classification of Fano threefolds of index greater than one and of Fano threefolds of index one and Picard number one ([I1,2 and 3]). *) As we have already said in the general introduction, in

*) Very recently Mori and Mukai have also classified Fano 3-folds of index one and Picard number greater than one, cf [M.-M]

these lectures we report on this work of Iskovskih and on some important contributions of Shokurov ([S1 and 2],[R]).

As to the literature we like to draw also the attention to the nice survey paper of Bonardi ([Bo]).

1.2. NOTATIONS AND CONVENTIONS

We always work over an algebraically closed field $\mathbf{k}$ of characteristic zero.

Let V denote a smooth, irreducible, projective variety. The *canonical class* of V is denoted by K_V. Let D be a divisor on V and $\mathcal{O}_V(D)$ its associated sheaf (cf.[H],p.144). We write $h^i(D) = \dim H^i(V,\mathcal{O}_V(D))$. Also we put $L(D) = H^0(V,\mathcal{O}_V(D))$ and we sometimes write $\ell(D) = h^0(D)$. The complete linear system determined by D is denoted by $|D|$, i.e. $|D| = \{D';D'\sim D, D'>0\}$, where $\sim$ means linear equivalence. The associated rational map is denoted by $\lambda_{|D|}$, i.e.

$$\lambda_{|D|}: V \dashrightarrow V' \subset \mathbb{P}_{\ell(D)-1}$$

where if $f_1,\ldots,f_{\ell(D)}$ is a k-basis for $L(D)$ then for $P \in V$ we have $\lambda_{|D|}(P) = (f_1(P),\ldots,f_{\ell(D)}(P))$. The divisor(class) D is called *very ample* if $\lambda_{|D|}: V \xrightarrow{\sim} V'$, and *ample* if nD is very ample for some $n > 0$.

If $W \subset V$ is a subvariety then $B_W(V) \to V$ denotes the blow up of the variety V with center W.

Furthermore: V_n^m means a variety of *dimension* n and *degree* m. Finally: by $V(n,m) \subset \mathbb{P}_N$ we mean the complete intersection in $\mathbb{P}_N$ of two hypersurfaces of degree n and m respectively.

1.3. DEFINITION

A variety V is called a *Fano variety* if $-K_V$ is ample (i.e. the anticanonical class is ample).

1.4. THE CASE dim $V = 1$

Clearly we have: a curve Γ is a Fano variety $\Leftrightarrow$ the genus $g(\Gamma) = 0$ $\Leftrightarrow$ Γ is a rational curve.

1.5. THE CASE dim $V = 2$

In the case of a surface S we have by definition (cf.[Ma],p.128): S is a Fano variety $\Leftrightarrow$ S is a *Del Pezzo* surface.

We recall here some properties of Del Pezzo surfaces (see [M],Chap.IV, [B1],Chap.IV and [S.-R.],Chap.VII). Let S be a Del Pezzo surface. Then:

i) S is a *rational* surface (but a rational surface of a very special type!).

ii) Put $d = (K_S.K_S)$, where $(-.-)$ denotes the intersectionnumber ($d=d(S)$ is called the *"degree"* of the Del Pezzo surface). Then $1 \leq d \leq 9$.

iii) $d = 9 \Rightarrow S \simeq \mathbb{P}_2$,

$d = 8 \Rightarrow S \simeq \mathbb{P}_1 \times \mathbb{P}_1$ or $S = B_{a_1}(\mathbb{P}_2)$ (the blow up of $\mathbb{P}_2$ in a point a_1).

$1 \leq d \leq 7 \Rightarrow S \simeq B_{a_1,a_2,\ldots,a_{9-d}}(\mathbb{P}_2)$ (the blow up of $\mathbb{P}_2$ in the points $a_1,a_2,\ldots,a_{9-d}$) .

Conversely for $d \geq 3$ the varieties of the above type are Del Pezzo surfaces provided no three of the points are on a line and no six are on a conic.

iv) Index of S . Write $-K_S = r.H$ with $H > 0$ and r maximal. $r = r(S)$ is called the *index* of S . In the above description $r(\mathbb{P}_2) = 3$, $r(\mathbb{P}_1\times\mathbb{P}_1) = 2$, the remaining Del Pezzo's have index 1. Note also that since $\mathrm{Pic}(S)$ is free we have that H is unique.

v) $3 \leq d \leq 9 \Rightarrow -K_S$ is *very* ample:

$$\lambda_{|-K_S|} : S \tilde{\to} S'^d \subset \mathbb{P}_d .$$

Moreover: S' is projectively normal (i.e. the hypersurfaces of fixed degree cut out a complete linear system). Conversely, smooth projectively normal surfaces of degree d in $\mathbb{P}_d$ are Del Pezzo surfaces (this is the classical definition of Del Pezzo surfaces). Example: $d = 3$, cubic surfaces in $\mathbb{P}_3$.

vi) Let $S = B_{a_1,\ldots,a_{9-d}}(\mathbb{P}_2)$. Then $|-K_S| = C'$, where C' is the strict (or proper) transform of *the cubics* of $\mathbb{P}_2$ *passing through* the points $a_1,\ldots,a_{9-d}$.

In particular: for $d \geq 2$ the system $|-K_S|$ has *no base points*. For $d = 2$ the $\lambda_{|-K_S|}$ gives a double covering of $\mathbb{P}_2$ (with ramification locus a quartic curve). For $d = 1$ the system $|-K_S|$ has one base point (coming from the "9th" point of intersection of the cubics going through the 8 fixed points $a_1,\ldots,a_8$ in the plane), and we get a rational map to $\mathbb{P}_1$.

vii) For $r = 1$ and $d \geq 3$ the Del Pezzo surfaces have, using the anticanonical embedding $\lambda_{|-K_S|}$, lines (for $r = 3$ and 2 there are also lines, but not in the anticanonical embedding).

1.6. EXAMPLES OF FANO VARIETIES OF DIMENSION 3 (= FANO 3-FOLDS)

In the following H denotes always a hyperplane section.

i) $V = \mathbb{P}_3$; $K_V = -4H$.

ii) $V = V(2) \subset \mathbb{P}_4$; $K_V = -3H$.

iii) $V = V(3) \subset \mathbb{P}_4$; $K_V = -2H$.

iv) $V = V(4) \subset \mathbb{P}_4 = K_V = -H$.

v) $V = V(3,2) \subset \mathbb{P}_5$; $K_V = -H$.

vi) $V = Gr(1,4).V(2).H.H' \subset \mathbb{P}_7$; $K_V = -H$. (Note: $K_{Gr(1,4)} = -5H$.)

vii) Let X be the Segre embedding of $\mathbb{P}_2 \times \mathbb{P}_2$ in $\mathbb{P}_8$, take

$V = X.H$; $K_V = -2H$.

viii) Let X be as above, take $V = X.V(2) \subset \mathbb{P}_8$; $K_V = -H$.

Note that in the examples i),-,vi) we have $Pic(V) \simeq \mathbb{Z}$, in examples vii) and viii) $Pic(V) \simeq \mathbb{Z} \oplus \mathbb{Z}$.

1.7. GENERAL PROPERTIES OF FANO THREEFOLDS

These general properties are obtained by applying three general theorems:

a) Serre duality theorem. Let D be a divisor on a variety V of dimension n then $h^i(D) = h^{n-i}(K_V - D)$, $0 \le i \le n$.

b) Riemann-Roch theorem on V with $\dim V = 3$. Then for a divisor D on V we have $\chi(V,D) = \sum_{i=0}^{3} (-1)^i h^i(D) = \frac{1}{6}D^3 - \frac{1}{4}D^2K_V + \frac{1}{12}D(K_V^2 + c_2) - \frac{1}{24}K_V \cdot c_2$, where $c_2 = c_2(V)$ is the second Chern class of V.

c) Kodaira vanishing theorem (here we use the fact that $\mathrm{char}(k) = 0$!) For an ample divisor D we have $h^i(-D) = 0$, $0 \le i < \dim V$.

In the following V is *always a Fano threefold*. Also we write often K instead of K_V when no confusion is possible.

LEMMA 1.

a) $h^i(mK) = 0$, $m > 0$, $i < 3$.

b) $h^i(-mK) = 0$, $m \ge 0$, $i > 0$.

PROOF.

a) immediate by Kodaira vanishing.

b) $h^i(-mK) = h^{3-i}((m+1)K) = 0$ by Serre duality and Kodaira vanishing.

COROLLARY 1.

a) $P_m(V) = h^0(mK) = 0$ *($m \ge 1$, mth plurigenus).*

b) $\kappa(V) = -\infty$ *(Kodaira dimension).*

COROLLARY 2.

$h^i(\mathcal{O}_V) = 0$ $(i>0)$. *In particular: the irregularity* $q(V) = h^1(\mathcal{O}_V) = 0$.

COROLLARY 3.

$\mathrm{Pic}^0(V) = 0$ *and* $\mathrm{Alb}(V) = 0$.

COROLLARY 4.

$Pic(V) \cong H^2(V,\mathbb{Z})$.

PROOF.

Look to the exponential sequence

$$0 \dashrightarrow \mathbb{Z} \dashrightarrow \mathcal{O} \overset{\exp}{\dashrightarrow} \mathcal{O}^* \dashrightarrow 1$$

and use $h^1(\mathcal{O}_V) = h^2(\mathcal{O}_V) = 0$.

COROLLARY 5.

$K_V . C_2(V) = -24$.

PROOF.

Apply Riemann-Roch to $D = 0$ and use Corollary 2.

LEMMA 2.

$h^0(-mK) = \frac{m(m+1)(2m+1)}{12}(-K^3) + 2m + 1 \quad (m \geq 0)$.

In paticular

$h^0(-K) = -\frac{1}{2}K^3 + 3$.

PROOF.

This follows immediately by applying Riemann-Roch to $D = -mK$ and using $K.C_2 = -24$.

1.8. CONSEQUENCES FOR $X \in |-K_V|$

Assume that $X \in |-K_V|$ is *smooth* (cf. with 1.10).

By the adjunction formula

$$K_X = (K_V+X).X = 0 .$$

Using the exact sequence

$$0 \longrightarrow \mathcal{O}_V(-X) \longrightarrow \mathcal{O}_V \longrightarrow \mathcal{O}_X \longrightarrow 0$$

and the fact that $H^1(\mathcal{O}_V) = H^2(\mathcal{O}_V(-X)) = 0$ (Lemma 1) we get $q(X) = \dim H^1(X,\mathcal{O}_X) = 0$. Hence

LEMMA 3.

X *is a K3-surface.*

LEMMA 4.

$|-K_V|.X \overset{\text{def}}{=} Tr_X|-K_V|$ *is complete.*

PROOF.

Look to the exact sequence

$$0 \longrightarrow \mathcal{O}_V \longrightarrow \mathcal{O}_V(X) \longrightarrow \mathcal{O}_X(X) \longrightarrow 0$$

and use $H^1(V,\mathcal{O}_V) = 0$ (Corollary 2 of Lemma 1).

1.9. CONSEQUENCES FOR THE CURVE SECTION

Assume that for $X_i \in |-K_V|$, $i = 1,2,$, we have that $\Gamma = X_1.X_2$ is a *a smooth irreducible curve* (cf. with 1.10).

LEMMA 5.

a) $|-K_V|.\Gamma \overset{\text{def}}{=} Tr_\Gamma|-K_V| = K_\Gamma$.

b) the genus $g(\Gamma) = -\frac{1}{2}(K_V)^3 + 1$ ($g=g(\Gamma)$ *is called the* <u>*genus of the Fano variety*</u> V).

PROOF.

By the adjunction formula we have

$$K_\Gamma = \Gamma._{X_1}\Gamma = X_1 ._V\Gamma = (-K_V).\Gamma .$$

Using the exact sequence

$$0 \longrightarrow \mathcal{O}_X \longrightarrow \mathcal{O}_X(\Gamma) \longrightarrow \mathcal{O}_\Gamma(\Gamma) \longrightarrow 0 ,$$

the fact that $q(X) = 0$ and Lemma 4 we get

$$H^0(V,-K_V) \twoheadrightarrow H^0(X,\mathcal{O}_X(X)) \twoheadrightarrow H^0(\Gamma,\mathcal{O}_\Gamma(\Gamma))$$

hence the anticanonical system $|-K_V|$ cuts out the complete canonical system K_Γ on Γ ; this gives a). As to b) this follows now by Lemma 2 since we have (as follows easily from the above used exact sequences) that

$$g(\Gamma) = \dim H^0(\Gamma,\mathcal{O}_\Gamma(K_\Gamma)) = \dim H^0(V,\mathcal{O}_V(-K_V)) - 2 = -\tfrac{1}{2}(K_V)^3 + 1 .$$

1.10. THE PRINCIPAL SERIES

DEFINITION. V is a Fano 3-fold of the *principal series* (or main series) if $-K_V$ is *very* ample.

Now if $-K_V$ is very ample then clearly the **assumptions** in 1.8 and 1.9 are fulfilled, by Bertini, for a sufficiently general $X \in |-K_V|$ and sufficiently general $X_1,X_2 \in |-K_V|$. Therefore we have

PROPOSITION 1.

Let V be a Fano 3-fold of the principal series embedded by means of $|-K_V|$. Then:

i) A sufficiently general hyperplane section is a K3*-surface.*

ii) (original Fano point of view) A sufficiently general curve section is a canonically embedded curve.

iii) Write $g = g(\Gamma)$ *, then* $g = -\tfrac{1}{2}(K_V)^3 + 1$.

iv) V spans $\mathbb{P}_{g+1}$ *and is of degree* $2g - 2$.

The fact iv) we shall often shortly indicate by writing

(1) $$V = V_3^{2g-2} \subset \mathbb{P}_{g+1} \quad (g \geq 3) .$$

EXAMPLES. 1) In case $V \simeq \mathbb{P}_3$ we have $g = \frac{1}{2}4^3 + 1 = 33$;
2) $V \simeq V(2) \subset \mathbb{P}_4 \Rightarrow g = 28$; 3) $V \simeq V(3) \subset \mathbb{P}_4 \Rightarrow g = 13$;
4) $V \simeq V(4) \subset \mathbb{P}_4 \Rightarrow g = 3$. Also note that in Examples 1), 2) and 3) the usual embedding is not via $|-K_V|$!

There is the following converse:

PROPOSITION 2.

Let $V \subset \mathbb{P}_N$ *be a smooth 3-fold such that for sufficiently general hyperplanes* H *and* H' *we have* $\Gamma = V.H.H'$ *is a canonically embedded curve of genus* g . *Then* V *is a Fano 3-fold of the principal series embedded by means of* $|-K_V|$.

PROOF.

Let X be a general hyperplane section. Then we have:

LEMMA 6.

i) $H^2(X,\mathcal{O}_X(n)) = 0$, $n > 0$.

$\dim H^2(X,\mathcal{O}_X) = 1$.

ii) $H^1(X,\mathcal{O}_X(n)) = 0$, $n \geq 0$.

PROOF.

For $n \geq 0$ consider the exact sequence

$$0 \longrightarrow \mathcal{O}_X(n-1) \longrightarrow \mathcal{O}_X(n) \longrightarrow \mathcal{O}_\Gamma(n) \longrightarrow 0$$

and the corresponding exact cohomology sequence

$$0 \longrightarrow H^0(\mathcal{O}_X(n-1)) \longrightarrow H^0(\mathcal{O}_X(n)) \xrightarrow{\alpha} H^0(\Gamma,\mathcal{O}_\Gamma(n)) \longrightarrow H^1(\mathcal{O}_X(n-1)) \longrightarrow$$

$$\xrightarrow{\beta} H^1(\mathcal{O}_X(n)) \longrightarrow H^1(\Gamma,\mathcal{O}_\Gamma(n)) \longrightarrow H^2(\mathcal{O}_X(n-1)) \longrightarrow H^2(\mathcal{O}_X(n)) \longrightarrow 0 .$$

For $n \geq 1$ we have that α is *onto* because Γ is a canonical embedded curve and hence is projectively normal (Noether-Enriques-Petri theorem), hence β is injective. Now assertion ii) follows by decreasing induction on n. Furthermore for $n > 1$ we have $H^1(\Gamma,\mathcal{O}_\Gamma(n)) = \check{H}^0(\Gamma,\mathcal{O}_\Gamma(1-n)) = 0$, this gives the first assertion of i) by decreasing induction and finally the second one follows since $\dim H^1(\Gamma,\mathcal{O}_\Gamma(1)) = 1$.

COROLLARY OF LEMMA 6.

X *is a* K3-*surface.*

PROOF.

By the adjunction formula we have

$$(K_X+\Gamma).\Gamma = K_\Gamma = \Gamma.\Gamma$$

hence $K_X.\Gamma = 0$, and hence by the equivalence criterium of Weil (cf. [Z1],p.120) we have $K_X = 0$ or X is a ruled surface. However $p_g(X) = 1$ by Lemma 6, hence X is not a ruled surface. Hence $K_X = 0$ and since, again by Lemma 6, $q(X) = 0$, we have X is a K3-surface.

RETURNING TO THE PROOF OF PROPOSITION 2 we apply again the adjunction formula, this time for X. We get

$$(K_V+X).X = K_X = 0 .$$

Hence $K_V + X = 0$ by Weil's equivalence criterium, hence $-K_V = X$ and this proves the proposition.

1.11. INDEX OF A FANO VARIETY

We have seen already (Corollary 3 of Lemma 1) that for a Fano variety V the $\mathrm{Pic}^0(V) = 0$ and $\mathrm{Pic}(V) = NS(V) = H^2(V,\mathbb{Z})$ (where $NS(-)$ is the Néron-Severi group). Write $-K_V = r.H$, with $H > 0$ and with r *maximal*. This $r = r(V)$ is called the *index* of V (cf. with the index of Del Pezzo surfaces, see 1.5). Note that H is *unique* up to torsion (and in fact unique as we shall see in a moment). Clearly H is *ample* and, similarly to Lemma 1 and 2, we have

LEMMA 7.

a) $h^i(-mH) = 0$, $i < 3$ *and* $m > 0$.

b) $h^i(mH) = 0$, $i > 0$ *and* $m \geq 1-r$ *(in particular for* $m \geq 0$ *).*

PROOF.

a) follows immediate from Kodaira vanishing theorem.

b) $h^i(mH) = h^{3-i}(K-mH) = h^{3-i}(-(m+r)H)$, hence $h^i(mH) = 0$ if $r+m \geq 1$, i.e. $m \geq 1-r$.

LEMMA 8.

$h^0(mH) = \frac{1}{12}\, m(m+r)(2m+r)H^3 + \frac{2m}{r} + 1$, *in particular:*

$h^0(H) = \frac{1}{12}\,(r+1)(r+2)H^3 + \frac{2}{r} + 1$.

PROOF.

Immediate by applying R.-R. to $D = H$ and using $K_V.C_2 = -24$ (Corollary 5 of Lemma 1).

1.12. CONSEQUENCES OF A THEOREM OF SHOKUROV

It may happen that the system $|H|$ has base points (see 1.15), nevertheless there is the important

THEOREM (SHOKUROV).

There exists a <u>*smooth and irreducible*</u> $H_0 \in |H|$.

We shall return to this theorem later but we start to draw conclusions from this fact.

COROLLARY 1.

For $r = 1$ H_0 is a K3-surface.

For $r > 1$ H_0 is a Del Pezzo surface with index $\geq (r-1)$.

PROOF.

For $r = 1$ we have seen this already (Lemma 3). For $r > 1$ we see from the adjunction formula

$$K_{H_0} = (K_V+H)H_0 = -(r-1)H.H_0 ,$$

hence H_0 is a Del Pezzo surface of index $\geq (r-1)$.

REMARK.

The idea is now: try to embed V by means of the linear system $|H|$, the fact that we have detailed knowledge about its hyperplane section (Del Pezzo or K3-surface) will yield information about V itself.

COROLLARY 2.

$1 \leq r \leq 4$.

PROOF.

Follows from 1.5 iv).

COROLLARY 3.

$\mathrm{Pic}(V)$ *is torsion free (and hence* H *unique).*

PROOF.

We have seen $\mathrm{Pic}(V) = H^2(V,\mathbb{Z})$ (Corollary 4 of Lemma 1); by Lefschetz $H^2(V,\mathbb{Z}) \hookrightarrow H^2(H,\mathbb{Z})$ and the latter is torsion free since H is Del Pezzo or K3-surface.

COROLLARY 4.

$\mathrm{Tr}_{H_0} |nH|$ *is complete* $(n \geq 1)$.

PROOF.

Use the exact sequence

$$0 \longrightarrow \mathcal{O}_V((n-1)H) \longrightarrow \mathcal{O}_V(nH) \longrightarrow \mathcal{O}_{H_0}(nH) \longrightarrow 0$$

and the fact that $H^1(V, \mathcal{O}_V((n-1)H) = 0$ if $n \geq 1$ (Lemma 6).

1.13. THE DEGREE OF A FANO VARIETY

Put

$$d = d(V) = H^3 = -\frac{K_V^3}{r^3} .$$

This integer is called the *degree* of V .

Now consider the rational map

(2) $$\lambda_{|H|}: V \dashrightarrow V' \subset \mathbb{P}_{\ell(H)-1} .$$

Question: What are the possible values of $\deg \lambda_{|H|}$? If $r = 1$ then we must (cf. Corollary 4 above) consider $\mathrm{Tr}_{H_0} |H|$ on the K3-surface H_0 . By *the theory of K3-surfaces* ([S.D],p.611 and 615) we have $\deg \lambda_{|H|}$ equals 1 or 2 or ∞ and the latter case happens only if $|H|$ has a *base locus*. If $r > 1$ then we have by the *theory of Del Pezzo surfaces* firstly that $\deg \lambda_{|H|} = \infty \Longleftrightarrow r = 2$ and $d = 1$ (see 1.5 vi)); furthermore again the only other possibilities are $\deg \lambda_{|H|}$ equals 1 or 2 (the latter case at most for $r = 2$ and $d = 2$). Also note that for $r > 1$ we have the relation

$$K_{H_0} . K_{H_0} = (r-1)^2 d$$

and hence, again by Del Pezzo theory (1.5 ii)) we have $1 \leq (r-1)^2 d \leq 9$, hence for $r > 1$ certainly $1 \le d \le 9$ (but in fact $1 \le d \le 7$, see later in §2).

1.14. Now we shall make a more precise analysis using the trivial:

LEMMA 9.

Let $X \subset \mathbb{P}_N$ *be an algebraic set spanning* $\mathbb{P}_N$ *. Then:*

(3) $$\deg X \ge \operatorname{codim} X+1 \ ;$$

PROOF.

Both sides do not change by taking hyperplane sections, so we can reduce to $\dim X = 0$ and then it is obvious.

Now apply this lemma to the variety V' from (2). Consider now the case $\deg \lambda_{|H|} \neq \infty \iff \dim V' = 3$; then using Lemma 8 and the fact that $H^3 = d$, the inequality $\deg V' \ge \operatorname{codim} V'+1$ takes the following form

(4) $$\frac{d}{\deg \lambda_{|H|}} = \deg V' \ge \frac{1}{12}(r+1)(r+2)d + \frac{2}{r} - 2 \ .$$

Note that both sides are integers. Using this, and examining the cases $r = 1,2,3$ or 4 separately we get the following result:

LEMMA 10.

i) $r = 4 \Rightarrow d = 1$, $\deg \lambda_{|H|} = 1$ *and* $V' = \mathbb{P}_3$.

ii) $r = 3 \Rightarrow d = 2$, $\deg \lambda_{|H|} = 1$ *and* $V' = V(2) \subset \mathbb{P}_4$.

iii) $r = 2 \Rightarrow 1 \le d \le 9$; *if* $d \neq 1,2$ *then* $\deg \lambda_{|H|} = 1$,
if $d = 2$, $\deg \lambda_{|H|} = 2$,
if $d = 1$, $\deg \lambda_{|H|} = \infty$.

iv) $r = 1 \Rightarrow \deg \lambda = 1,2$ *or* ∞ *(the latter case happens only if* $|H|$ *has a base locus).*

PROOF.

The cases i) and ii) follow immediately from (4). For $r = 2$ (4) takes the form

$$\frac{d}{\deg \lambda} \geq d-1$$

and the rest follows immediately from Del Pezzo theory. Finally the case $r = 1$ has been considered already in 1.13 (note that (4) now takes the form

$$\frac{d}{\deg \lambda} \geq \tfrac{1}{2}d .$$

1.15. REMARK ABOUT POSSIBLE BASE POINTS IN $|H|$

The linear system $|H|$ has *no fixed component* by the theorem of Shokurov (1.12). Take a smooth, irreducible $H_0 \in |H|$; since $Tr_{H_0}|H|$ is complete (Corollary 4 in 1.12) we have:

i) base curves in $|H| \leftrightarrow$ fixed components in $Tr_{H_0}|H|$.

ii) base points in $|H| \Leftrightarrow$ base points in $Tr_{H_0}|H|$.

Hence:

a) if $r \geq 2$ then (by the Del Pezzo theory): there is one and only one base point iff $r = 2$, $d = 1$,

b) if $r = 1$ then H_0 is a K3-surface and a linear system $|D|$ on a K3-surface has *no base points outside its fixed components* ([S.D],p.611); moreover if fixed components occur then there is detailed information on the shape of D (ibid,p.611). Using this Iskovskih obtains the following results ([I3],Theorem 6.3): If $r = 1$ and if $|H|$ has a base locus then $Tr_{H_0}|H| = |mE+Z|$ with Z a rational curve, and Z is the unique base locus of $|H|$. Moreover this can occur only if $m = 3$ or 4 and a precise description for these varieties is given.

1.16. PROJECTIVE NORMALITY

LEMMA 11.

Assume:

i) $|H|$ *has no base locus.*

ii) $\deg \lambda_{|H|} = 1$.

Then:

$$(5) \qquad S^nH^0(V,\mathcal{O}_V(H)) \twoheadrightarrow H^0(V,\mathcal{O}_V(nH)) \quad (n\geq 1) ,$$

where $S^nH^0(-)$ *are the terms of degree* n *in the symmetric algebra* $S^*H^0(-)$ *generated by* $H^0(-)$.

PROOF.

We proceed by induction on n .

STEP I. Reduction from V to H_0 .

Consider the exact sequence

$$0 \longrightarrow \mathcal{O}_V((n-1)H) \xrightarrow{\cdot H} \mathcal{O}_V(nH) \longrightarrow \mathcal{O}_{H_0}(nH) \longrightarrow 0 .$$

The step follows by considering the corresponding exact cohomology sequence and an easy diagram chase:

$$\begin{array}{ccccccc}
H^0(V,(n-1)H) & \xrightarrow{\cdot H} & H^0(V,nH) & \longrightarrow & H^0(H_0,\mathcal{O}_{H_0}(nH)) & \longrightarrow & H^1(V,(n-1)H) \\
\uparrow\!\!\uparrow & & \uparrow & & \uparrow\!\!\uparrow & & \| \\
S^{n-1}H^0(V,H) & & S^nH^0(V,H) & \twoheadrightarrow & S^nH^0(H_0,\mathcal{O}_{H_0}(H)) & & 0
\end{array}$$

Note that $Tr_{H_0}|H|$ is complete (Corollary 4 of 1.12), hence the surjectivity in the lower row.

STEP II. Reduction from H_0 to $C = H_0.H$.

Note that due to our assumptions C is a smooth irreducible curve. This step is now entirely similar to Step I.

STEP III. The case of curves.

r = 1 : Then C is a curve such that $Tr_C|H| = K_C$ (see Lemma 5). Moreover $\deg \lambda_{|H|} = 1$ implies that the restriction of $\lambda_{|H|}$ to C has also degree 1 and hence gives the canonical embedding. Using this embedding (and hence $|H|$ is now the hyperplane system on C), the assertion becomes $S^nH^0(C,\mathcal{O}_C(1)) \twoheadrightarrow H^0(C,\mathcal{O}_C(n))$ for the (now) cannonical embedded curve; this is true since by the Noether-Enriques-Petri theorem C is projectively normal.

r > 1 : Put $g = g(C)$. Using the adjunction formula for C on H_0 we get: $2g - 2 = \deg(K_C) = \deg\{(C+K_{H_0}).C\} = -(r-2)H^3$. Hence we have: $r = 2 \Rightarrow g = 1$ and $r > 2 \Rightarrow g = 0$. Moreover, put $D = C.H$, then $\deg(D) = H^3 = d$, hence (by our assumption that λ_H is birational) for $r = 2$ we have $\deg(D) \geq 3$ and for $r > 2$ we have $\deg(D) > 1$. The assertion follows now from the following lemma in Mumford's Varenna lectures of 1969:

LEMMA ([M],p.55):

If C *is a curve of genus* g *and* D *a divisor on* C *with* $\deg(D) \geq 2g + 1$, *then*

$$S^nH^0(C,\mathcal{O}_C(D)) \twoheadrightarrow H^0(C,\mathcal{O}_C(nD)) .$$

COROLLARY 1.

Let the assumptions be as in Lemma 11. *Then* $|H|$ *is* *very ample (and hence we can identify* $\lambda_{|H|}: V \tilde{\to} V' \subset \mathbb{P}_N$).

PROOF.

Again this follows from a lemma in Mumford's Varenna lectures:

LEMMA ([M],p.38).

Let D *be an* <u>*ample*</u> *divisor on a smooth variety* V *such that*

$$(*) \qquad S^n H^0(V,\mathcal{O}_V(D)) \twoheadrightarrow H^0(V,\mathcal{O}_V(nD)) \quad (\textit{all } n>0) .$$

Then D *is* <u>*very*</u> *ample.*

PROOF.

Look to the commutative diagram (and write for abbreviation $H^0(V,\mathcal{O}_V(D)) = \Gamma(D)$, etc):

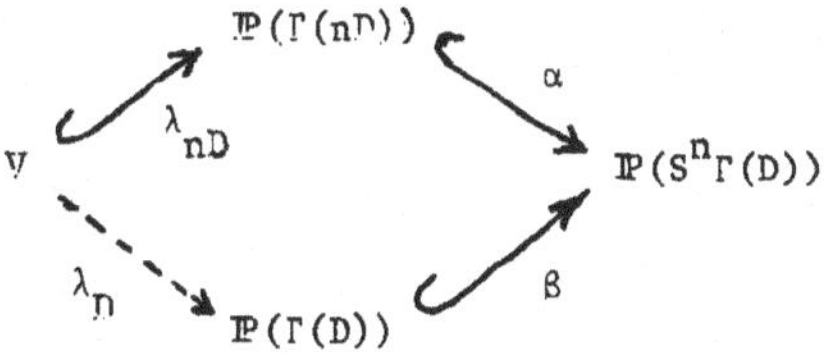

Now λ_{nD} is a closed immersion for some $n >> 0$, α is the closed immersion coming from the assumption (*), β is the standard Veronese immersion. Hence also λ_D must be a closed immersion.

Hence, for explicit reference, we have:

COROLLARY 2.

If $|H|$ *has no base points and* $\lambda_{|H|}$ *is birational then* $|H|$ *is* <u>*very*</u> *ample. This is in particular the case for* $r > 2$ *and* $r = 2$, $d > 2$.

COROLLARY 3.

Let the assumptions be as in Lemma 11. *Then* $\lambda_{|H|} : V \xrightarrow{\sim} V' \subset \mathbb{P}_N$ *with* V' <u>*projectively normal*</u>. *In particular: if* V *is embedded by means of* $|H|$ *then* V *is projectively normal.*

PROOF.

We have now $V \simeq V'$ by Corollary 2, hence V' is smooth and hence certainly normal; the projective normality follows now from (5).

1.17. EQUATIONS FOR V

For use later on we state a general lemma:

LEMMA 12.

Let X *be an irreducible variety, spanning* $\mathbb{P}_N$ *and with* $\dim X \geq 2$. *Let the hyperplanes of* $\mathbb{P}_N$ *cut out a complete system* $|H|$. *Let* $Y = X.H_0$ *for some hyperplane* H_0 . *Assume:*

$$(*) \qquad S^{m-1}H^0(X,\mathcal{O}_X(H)) \twoheadrightarrow H^0(X,\mathcal{O}_X((m-1)H)) \quad (m\geq 2) .$$

Let $I(X)$, *resp.* $I(Y)$, *be the ideals defined by* X *and* Y *respectively, and let* $F_m \in I_m(Y)$ *(the homogeneous part of degree* m *). Then there exists* $F_m^* \in I_m(X)$ *such that* $F_m^* \cap H_0 = F_m$. *Moreover in case* $m = 2$ *the* F_m^* *is uniquely determined by* F_m .

REMARK. Note that (*) certainly is fulfilled if X is projectively normal.

PROOF.

Look to the following commutative diagram with exact rows and columns in the way as indicated (note that the exactness of the first column is precisely the assumption(*)).

$$\begin{array}{ccccccccc}
 & & 0 & & 0 & & 0 & & \\
 & & \downarrow & & \downarrow & & \downarrow & & \\
0 & \longrightarrow & I_{m-1}(X) & \xrightarrow{.H_0} & I_m(X) & \xrightarrow{\alpha_m} & I_m(Y) & & \\
 & & \downarrow & & \downarrow & & \downarrow & & \\
0 & \longrightarrow & H^0(\mathbb{P}_N,\mathcal{O}(m-1)) & \xrightarrow{.H_0} & H^0(\mathbb{P}_N,\mathcal{O}(m)) & \longrightarrow & H^0(\mathbb{P}_{N-1},\mathcal{O}(m)) & \longrightarrow & 0 \\
 & & \downarrow & & \downarrow & & \downarrow & & \\
0 & \longrightarrow & H^0(X,\mathcal{O}(m-1)) & \xrightarrow{.H_0} & H^0(X,\mathcal{O}(m)) & \longrightarrow & H^0(Y,\mathcal{O}(m)) . & & \\
 & & \downarrow & & & & & & \\
 & & 0 & & & & & &
\end{array}$$

The required surjectivity of α_m follows by an easy diagram chase (cf. with the lemma of the snake). As to the uniqueness of F^*_m in case $m = 2$, this follows since by assumption $I_1(X) = 0$ since X spans $\mathbb{P}_N$.

COROLLARY 1.

Let the assumptions be as in Lemma 12 *with* (*) *fulfilled for all* $m \geq 2$ *(OK if* X *is projectively normal). Then: if* $I(Y)$ *is generated by quadrics, so is* $I(X)$. *Similarly if* $I(Y)$ *is generated by quadrics and cubics so is* $I(X)$.

PROOF.

If $F_{2,1},\dots,F_{2,s}$ are generators of $I(Y)$ of degree 2 (say); lift then, by Lemma 12, to $F^*_{2,j} \in I(X)$.

Now let $G^* \in I_m(X)$, we apply induction on m. Consider $G = G^*|H_0 \in Im(Y)$ and write $G = \sum_j A_j F_{2,j}$ where the A_j are forms of degree $(m-2)$. Lift A_j arbitrary to forms A^*_j of degree $(m-2)$ in $\mathbb{P}_N$, then we have $G^* - \sum_j A^*_j F^*_{2,j} = H_0.B^*$ with $B^* \in I_{m-1}(X)$, and apply the induction assumption.

We shall apply this lemma later on in the cases $r = 2$ and $r = 1$.

1.18. THEOREM OF SHOKUROV

Finally we want to look a little into the proof of this theorem which we mentioned and used already in 1.12.

THEOREM.

There exists a smooth and irreducible divisor $H_0 \in |H|$.

1.19. We shall only consider the case $r = 1$, i.e. the case $|H| = |-K_V|$ and we shall give only some indications of the-complicated-proof.

STEP 1:

If $D \in |-K_V|$, then D is *connected* because from the exact sequence $0 \to \mathcal{O}_V(-D) \to \mathcal{O}_V \to \mathcal{O}_D \to 0$ and the fact (Corollary 2 of Lemma 1) $H^1(V,\mathcal{O}_V) = 0$, follows $\dim H^0(D,\mathcal{O}_D) = 1$.

STEP 2:

LEMMA 13.

Let $\lambda_{|-K_V|}: V \dashrightarrow W \quad \mathbb{P}_{g+1}$ *be the rational map determined by* $|-K_V|$ *and* W *the image of* V. *Then* $\dim W > 1$.

PROOF ([S1],p.399,Lemma 3.2).

Suppose $\dim W = 1$. Write $|-K_V| = |D_0+D|$ with D_0 the *fixed* part and D the movable part. Since $q(V) = 0$ we have for the genus $g(W) = 0$ and W is a smooth rational curve of degree $(g+1)$ spanning $\mathbb{P}_{g+1}$ (i.e. a "rational normal" curve in classical terminology). Hence we have $D \sim (g+1)E$ with $|E|$ a *linear pencil*. Consider the rational map $\pi: V \dashrightarrow W \simeq \mathbb{P}_1$ determined by $|E|$. We have the following relation

$$(6) \qquad 2g - 2 = -K_V^3 = ((D_0+(g+1)E)^2.-K_V) =$$

$$= (g+1)^2(E^2.-K_V) + (g+1)(E.D_0.-K_V) + D_0.K_V^2 .$$

Due to the movability of E and the ampleness of $-K_V$, all terms are non-negative. If $(E^2.-K_V) > 0$ then we get $2g - 2 \geq (g+1)^2$ which is a contradiction, hence $E^2 = 0$ and $\pi: V \to W$ is a *morphism*. From Step 1 it follows now that $D_0 \neq 0$ and $D_0 \cap E \neq \emptyset$. Using the ampleness of $-K_V$ we see that the remaining two terms in (6) are strictly positive and we get the following

$$(7) \qquad (E.D_0.-K_V) = 1 , \quad (D_0.K_V^2) = g-3 > 0 , \quad E^2 = 0 , \quad (E.K_V^2) = 1 .$$

Next we claim: D_0 is a *reduced and irreducible* divisor. In order to see this first note that $(E.K_V^2) = 1$ implies (again using the ampleness of $-K_V$) that the fibers E of $\pi: V \to W$ are all reduced and irreducible, hence no component of D_0 is contained in a fibre and then $(E.D_0.-K_V) = 1$ implies that D_0 is reduced and irreducible.

Unfortunately the final part of the proof of [S1], Lemma 3.2 contains a gap as has been pointed out to me by Mori. Using *his theory of extremal rays* Mori suggests to complete the proof of the lemma as follows:

Since $-K_V$ is ample, V contains an *extremal rational curve* ℓ ([Mo], Theorem 1 and Corollary 2). So we have $(\ell.-K_V) \leq 4$ and ℓ spans an *extremal ray* $R = \mathbb{R}_+[\ell]$ (where [C] denotes the numerical equivalence class of a curve C). Moreover since the cone $NE(V)$ is spanned by such extremal rays (ibid, Corollary 2) we can choose ℓ such that $(\ell.E) \neq 0$ and then (since $|E|$ is free) we have $(\ell.E) > 0$. There exists ([Mo], Theorem 4) a morphism $\phi: V \to Y$ to a projection variety Y such that for an irreducible curve C on V we have that $[C] \in R \Longleftrightarrow \dim \phi(C) = 0$.

Using the relation $-K_V = D_0 + (g+1)E$, the fact that $g \geq 4$ (which follows from (7)) and $(\ell.E) > 0$, we get

$$4 \geq (\ell.-K_V) = (\ell.D_0) + (g+1)(\ell.E) \geq (\ell.D_0) + 5 ,$$

hence $(\ell.D_0) < 0$. Hence the extremal ray R is *not* numerical effective and we have (ibid, Theorem 5) $\dim Y = 3$ and there exists an irreducible divisor D such that $\phi: V - D \xrightarrow{\sim} Y - \phi(D)$ and $\dim \phi(D) \leq 1$. Moreover (always by ibid, Theorem 5) the following cases are possible:

a) $\phi(D)$ is a curve and D a $\mathbb{P}_1$-bundle over $\phi(D)$. Let ℓ_0 denote a fibre of D over $\phi(D)$, then $(\ell_0.D) = -1$. Moreover using the adjunction formula for K_D and the fact that on D we have $K_D = -2s +$ fibers (where s is a section) we easily find $(\ell_0.-K_V) = 1$.

b) $\phi(D) = Q$ is a non-singular point on Y, $D \simeq \mathbb{P}_2$ and $\mathcal{O}_D(D) \simeq \mathcal{O}_{\mathbb{P}_2}(-1)$. Take for ℓ_0 a "line" in D. Then we get $(\ell_0.D) = -1$ and, again using the adjunction formula, $(\ell_0.-K_V) = 2$.

c) $\phi(D) = Q$ ia a double point of Y, $D \simeq \mathbb{P}_1 \times \mathbb{P}_1$ and $\mathcal{O}_D(D) \simeq \mathcal{O}(-1,-1)$. Take $\ell_0 = a \times \mathbb{P}_1$. We have $(\ell_0.D) = -1$ and, always proceeding as above $(\ell_0.-K_V) = 1$.

d) $\phi(D) = Q$ is a double point of Y, $D \simeq$ an irreducible reduced singular quadric surface in $\mathbb{P}_3$ and $\mathcal{O}_D(D) \cong \mathcal{O}_{\mathbb{P}_3}(-1) \otimes \mathcal{O}_D$. Take for ℓ_0 a "line in D through the vertex"; we have $(\ell_0.D) = -1$ and $(\ell_0.-K_V) = 1$.

e) $\phi(D) = Q$ is a quadruple point of Y, $D \simeq \mathbb{P}_2$ and $\mathcal{O}_D(D) \simeq \mathcal{O}_{\mathbb{P}_2}(-2)$. Take for ℓ_0 a "line" in D, then $(\ell_0.D) = -2$ and $(\ell_0.-K_V) = 1$.

In all cases $[\ell_0] \in R$ by (ibid, Theorem 4). Replacing now the curve ℓ from above by ℓ_0, we get $(\ell_0.D_0) < 0$. However from the explicit description a),-,e) above we see that then necessarily $D_0 = D$. Finally, using this, and the results above we get

$$2 \geq (\ell_0.-K_V) = (\ell_0.D_0) + (g+1)(\ell_0.E) \geq$$

$$\geq -2 + (g+1) = g-1 .$$

Hence $g \leq 3$ contradicting $g \geq 4$ from (7). Hence $\dim W > 1$, proving Lemma 13 and completing Step 2.

STEP 3:

$|-K_V|$ has no fixed components and the general number $D \in |-K_V|$ is irreducible and has at most double points.

For the proof we refer now to the original paper [S1], p.400-403. Here we give only some indications. Write, as before in Step 2, $|-K_V| = |D_0+D|$ with D_0 fixed and now by Bertini the general D is irreducible. Now resolve the singular locus of such a general D by monoidal transformations:

$$\begin{array}{ccc} V & \xleftarrow{\ \sigma\ } & \widetilde{V} \\ \lambda \vdots & \swarrow & \\ W & & \end{array}$$

One has $\sigma^*(D) = \widetilde{D} + \Sigma n_i E_i$ ($\widetilde{D}$ strict transform, $n_i \geq 2$) with E_i the blown up centers, and also $-K_{\widetilde{V}} = \sigma^*(K_V) + \Sigma \delta_i E_i$ ($\delta_i = 1$ or 2 according as the center is a curve or a point). From this, and the adjunction formula one gets

$$(8) \qquad K_{\widetilde{D}} = -(\widetilde{D}.(\sigma^*(D_0)+\Sigma(n_i-\delta_i)E_i)) .$$

Since $n_i \geq \delta_i$ one has $K_{\widetilde{D}} \leq 0$. On $\widetilde{D}$ one studies two divisor (classes):

$$F = \widetilde{D}.\sigma^*(-K_V) \quad \text{and} \quad G = \widetilde{D}.(\widetilde{D}+\Sigma\delta_i E_i) .$$

Writing p_g (resp. q) for the geometric genus (resp. the irregularity) of $\widetilde{D}$, a study of this divisor classes leads, using Riemann-Roch for G and vanishing criteria for F , to the inequality

$$p_g - q - 1 \geq 0 .$$

Since $K_{\widetilde{D}} \leq 0$ this is only possible if $K_{\widetilde{D}} = 0$, $p_g = 1$, $q = 0$. Hence $\widetilde{D}$ is a K3-surface and $n_i = 2 = \delta_i$; i.e. there are only isolated singular points (since $\delta_i \neq 1$) and they are double points ($n_i = 2$) . Also $0 = K_{\widetilde{D}} = -\widetilde{D}.\sigma^*(D_0)$; projecting to V we get $D.D_0 = 0$, but then using connectedness (Step 1) we have $D_0 = 0$.

STEP 4.

In fact D is smooth.

This last step is done via a detailed study of linear systems on K3-surfaces.

1.20. THE CASE $r \geq 2$

For this we refer to the original paper [S1],p.403.

§2. FANO THREEFOLDS OF INDEX 2

2.1. INDEX ≥ 2

We have seen (Corollary 2 in 1.12) that for the index $r = r(V)$ we have $1 \leq r \leq 4$. In this lecture we consider the cases $r \geq 2$. In the following H always denotes the divisor class such that $-K_V = rH$ and we consider

$$\lambda_{|H|}: V \dashrightarrow V' \subset \mathbb{P}_N$$

with $N = h^0(H) - 1 = \frac{1}{12}(r+1)(r+2)H^3 + \frac{2}{r}$.

From Lemma 10 (in 1.14) and Corollary 2 and 3 in 1.16 we have:

$r = 4 \Rightarrow \lambda_{|H|}: V \tilde{\to} \mathbb{P}_3$,

$r = 3 \Rightarrow \lambda_{|H|}: V \tilde{\to} V(2) \subset \mathbb{P}_4$.

2.2. THEOREM 1.

Let V *be a Fano threefold (always smooth) with index* $r = 2$; *let* H *be the divisor class such that* $-K_V = 2H$. *Put* $d = H^3$. *Then we have*

i) $1 \leq d \leq 7$

ii) For $3 \leq d \leq 7$ *the* H *is* <u>*very*</u> *ample and* $\lambda_H: V \tilde{\to} V' \subset \mathbb{P}_{d+1}$ *(spanning it) with* $\deg V' = d$.

Moreover: V' *is projectively normal and for* $d \geq 4$ *the* V' *is an intersection of quadrics.*

iii) Conversely: if $W = W_3^d \subset \mathbb{P}_{d+1}$ *is smooth and projectively normal then* $3 \leq d \leq 8$. *For* $3 \leq d \leq 7$ *the* W *is a Fano threefold of index* 2, *embedded by means of* $|H|$. *For* $d = 8$ *the* W *is the Veronese image of* $\mathbb{P}_3$.

PROOF (some parts).

For $r = 2$ we have that a smooth $H_0 \in |H|$ is a Del Pezzo surface with $-K_{H_0} = H.H_0$ and the well-known relation $1 \leq K_{H_0}.K_{H_0} \leq 9$ (1.5 ii)) gives $1 \leq d \leq 9$.

Also we have seen (in 1.15) that for $r = 2$, $d > 1$ the $\lambda_{|H|}$ is a morphism which is birational as soon as $d > 2$. Hence for $r = 2$, $d \geq 3$ we have (Corollary 2 in 1.16) that $\lambda_{|H|}: V \xrightarrow{\sim} V' \subset \mathbb{P}_N$ with $N = d+1$ and $\deg(V') = d$. Moreover V' is projectively normal by Corollary 3 in 1.16.

In order to see that for $d \geq 4$ the V' is an intersection of quadrics it suffices, by Corollary 1 in 1.17, to see this for the curve section C of V'. C is the hyperplane section of the Del Pezzo surface H_0', with H_0' itself the hyperplane section of V'. Hence C is an elliptic curve of degree d, and it is well-known that, for $d \geq 4$, C is an intersection of quadrics (see [M],p.58). This completes the proof of ii).

In order to complete the proof of i) we still have to see that the cases $d = 8$ and $d = 9$ are impossible. For simplicity of notations we identify $V \xrightarrow{\sim} V'$.

For $d = 9$, take a general hyperplane L and consider $V.L = H_0$. H_0 is now a Del Pezzo surface of degree 9, hence the Veronese image of $\mathbb{P}_2$ by $\mathcal{O}_{\mathbb{P}_2}(3)$. By Lefschetz we have a *surjective* map $H_2(H_0,\mathbb{Z}) \to\to H_2(V,\mathbb{Z})$, and since $H_2(H_0,\mathbb{Z})$ is free of rank 1 and the image does not consists entirely out of torsion, this map is an *isomorphism*. Hence we have also $\mathrm{Pic}(V) \cong H^2(V,\mathbb{Z}) \xrightarrow{\sim} H^2(H_0,\mathbb{Z}) \cong \mathrm{Pic}(H_0)$. So both are of rank one; also note that the map is: $D \mapsto D.L$. Since $r = 2$ we see that the generator of $\mathrm{Pic}(V)$ is H $(=VL)$. However $H_0.L$ is *not* the generator of $\mathrm{Pic}(H_0)$ because, since H_0 is the Veronese image of $\mathbb{P}_2$ by $\mathcal{O}(3)$ we have $H_0.L = 3\ell$, where ℓ is the image of the lines in $\mathbb{P}_2$. Hence we have a contradiction and $d = 9$ is impossible.

Also $d = 8$ is impossible but we do not give the proof here (which is more difficult than for $d = 9$).

Also to iii), take hyperplane sections, then we get surfaces $S^d \subset \mathbb{P}_d$. It is well-known that such surfaces are rational ([N],p.366). Now $H^1(W,\mathcal{O}(m-1)) = 0$ (for $m>1$ use Kodaira Vanishing; for $m=1$ use $\mathrm{Pic}^0(W) \hookrightarrow \mathrm{Pic}^0(S) = 0$ since S is rational). Using this one gets

$H^0(W,\mathcal{O}(n)) \twoheadrightarrow H^0(S,\mathcal{O}_S(n))$ for all $n > 0$, hence S is, like W, projectively normal. However then S is a Del Pezzo surface embedded by $|-K_S|$ (1.5 v)). Hence by the adjunction formula

$$(K_W+S).L = K_S = -L.S$$

we get, using $\mathrm{Pic}(W) \hookrightarrow \mathrm{Pic}(S)$ (by the equivalence criterion of Weil, [Z1], p. 120), that $K_W = -2L.W$ (L always a hyperplane section). Hence W is Fano, of index $r \geq 2$. Let $L.W = sD$ (maximal s), then $r = 2s$. So only $r = 4$ or $r = 2$; if $r = 4$ then $W \simeq \mathbb{P}_3$ embedded by means of $\mathcal{O}(2)$ and this occurs iff $\deg(W) = 8$, i.e. only for $W^8 \subset \mathbb{P}_9$. Hence we have $3 \leq d \leq 8$ and for $3 \leq d \leq 7$ the W is a Fano of index 2 and $|H| = |L.W|$.

2.3. THEOREM 2.

Let V *be a Fano threefold of index 2, embedded by means of* $|H|$ *, and let* $d = H^3$. *Then:*

$d = 3 \Rightarrow V = V(3) \subset \mathbb{P}_4$;

$d = 4 \Rightarrow V = V(2,2) \subset \mathbb{P}_5$;

$d = 5 \Rightarrow V \simeq Gr(1,4).L_1.L_2.L_3 \subset \mathbb{P}_6$ *(with* L_i *hyperplanes in the ambient space);*

$d = 6 \Rightarrow V \simeq \mathbb{P}_1 \times \mathbb{P}_1 \times \mathbb{P}_1 \subset \mathbb{P}_7$ *via the Segre embedding or* $V \simeq L.S(\mathbb{P}_2 \times \mathbb{P}_2)$ *where* S *is again the Segre embedding (see examples* vii) *of* 1.6 *)*);*

$d = 7 \Rightarrow V = V^7 \subset \mathbb{P}_8$ *obtained by projecting the Veronese image of* $\mathbb{P}_3$ *from one of its points.*

Conversely, varieties of the above type are Fano 3-folds of index 2, *embedded via* $|H|$, *and of the corresponding degree* d .

*) This last case is missing in the list [I3], see [Se].

PROOF (only indications of some cases).

We have already seen in Theorem 1 that Fano's of index 2, of degree $3 \leq d \leq 7$, and embedded via $|H|$ give varieties $V^d \subset \mathbb{P}_{d+1}$.

The case $d = 3$ is then immediate, the case $d = 4$ follows from Theorem 1 ii).

Consider now the case $d = 5$. First a general remark.

2.4. REMARK.

A Fano 3-fold of index 2 and $d \geq 3$ contains lines because the smooth surface $H_0 \in |H|$ is Del Pezzo surface of degree $d \geq 3$ and hence H_0 contains lines. Using deformation theory it is not difficult to see that (always $d \geq 3$) the family of lines has dimension 2, covers V and for the normal sheaf there are only two possibilities:

a) $N_{\ell|V} \simeq \mathcal{O}_\ell \oplus \mathcal{O}_\ell$

b) $N_{\ell|V} \simeq \mathcal{O}_\ell(-1) \oplus \mathcal{O}_\ell(1)$

and a) is the "general" case.

2.5. THE CASE $d = 5$

Take a lim $Z \subset V$ and such that $N_{Z|V} \simeq \mathcal{O}_Z \oplus \mathcal{O}_Z$. Now *project* V *from* Z:

$$\pi_Z \colon V \dashrightarrow W \subset \mathbb{P}_4 .$$

Blow up V along Z, write $Z' = \sigma^{-1}(Z) \subset V' = B_Z(V)$. We have now a diagram:

(*)

$$\begin{array}{ccccc}
Z' & \hookrightarrow & V' = B_Z(V) & & \\
\downarrow \sigma & & \downarrow \sigma & \searrow^{\phi_Z} & \\
Z & \longrightarrow & V & \overset{}{\underset{\pi_Z}{\dashrightarrow}} & W \subset \mathbb{P}_4
\end{array}$$

Write $H' = \sigma^*(H)$ for the total transform; clearly $\phi_Z = \pi_Z \cdot \sigma$ is given by the linear system $|H'-Z'|$. Also $Z' = \mathbb{P}(\check{N}_{Z|V}) = \mathbb{P}_1 \times \mathbb{P}_1$.

STEP 1: ϕ_Z is a *birational* morphism and W is a quadric. One computes (via standard formulas by blowing-ups) $(H'-Z')^3 = 2$; now $(H'-Z')^3 =$ $= \deg W . \deg(\phi_Z)$ and since $\deg(W) = 1$ is impossible (W spans $\mathbb{P}_4$ because $\dim H^0(V', \mathcal{O}_{V'}(H'-Z')) = \dim H^0(V, \mathcal{O}_V(1)) - 2 = 5$), one gets $\deg W = 2$ and $\deg(\phi_Z) = 1$.

STEP 2: $\phi_Z|Z' : Z' \xrightarrow{\sim} Q \subset W$, where Q is a smooth quadric in $\mathbb{P}_3$.

For this one computes $\mathrm{Tr}_{Z'}|H'-Z'|$; this turns out to be the system $|f_1+f_2|$ on $\mathbb{P}_1 \times \mathbb{P}_2$, where $f_1 = a \times \mathbb{P}_1$ and $f_2 = \mathbb{P}_1 \times b$.

STEP 3: Consider the surface E on V which is the union of all the lines $\ell \subset V$ *meeting* Z, and let E' be the strict transform in V'. E' (or E) is *contracted* by ϕ_Z (resp. by π_Z) into a curve $\phi_Z(E') = Y \subset Q$. Now one proves (again using, among other things, Del Pezzo theory) that Y is a *smooth rational curve of degree* 3.

So starting with V we have now a triple $(W \supset Q \supset Y)$ in $\mathbb{P}_4$, and in fact Q is determined by Y because it is the hyperplane section of W spanned by Y. Moreover one shows that W is *smooth*.

STEP 4: *The couple* (W,Y) *determines* V, as follows:
Consider the linear system $|\mathcal{O}_W(2)-Y|$; let V'' be the blow up of W along Y and let Q'' be the strict transform of Q, then it turns out that Q'' can be contracted into a line Z and in this way one obtains V :

$$\begin{array}{ccc} V' & & V'' = B_Y(W) \supset Q'' \\ \sigma \downarrow & \overset{\psi}{\swarrow} & \downarrow \\ V & \dashrightarrow & W \supset Q \end{array} .$$

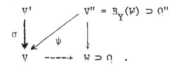

STEP 5: If V_1 and V_2 are two Fano's of the considered type then one can show that the corresponding couples (W_1,Y_1) and (W_2,Y_2) are projectively equivalent, and hence also V_1 and V_2 are isomorphic.

STEP 6: $V = Gr(1,4)L_1.L_2.L_3$ is of the required type.

2.6. For the cases $d = 6$ and 7 see ([I1],§6). However we want to remark that for $d = 6$ one proceeds very similarly as for $d = 5$. Projecting from a line Z one gets again a diagram (*) and a variety $W \subset \mathbb{P}_5$. Now V is determined via W (as above) and for determining W the crucial fact is that we have the *equality* (cf. 1.14):

(3') $\qquad \deg W = \operatorname{codim} W + 1$

and for such varieties there is the classical:

2.7 THEOREM (Enriques) ([S.D],p.607;[I1],2.8):

If $W \subset \mathbb{P}_N$ spans $\mathbb{P}_N$ and satisfies (3') then W is one of the following varieties:

1) $\mathbb{P}_N$ itself,
2) a quadric in $\mathbb{P}_N$,
3) a rational scroll,
4) a cone over a rational scroll,
5) the Veronese surface in $\mathbb{P}_5$,
6) a cone over the Veronese surface,
7) a cone over a rational normal curve.

2.8. Finally we mention also the cases $r = 2$, $d = 2$ or $d = 1$.

$\underline{d = 2}$: $\lambda_{|H|}: V \to \mathbb{P}_3$ is a double covering with ramification locus a smooth quartic surface D ; conversely such a covering is such a Fano.

These varieties have been studied now extensively by Clemens (see his lectures), Welters and Tikhomirov.

<u>d = 1</u> : $\lambda_{|H|}: V \dashrightarrow \mathbb{P}_2$, with one point of indeterminancy; the fibres are elliptic curves. Moreover $\lambda_{|-K_V|}: V \longrightarrow W \subset \mathbb{P}_6$ is a double cover over W , where W is a cone over the Veronese surface in $\mathbb{P}_5$, the ramification locus D is cut out on W by a cubic hypersurface not passing through the vertex of W ; conversely, such a covering is such a Fano.

§3. ON THE EXISTENCE OF LINES OF FANO THREEFOLDS

3.1. INTRODUCTION

In the classification theory of Fano 3-folds a crucial rôle is played by the lines on the threefold. The lack of a proof for the existence of lines by Fano is one of the most unsatisfactory parts of his theory. Also Iskovskih was unable to overcome this difficulty and stated explicitely the existence as a hypothesis (the so-called "Fano-conjecture"). It was only in 1979 that Shokurov succeeded in proving the existence of lines [S2]. Shokurov's proof has been discussed and at some points modified by Miles Reid [R]. Our lecture is basèd upon this Mittag-Leffler Report of Reid. We can give only some indications of the long and complicated proof.

3.2. SOME DEFINITIONS

As always V denotes a smooth Fano 3-fold.

DEFINITION. An irreducible curve $C \subset V$ is called a *quasi-line* (or a "line on V") if the intersection number $(C.-K_V) = 1$. Similarly C is called a *quasi-conic* (or a "conic on V") if $(C.-K_V) = 2$.

Note that a quasi-line is not the same thing as a "usual" line (or "straight" line). For instance a cubic 3-fold $V(3) \subset \mathbb{P}_4$ contains ∞^2 lines, but, since $-K_V = 2H$, it does not contain quasi-lines. Clearly a V of index $r \geq 2$ never contains quasi-lines.

On the other hand if V is embedded by means of $|-K_V|$ then the notion quasi-line coincides with the notion of straight line (simlarly for conics).

3.3. THEOREM (Shokurov).

If V *is a smooth Fano 3-fold then*

either i) V *contains a quasi-line,*

or ii) index $(V) \geq 2$,

or *iii)* $V \simeq \mathbb{P}_1 \times \mathbb{P}_2$.

3.4. COROLLARY ("Fano-conjecture")

Let V *be a smooth Fano* 3*-fold of index* 1 *embedded by means of* $|-K_V|$ *and not isomorphic to* $\mathbb{P}_1 \times \mathbb{P}_2$. *Then* V *contains a (usual) line.*

3.5. REMARK.

It is an easy excercise (left to the reader) to verify that $\mathbb{P}_1 \times \mathbb{P}_2$ has index 1 but contains no quasi-lines.

3.6. ABOUT THE PROOF

Assume that the theorem is *not* true. Take *among the Fano* 3*-folds contradicting the theorem a threefold* V with *minimal* $\rho(V) = \mathrm{rk}\ \mathrm{Pic}(V)$. Now we consider *two cases:*

3.7. CASE A: Assume there exist *positive* divisors D_1, D_2 such that $D_1 + D_2 \in |-K_V|$ (we say shortly: $|-K_V|$ *has splitting).*

The proof of this case is based upon

KEY LEMMA I.

Let V *be a Fano* 3*-fold. Assume there exist* $D_1 > 0$, $D_2 > 0$ *such that* $D_1 + D_2 \in |-K_V|$. *Then* V *satisfies the theorem or* V *contains a "Veronese surface".*

DEFINITION.

A surface $S \subset V$ is called a "*Veronese surface in* V " if:

a) $S \simeq \mathbb{P}_2$

b) $\mathcal{O}_S(-K_V) \simeq \mathcal{O}_{\mathbb{P}_2}(2)$.

REMARKS.

1) If V is embedded via $|-K_V|$ then the notion Veronese surface in V coincides with the usual notion of Veronese surface (in $\mathbb{P}_5$).

2) If $S \subset V$ is a Veronese surface then it follows from the adjunction formula that $\mathcal{O}_S(S) \simeq \mathcal{O}_{\mathbb{P}_2}(-1)$.

3.8. KEY LEMMA I $\Rightarrow$ THEOREM IN CASE A

In order to see this one has only to take care of the case that the Fano threefold V , contradicting the theorem and *among such with minimal* $\rho(V)$, contains a Veronese surface S . Since $\mathcal{O}_S(S) \simeq \mathcal{O}_{\mathbb{P}_2}(-1)$ one can contract S to a point (via $|-K_V+2S|$) :

$$\begin{array}{ccc} S & \subset & V \\ \downarrow & & \downarrow \sigma \\ P_1 & \in & V_1 \end{array} .$$

Using the relation $K_V = \sigma^*(K_{V_1}) + 2S$ one shows, via Nakai, that $-K_{V_1}$ is again ample, i.e. V_1 again Fano. Also it is not difficult to check that V_1 *contradicts again* the theorem. (Note: one has to take care that the index may change under σ however the parity of the index is preserved and so one has, concerning this point, only to worry about V_1 with index 3, but then one easily produces a quasi-line on V). However now we have $\rho(V_1) = \rho(V) - 1$ contradicting the choice of V (see beginning of 3.6). Hence Key Lemma 1 $\Rightarrow$ Theorem in Case A .

3.9. ABOUT THE PROOF OF KEY LEMMA I

First of all one remarks that one *can make some extra assumptions* at the cost of omitting a "number of well-known cases". Therefore we make:

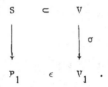

EXTRA ASSUMPTION 1: $|-K_V|$ is *very* ample.

REMARKS.

1) We use now $|-K_V|$ for the embedding, hence $V^{2g-2} \subset \mathbb{P}_{g+1}$, and also quasi-lines are now usual lines so we speak from now on simply about lines.

2) We have lost now the following types:

a) $|-K_V|$ with base points,

b) the "hyperelliptic cases", i.e. those V's for which $\lambda_{|-K_V|}: V \to V'$ is a morphism of degree 2.

Both types are "well-known" (cf. 1.15 and [I3],Theorem 2.2).

EXTRA ASSUMPTION 2: V is an intersection of quadrics.

REMARK.

Again we are loosing well-known cases:

a) for index 2 some cases, see 2.2.

b) for index 1 the "trigonal" ones (see 4.3) and some complete intersections (see 4.2).

3.10. LEMMA 1.

Suppose V *is a Fano 3-fold (always smooth) with a* $|-K_V|$ *which has splitting (i.e. case* A, *see* 3.7). *Assume moreover that* V *satisfies the extra assumptions* 1 *and* 2 . *Then* V *is among the following types:*

1) V *contains a free (= no fixed components, no base points) linear pencil of Del Pezzo surfaces.*

2) V *contains a* $S_2^{n-1} \subset \mathbb{P}_n$ *(a surface of degree* (n-1) *spanning* $\mathbb{P}_n$ *).*

3) V *is a conic bundle:* $V \to \mathbb{P}_2$.

4) the index $r(V) \geq 2$.

5) V *is the blow up of* $\mathbb{P}_3$ *or of* $V(2) \subset \mathbb{P}_4$ *in a smooth (possibly reducible) curve.*

3.11. LEMMA 1 ➡ KEY LEMMA I (almost!)

Consider the cases 1,...,5 separately:

1) $V \supset S$ a Del Pezzo, general member of the free pencil. We have $K_S = K_V.S$, hence S is embedded by means of $|-K_S|$ (V being embedded by $|-K_V|$). So we have lines on the Del Pezzo surfaces except in the cases $K_S^2 = 8$ or 9 which have to be treated separately (see 1.5 viii)).

2) Here we are in the case $\deg S = \operatorname{codim} S+1$, so the theorem of Enriques (see 2.7) can be applied and this suffices.

3) Here we get lines from the degenerated conics. Hence only the case that all conics are smooth has to be considered separately.

4) Clearly OK .

5) This very explicite case is easy to handle.

We shall not go into the "still to be treated special cases" in 1) and 3).

3.12. ABOUT THE PROOF OF LEMMA 1

Now we come to the hard work of proving Lemma 1. We shall not do this, but - in order to give some feeling about the intricacy fo the proof - we shall do *one* step! First again some remarks.

REMARK 1. If $|D|$ is a *free linear pencil* on V then $|D|$ is a pencil of Del Pezzo surfaces. For: $K_D = (K_V+D).D = K_V.D$, hence $-K_D$ is ample.

REMARK 2. We have now always the embedding via $|-K_V|$ and use the notations

$$V \supset X = V.H \supset \Gamma = V.H.H'$$

for the general hyperplane section and the general codim. 2 section.

Suppose now that $|D|$ is a linear system of divisors on V such that:

a) there are no fixed components,

b) $\operatorname{Tr}_X|D|$ is complete.

Then $|D|$ is (also) base point free.

For: if $P \in V$, take a general $X_0 = V.H_0$ through P (hence avoiding possible base curves of $|D|$), then it follows from the theory of K3-surfaces that $Tr_{X_0}|D|$ has no base points.

REMARK 3. By standard exact sequence arguments we see that $Tr_X|D|$ is complete if $H^1(\mathcal{O}_V(D-X)) = 0$.

3.13. Let V now be as in Lemma 1. Now if $D_1 + D_2 \in |-K_V|$ with $D_i > 0$ $(i=1,2)$, the (very) rough idea is to study $|D_1|$ and $|D_2|$ and to try to break it into smaller linear systems and to end up with a free pencil.

STEP 1. Suppose V has *no* free pencil. Let $H_0 \in |-K_V|$ and $A+B = H_0$ with $A > 0$, $B > 0$, A reduced and irreducible and of *minimal* degree among the components of H_0 . Then: either $|A|$ is free, or $A = S_2^{n-1} \subset \mathbb{P}_n$ $(n \leq g = g(\Gamma))$.

PROOF.

By remarks 2 and 3 above we are done if $H^1(\mathcal{O}_V(-B)) = 0$, and using the exact sequence $0 \to \mathcal{O}_V(-B) \to \mathcal{O}_V \to \mathcal{O}_B \to 0$ and $H^1(\mathcal{O}_V) = 0$, this is OK if B is connected. Therefore we start to study $|B|$. Suppose first that $|B|$ has a *fixed component* B_0 . Applying the argument the other way around, since A is irreducible we have $H^1(\mathcal{O}_V(-A)) = 0$, and hence $Tr_X|B|$ complete and similarly $Tr_\Gamma|B|$ complete. In our situation $Tr_\Gamma|B| = |K_\Gamma - A.\Gamma|$ (because $K_\Gamma = Tr_\Gamma|-K_V|$, see 1.10). Now we use the following

FACT ABOUT CURVES ([S2], 3.23 and [R],p.10):

If C is a canonically embedded curve, of genus g , and D a special divisor on C (i.e. $K_C-D>0$). Then: deg (fixed part of $|K_C-D|$) $\leq$ deg D . Moreover we have *equality* iff $\deg(D) = g-1$ and $h^0(\mathcal{O}_C(D)) = 1$.

We apply this in our situation to $C = \Gamma$ and $D = Tr_\Gamma|A|$. Now

$|K_\Gamma - A.\Gamma| = Tr_\Gamma|B|$ has (by our assumption) certainly the fixed component $B_0.\Gamma$. By the minimality of $\deg(A)$ in H_0 we are in the case of *equality*. Hence $\deg A = \deg(A.\Gamma) = g-1$ hence $A = A_2^{g-1} \subset \mathbb{P}_g = \mathbb{P}_{g+1}.H$.

Therefore in order to prove Step 1 we can assume that $|B|$ has no fixed component. Then by Remarks 2 and 3 and the fact that $H^1(\mathcal{O}_V(-A)) = 0$ (see above) we get that $Tr_X|B|$ is complete and that B has no *base points* (Remark 2). Now consider the *morphism* $\lambda_{|B|}: V \to W$. Now we can assume that $|B|$ is *not composed* with a pencil, for then (cf. with 1.18 Step 2) we would have W a rational curve (for $q(V)=0$) and then the fibers of $\lambda_{|B|}$ should give a *free pencil* contrary to our assumption. However if $|B|$ is not composed with a pencil then a general member of $|B|$ is irreducible, hence all members connected and - as we have seen above - this suffices to prove that $|A|$ is free, which completes Step 1.

3.14. Several more steps are needed for the proof of Lemma I but we refer to [R],p.8-17.

3.14 CASE B. Assume $|-K_V|$ has *no splitting* (see 3.7).

Again we can make some extra assumptions (cf. with 3.9) namely we assume now the following:

1) $|-K_V|$ very ample and V embedded via $|-K_V|$ (hence $V=V^{2g-2} \subset \mathbb{P}_{g+1}$) ,
2) $|-K_V|$ has no splitting (i.e. we are in Case B) ,
3) V is an intersection of quadrics.

Now we need:

KEY LEMMA II.

Let V *satisfy* 1, 2 *and* 3 *from above. Let* $P \in V$ *not lying on a line or a conic. Let* $\sigma: W = B_P(V) \to V$ *be the blow up of* V *in* P *and* $S = \sigma^{-1}(P)$. *Then:* $|-K_W|$ *is again very ample.*

KEY LEMMA III.

Same assumptions about V . *Let* C *be a conic on* V *meeting no lines such that the normal bundle of* C *in* V *is of type* $N_{C|V} \simeq \mathcal{O}_{\mathbb{P}_1} \oplus \mathcal{O}_{\mathbb{P}_1}$ *or* $N_{C|V} \simeq \mathcal{O}_{\mathbb{P}_1}(1) + \mathcal{O}_{\mathbb{P}_1}(-1)$. *Let* $\sigma\colon W = B_C(V) \to V$ *be the blow up of* V *along* C *and* $S = \sigma^{-1}(C)$. *Then:* $|-K_W|$ *is again very ample.*

The proof of these two lemmas (KL II and KL III) is the most delicate part of the entire proof; we cannot enter into them and refer to [R],p.25-38.

3.15. THE SURFACE S IN KL II AND KL III.

In KL II we clearly have $S \simeq \mathbb{P}_2$ and one easily sees that $\mathcal{O}_S(S) \simeq \mathcal{O}_{\mathbb{P}_2}(-1)$ hence, since $K_W = \sigma^*(K_V) + 2S$, we get $\mathcal{O}_S(K_W) \simeq \mathcal{O}_{\mathbb{P}_2}(-2)$ and S is a Veronese surface in W (see 3.7).

For the case of KL III recall that the standard surface $\mathbb{F}_n = \mathbb{P}_{\mathbb{P}_1}(\mathcal{O}\oplus\mathcal{O}(-n))$ has $\mathrm{Pic}(\mathbb{F}_n) = \mathbb{Z}b \oplus \mathbb{Z}f$ with b the unique section with $b^2 = n$ and f the fibre, and the linear system $|(n+r)f+b|$ embeds $\mathbb{F}_n$ in $\mathbb{P}_{n+2r+1}$ as a surface of degree (n+2r) .

DEFINITION. A surface $S \subset V$ is called a *rational scroll of type* $\mathbb{F}_{n,r}$ *in* V *if:*

a) $\mathbb{F}_{n,r} \simeq \mathbb{F}_n$

b) $\mathcal{O}_S(-K_V) \cong \mathcal{O}_{\mathbb{F}_n}((n+r)f+b)$.

Now in KL III in case $N_{C|V} \simeq \mathcal{O} + \mathcal{O}$ we clearly have $S \simeq \mathbb{F}_0$ and an (easy) computation shows that S is of type $\mathbb{F}_{0,2}$ in W . On the other hand if $N_{C|V} \simeq \mathcal{O}(1) \oplus \mathcal{O}(-1)$ we clearly have $S \simeq \mathbb{F}_2$ and we find that S is of type $\mathbb{F}_{2,1}$ in W .

We refer to both types together as *quartic scrolls* in V .

3.16 KEY LEMMA IV.

Let W *be a Fano 3-fold (always smooth) with very ample* $|-K_W|$ *and containing a surface* S *which is* *Veronese* *or a* *quartic scroll*. *Then there exist positive divisors* D_1, D_2 *such that* $S \not\subset D_i$ $(i=1,2)$ *and* $D_1 + D_2 + nS \in |-K_W|$ $(n \in \mathbb{Z})$ *(splitting modulo* S *).*

3.17. KL II + III + IV $\to$ THEOREM IN CASE B

PROOF.

Assume that V does not satisfy the theorem and that we are in Case B. Moreover we can make the supplementary assumptions 1) and 3) of 3.14 (and we have also 2)). In particular there are no lines on V. In case V is *not* covered by conics we can apply KL II. In case V *is covered by conics* we can apply KL III because it is well known ([12],§4) that for $N_{C|V}$ there are only the possibilities $\mathcal{O} \oplus \mathcal{O}$, $\mathcal{O}(1) \oplus \mathcal{O}(-1)$, $\mathcal{O}(2) \oplus \mathcal{O}(-2)$ and $\mathcal{O}(4) \oplus \mathcal{O}(-4)$, however conics of the two last types *do not cover* V. Therefore by blowing up we get a 3-fold W satisfying (see 3.15) the assumptions of KL IV. Therefore we have divisors D_i on W $(i=1,2)$ with $D_i > 0$, $S \not\subset D_i$ and such that $D_i + D_2 + nS \in |-K_W|$ $(n \in \mathbb{Z})$. Now $|-K_W| = |\sigma^*(-K_V)-2S|$ or $|-K_W| = |\sigma^*(-K_V)-S|$. Therefore applying the map $\sigma_*: \mathrm{Pic}(W) \to \mathrm{Pic}(V)$ we get $\sigma_*(D_1) + \sigma_*(D_2) \in |-K_V|$, contradicting our assumption that we are in Case B, i.e. that $|-K_V|$ has no splitting.

3.18. ABOUT THE PROOF OF KL IV

The proof of KL IV is easily reduced to

KEY LEMMA IV'

Same assumptions as in KL IV. *Assume moreover that there is no splitting* modulo S. *Then the linear system* $|E| = |-K_W-2S|$ *is free and* $\dim \lambda_{|E|}(W) = 3$.

3.19. IV' $\Rightarrow$ IV: The assumption that there is no splitting modulo S leads to a contradiction in the following way. The conclusion of IV' precisely guarantees that we can apply the Ramanujam vanishing theorem ([Ra],p.49,Theorem 3) to $|E|$ and hence $H^i(W,\mathcal{O}_W(-E)) = 0$, $0 \leq i < 3$. Therefore, taking a general member $E \in |E|$ we get from the exact sequence $0 \to \mathcal{O}_W(-E) \to \mathcal{O}_W \to \mathcal{O}_E \to 0$ that $H^1(E,\mathcal{O}_E) = 0$ and hence $\chi(E,\mathcal{O}_E) > 0$. On the other hand using that S is Veronese or a quartic scroll we get by an easy computation (proceeding in two steps: first looking to $D \in |-K_W-S|$, next to $E \in |D-S|$) that $\chi(E,\mathcal{O}_E) = -2$ (resp. -1).

3.20. ABOUT THE PROOF OF IV'

This is again difficult. As could be expected one studies first $D = |-K_W-S|$, next $|E| = |D-S|$. We must refer to [R],p.18-24. We *mention only the major steps*. One shows:

1) $\dim|D| \geq 1$,

2) $|D|$ is free and all members irreducible,

3) $\dim \lambda_{|D|}(W) = 3$,

4) λ_D birational and $g \geq 16$ (resp. 15).

Next turning to E , let $Y \in |D|$ be a general member. Then:

5) $\dim|E| \geq 1$,

6) the members of $Tr_Y|D|$ are numerically 2-connected,

7) through every point $P \in V$ there passes a non-singular element of $|D|$.

§4. FANO THREEFOLDS OF INDEX ONE AND PICARD NUMBER ONE

4.1. ASSUMPTIONS

As always V denotes a smooth Fano threefold. We assume now throughout the rest of the lectures that:

a) index $r = r(V) = 1$,

b) V is of the *principal series*, i.e. $-K_V$ is *very* ample (cf. 1.10),

c) V is embedded by means of $|-K_V|$; so we have (cf. 1.10):

$$V = V^{2g-2} \subset \mathbb{P}_{g+1} \quad (g\geq 3) \ . \tag{1}$$

As a matter of notation we use $|H|$ for the hyperplane system in the ambient projective space and also, by abuse of language, for the restriction to V itself (i.e. then $|H| = |-K_V|$).

Let $V \supset X = V.H \supset \Gamma = V.H.H'$, with X (resp. Γ) a general hyperplane section (i.e. a general codim. 2 section). So X is a K3-surface, and Γ a canonically embedded curve of genus g .

REMARK.

In making the supplementary assumption b) we have lost:

i) those V for which $|-K_V|$ has base points; see 1.5,

ii) those V with free $|-K_V|$ but for which $\lambda = \lambda_{|-K_V|}$ is *not* birational; then (see 1.14 Lemma 10) λ is a *morphism of degree* 2 and these are the so-called *hyperelliptic* Fano's. For a description of those see [I1],§7.

4.2. EQUATIONS FOR V

LEMMA 1.

Let V *be as above. Then:*

i) V is projectively normal,

ii) If $g > 3$ *then* V *is an intersection of cubics and quadrics,*

iii) If $g > 4$ *then* V *is an intersection of quadrics provided* Γ *is* <u>*non-*</u>

trigonal (recall that a curve is trigonal if it has a 1-dimensional linear system g_3^1 *of degree* 3*).*

PROOF.

We have seen already i) as Corollary 3 of Lemma 11 in 1.16. In fact it was a consequence, via the corresponding statement for Γ , of the Noether-Enriques-Petri theorem for the canonical curve Γ . Using Corollary 1 of 1.17 the same is true for ii) and iii), except that for iii) we have to rule out also the case of a Γ with $g(\Gamma) = 6$ and isomorphic to a plane curve of degree 5. However such a Γ does not occur as linear section on V because it does not occur on the K3-surface X ([SD],p.633).

COROLLARY 1.

$g = 3 \iff V = V(4) \subset \mathbb{P}_4$

$g = 4 \iff V = V(3,2) \subset \mathbb{P}_5$

$g = 5 \iff V = V(2,2,2) \subset \mathbb{P}_6$.

4.3. ABOUT THE TRIGONAL ONES

V is called *trigonal* if Γ is trigonal. Consider all the quadrics through V ; they determine a variety W . By considering W.H.H' we are again back in the situation of Γ and the Noether-Enriques-Petri theorem gives a description of the surface W.H.H' . Using this we can get information over W and finally over V itself. It turns out, except for $g = 3$ and 4 where all Γ are trigonal, that trigonal V occurs only for $g = 6$, 7 and 10 and such V can be described ([I2], 2). Moreover one can show that for such V we have $Pic(V) = \mathbb{Z} \oplus \mathbb{Z}$.

4.4. RESTRICTION TO $\rho(V) = 1$. LINES ON V .

From now on we shall restrict to the case the *Picard number* $\rho(V) = rk\ Pic(V) = 1$. The classification theorem of Iskovskih (and Fano) deals only with the case. Very recently Mori and Mukai have classified the

Fano's with $r(V) = 1$ and $\rho(V) > 1$.

If $\rho(V) = 1$ then, since $r = 1$, we have $\mathrm{Pic}(V) = \mathbb{Z}$ with *generator* $-K_V = H$.

By the theorem of Shokurov (3.4) we know that V *contains lines*. Using deformation theory one can show then rather easily that the dimension of the family of lines is 1 or 2; however dim 2 can occur only if V contains planes and that case is now ruled out by our assumption $\rho(V) = 1$. Therefore our V *contains a one-dimensional* family $F(V)$ of lines. Moreover for the normal sheaf of a line $Z \subset V$ only the following possibilities occur:

a) $N_{Z|V} \simeq \mathcal{O}(-1) \oplus \mathcal{O}$

b) $N_{Z|V} \simeq \mathcal{O}(-2) \oplus \mathcal{O}(1)$.

Furthermore case a) is the *general* case.

Finally if we take a sufficiently general line $Z \subset V$ (so we are in case a)) then there exist $(m+1)$ - other lines $Z_i \subset V$ meeting Z and all are in case a) and $m > 0$. (In fact m is the degree of th hypersurface from the ambient space cutting out on V the ruled surface consisting of the lines on V). For all of this see [I2],§3 or [I3], Chapter III,§2.

4.5. PROJECTION FORM A LINE

One of the ideas of Fano is to study V by projecting it from a line $Z \subset V$ into $\mathbb{P}_{g-1}$ (and "doubly-projecting" it to $\mathbb{P}_{g-6}$, see later in 4.9).

Assume from now on that $g \geq 5$ ($g = 3$ and 4 we know already, see Corollary in 4.2). Note that, due to $\rho(V) = 1$, the V is now non-trigonal (see 4.3) and an intersection of quadrics. Take a line $Z \subset V$ sufficiently general as described in 4.4. Let $\pi_Z : V \dashrightarrow V^* \subset \mathbb{P}_{g-1}$ be the projection from Z ; if $|H| = |-K_V|$ then we write (by abuse of language) $|H-Z|$ for the sublinear system of those H's going through Z .

Let $\sigma : V' = B_Z(V) \to V$ be the blow up of V along Z ; write $Z' = \sigma^{-1}(Z)$ and $H' = \sigma^*(H)$. Let $\phi_Z = \pi_Z \cdot \sigma$. Finally for $Z_i \subset V$

$(i=1,\dots m+1)$, the lines on V meeting Z , let $Z_i^0 \subset V'$ be their proper transforms (hence *curves* on V').

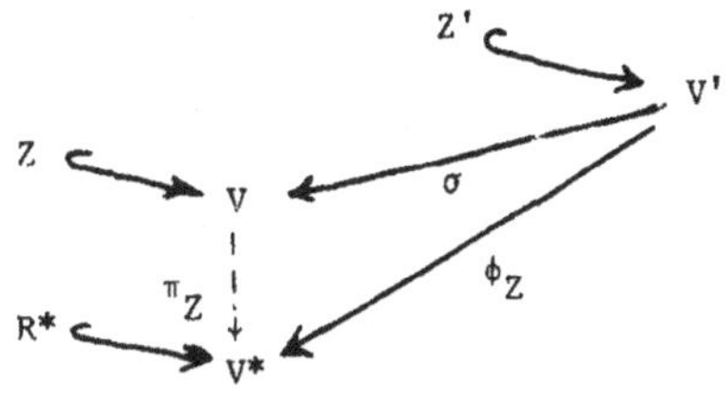

Clearly ϕ_Z is a *morphism* and it is given by the linear system $|H'-Z'|$. Furthermore since $N_{Z|V} \cong \mathcal{O}(-1) \oplus \mathcal{O}$ we have that $Z' = \mathbb{P}_{\mathbb{P}_1}(\mathcal{O}+\mathcal{O}(1)) = \mathbb{F}_1$ and hence $\mathrm{Pic}(Z') = \mathbb{Z}b \oplus \mathbb{Z}f$, where f is a fibre and $b^2 = -1$.

Let $R^* = \phi_Z(Z')$, then geometrically it is clear that R^* is a ruled surface obtained from projecting the "tangentlines" Z , or, more explicitely if we take $\mathbb{P}_{g-1} \subset \mathbb{P}_{g+1}$ with $\mathbb{P}_{g-1} \cap Z = \emptyset$ and $v \in Z \subset V$ and $T_{Z,v}$ is the tangentspace to V in v , then R^* is the union of the lines $(T_{V,v} \cap \mathbb{P}_{g-1})$ by varying v along Z .

4.6. LEMMA 2.

Let $g \geq 5$. *Then:*

i) ϕ_Z *is a birational morphism, determined by* $|H'-Z'|$

ii) $\phi_Z \colon V' - Z' - \bigcup_i Z_i^0 \xrightarrow{\sim} V^* - R^*$

iii) ϕ_Z *contracts the* Z_i^0 *to points* $z_i^* \in R^*$ *(*$i=1,\dots,m+1$*)* .

PROOF.

To see that ϕ_Z is *birational*, we can look to π_Z . Since $g \geq 5$ the V is an intersection of quadrics Q_α . Now if $v \in V$, the span $\langle Z,v \rangle$ has intersection

$$\langle Z,v \rangle \cap Q_\alpha = Z + \ell_\alpha$$

with a line ℓ_α and if $v \notin Z_i$ then $\bigcap_\alpha \ell_\alpha = v$ and hence v is uniquely determined by $v^* = \pi_Z(v)$.

The ohter assertions are clear.

4.7. Now we study $Tr_{Z'} |H'-Z'|$. We have

LEMMA 3.

i) $Tr_{Z'} |H'-Z'| = |b+2f|$,

ii) $H'.Z' = f$, $-Z'.Z' = b + f$,

iii) $\phi_Z(Z') = R^*$ *is a smooth cubic scroll, spanning a* $\mathbb{P}_4$.

PROOF (indication).

An easy computation gives first $Tr_{Z'} |H'-Z'| \subset |b+2f|$, namely clearly $(H'-Z').Z' = b + qf$ and

$$-1 + 2q = (b+qf.b+qf)_{Z'} = (H'-Z')^2.Z' = 3 ,$$

hence $q = 2$. Next $H'.Z' = (H.Z).f$ by standard formulas for blowing-ups. Next one checks that $Tr_{Z'} |H'-Z'|$ contracts nothing on Z' and gives an injective map, this gives *equality* in i). Finally $|b+2f|$ maps Z' onto a smooth cubic scroll spanning $\mathbb{P}_4$.

4.8. STUDY OF V^*

LEMMA 4.

i) V^* *is smooth, except in* z_i^* $(i=1,\ldots,m+1)$,

ii) z_i^* *are ordinary double points (hence* V^* *normal and Govenstein),*

iii) $\phi_Z: V' - \bigcup_i Z_i^0 \xrightarrow{\sim} V^* - \bigcup_i z_i^*$,

iv) V' *projectively normal,*

v) *put* $g_1 = g - 2$, *then* $\deg(V^*) = 2g_1 - 2$ *and* V^* *spans a* $\mathbb{P}_{g_1+1}$,

vi) *the dualising sheaf* $\omega_{V^*} \simeq \mathcal{O}_{V^*}(-1)$,

vii) $\mathrm{Pic}(V^*) = \mathbb{Z}.\mathcal{O}_{V^*}(1)$.

REMARK: Now the idea of Fano becomes clear: by projecting from a line on V we get *another* Fano V^* with a *lower* g (but with double points).

PROOF (some points).

ii): Blow up V' in Z_i^0, this gives a resolution $\alpha: B_{Z_i^0}(V') \to V' \to V^*$ for z_i^* with inverse image $\tilde{Z}_i^0$ (= inverse image of Z_i^0). Now we have $N_{Z_i/V} \simeq \mathcal{O} \oplus \mathcal{O}(-1)$, then $N_{Z_i^0/V'} = \mathcal{O}(-1) + \mathcal{O}(-1)$; hence $\tilde{Z}_i^0 \simeq \mathbb{P}_1 \times \mathbb{P}_1$, hence z_i ordinary double point. Hence in particular V^* normal and Gorenstein and ω_{V^*} invertible.

v): $\deg V^* = (H'-Z')^3 = H'^3 - 3H'^2Z' + 3H'Z'^2 - H'^3 = (2g-2) - 0 - 3 - 1 =$
$= 2g - 6 = 2g_1 - 2$.

vi) and vii): we have a commutative diagram

$$\begin{array}{ccc} \mathrm{Pic}(V^*) & \hookrightarrow & \mathrm{Pic}_{\mathrm{Weil}}(V^*) \\ \downarrow & & \downarrow \simeq \\ \mathrm{Pic}(V') & = & \mathrm{Pic}_{\mathrm{Weil}}(V') \end{array}$$

From this, and the relation between ω_{V^*} and the canonical class of V^* in the sense of Zariski ([Z2],p.32) we get vi). Finally for vii) we use that $\mathrm{Pic}(V') = \mathbb{Z}.(H'-Z') \oplus \mathbb{Z}.Z'$ and hence $\mathrm{Pic}_{\mathrm{Weil}}(V^*) = \mathbb{Z}.\mathcal{O}(1) + \mathbb{Z}.R^*$, but one can show that *no multiple* of R^* is a Cartier divisor. Namely if f and s are the standard generators of $\mathrm{Pic}(\tilde{Z}_i^0)$ then $\alpha^*(R^*).\tilde{Z}_i^0 = f$ whereas for a Cartier divisor D^* on V^* we get $\alpha^*(D^*).\tilde{Z}_i^0 = nf + ns$.

4.9. DOUBLE PROJECTION

Consider in the ambient space $\mathbb{P}_{g-1}$ of V^* the space $\langle R^*\rangle = \mathbb{P}_4$ and project from this $\mathbb{P}_4$ to $\mathbb{P}_{g-6}$

$$\pi_{\mathbb{P}_4} : \mathbb{P}_{g-1} \dashrightarrow \mathbb{P}_{g-6} .$$

Write $\pi_{2Z} = \pi_{\mathbb{P}_4} \cdot \pi_Z$ and $\phi_{2Z} = \pi_{\mathbb{P}_4} \cdot \phi_Z$ and note (as is easily seen) that ϕ_{2Z} is the rational map determined by the linear system $|H'-2Z'|$ (whereas ϕ_Z was given by $|H'-Z'|$). Let W be the image of V by π_{2Z} .

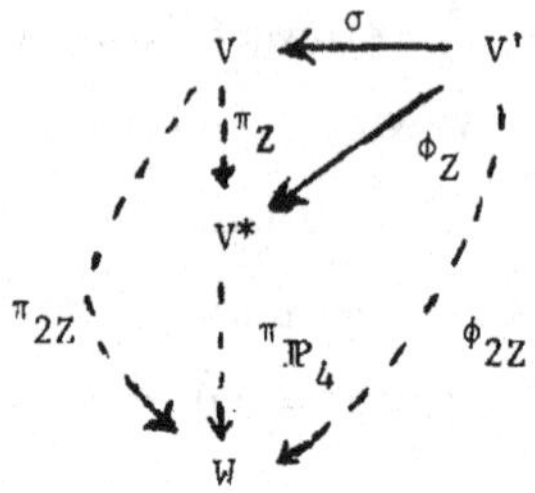

4.10. LEMMA 5.

Let $g \geq 5$. *Then:*

i) $h^i(\mathcal{O}_{V'}(-Z')) = 0$, $i > 0$

ii) $h^i(\mathcal{O}_{V'}(H'-Z')) = 0$, $i > 0$

iii) $h^i(\mathcal{O}_{V'}(H'-2Z')) = 0$, $i > 0$

iv) $h^0(\mathcal{O}_{V'}(H'-2Z')) = g - 5$ *(hence* $\langle W \rangle = \mathbb{P}_{g-6}$ *)*

v) $h^0(\mathcal{O}_{V'}(H'-3Z')) \leq 1$.

PROOF (some indications).

Assertions i)-iv) are easily obtained from the standard exact sequences (j=1,2) :

$$0 \to \mathcal{O}_{V'}(H'-jZ') \to \mathcal{O}_{V'}(H'-(j-1)Z') \to \mathcal{O}_{Z'}(H'-(j-1)Z') \to 0$$

due to the fact that we know $Tr_{Z'}|H'-(j-1)Z'|$ (see Lemma 3). For the important assertion v) we have *to use the conics on* V . From the existence of lines (§3) we get first the existence of reducible conics and then, since moreover $\rho(V) = 1$, by deformation theory also of smooth conics. Furthermore through a genearl point pass only finitely many conics (see [I2],§4 or [I3], Chapter III,§3). Now returning to the proof of v) consider the surface Q which is the union of all conics C meeting Z ; let Q' (resp. C') be the proper transform of Q (resp. C) on V' . For a general such C we have $(H'-3Z').C' = -1$. Now if $h^0(\mathcal{O}_{V'}(H'-3Z')) > 1$ then $|H'-3Z'|$ has a movable part, but from the above we see that C' is on a fixed component, hence Q' is a fixed component. Hence for some hyperplane H_0 we have $H_0'.V' = 3Z' + Q' + D'$, with $D' > 0$ the movable part; hence going back to V we get $H_0.V = Q + D$, $D > 0$, but this contradicts $\mathrm{Pic}(V) = \mathbb{Z}.H$.

4.11. FROM NOW ON WE ASSUME ALWAYS $g \geq 7$

LEMMA 6.

i) $|H'-2Z'|$ has no fixed component,

ii) the base locus $\mathcal{B}(H'-2Z') = \bigcup_i Z_i^0$ (i=1,...,m+1) .

PROOF (indication).

Since $\dim|H'-2Z'| = g-6$ (Lemma 5) and $g \geq 7$, there is a movable part. From $\mathrm{Pic}(V) = \mathbb{Z}.H$ we get as only candidate for fixed component Z'; however then $|H'-2Z'| = |H'-3Z'|$ contradicting the dimensions of these linear systems (Lemma 5 and $g \geq 7$).

From $Z_i^0.(H'-3Z') = -1$ we get that the Z_i^0 are in the base locus; that there are no other base points follows via hyperplane sections from the theory of linear systems on K3-surfaces.

4.12. LEMMA 7.

Put $L = \mathrm{Tr}_{Z'}|H'-2Z'|$. *Then:*

i) $L \subset |2b+3f| = |-K_{Z'}|$,

ii) L *has no fixed components and the base locus* $\mathcal{B}(L)$ *consists of the points* $z_i' = Z_i^0 \cap Z'$ (i=1,...,m+1) .

PROOF.

i) follows directly from $H'.Z' = f$ and $Z'.Z' = -(b+f)$ (see Lemma 3),

ii) folllows from Lemma 6.

COROLLARY 1.

$m \leq 7$.

PROOF.

First recall that $(m+1)$ is the number of lines Z_i on V meeting the general line Z .

The surface $Z' \cong \mathbb{F}_1$ can also be considered as the blow up of $\mathbb{P}_2$ in a point a

$$\mu: Z' = \mathbb{F}_1 \tilde{\to} B_a(\mathbb{P}_2) \to \mathbb{P}_2$$

and then for the section b with $b^2 = -1$ we have simply $b = \mu^{-1}(a)$; put $c_i = \mu(z_i')$ where $z_i' = Z_i^0 \cap Z'$. Since $|2b+3f| = |-K_{Z'}|$ this is simply the system of the *strict* transforms under μ of the *cubics in* $\mathbb{P}_2$ *passing through the point* a and L is the system of the strict transforms of the *cubics in* $\mathbb{P}_2$ *passing through* a *and the points* c_i $(i=1,\dots,m+1)$. Due to the fact that Z is *general* one can show that the $z_i' \notin b$ ([I2], 3.7 (iii)) hence we get $(m+2)$ different points in $\mathbb{P}_2$ and so we have $m + 2 \leq 9$, hence $m \leq 7$.

4.13. LEMMA 8.

Always $g \geq 7$. *Then:* $g = 14 + h^0(H'-3Z') - h^1(H'-3Z')$.

PROOF.

Consider the exact sequence

$$0 \to \mathcal{O}_{V'}(H'-3Z') \to \mathcal{O}_{V'}(H'-2Z') \to \mathcal{O}_{Z'}(H'-2Z') \to 0 .$$

This gives the exact sequence (using the vanishing from Lemma 5):

$0 \to H^0(\mathcal{O}_{V'}(H'-3Z')) \to H^0(\mathcal{O}_{V'}(H'-2Z')) \to H^0(\mathcal{O}_{Z'}(H'-2Z')) \to H'(\mathcal{O}_{V'}(H'-3Z')) \to 0$.

Now $\dim H^0(\mathcal{O}_{V'}(H'-2Z')) = g-5$ (Lemma 5) and since $\mathcal{O}_{Z'}(H'-2Z') = \mathcal{O}_{Z'}(2b+3f)$

we have $\dim H^0(\mathcal{O}_{Z'}(H'-2Z')) = 9$ and this proves the lemma.

COROLLARY 1.

We have $g \leq 15$. *(Hence: the sequence of Fano's* $V^{2g-2} \subset \mathbb{P}_{g+1}$ *with* $\rho(V) = 1$ *stops!)*

PROOF.

Immediate from Lemma 8 and Lemma 5 (v).

COROLLARY 2.

i) $h^1(H'-3Z') = \dim |-K_{Z'}| - \dim L$,

ii) *if* $m \neq 7$ *then* $h^1 \geq (m+1)$,

iii) $g \leq 13$.

PROOF.

Assertion i) follows immediately from the exact sequence of cohomology groups considered in the proof of Lemma 8. Next if $m \neq 7$ we have $(m+2) < 9$ points in $\mathbb{P}_2$ through which the cubics have to go (see the proof of Corollary 1 of Lemma 7), hence we get at least $(m+1)$ conditions from the points c_i , hence $h^1 \geq (m+1)$ (note: we have only an inequality because there may be also "infinitesimal" conditions). Using this we get, for $m \neq 7$, from the cohomolgy sequence that $g \leq 13 + h^0(H'-3Z') - m$ and hence $g \leq 13$. On the other hand if $m = 7$ then $g \leq 7 + h^0(H'-3Z')$.

4.14. It is impossible to report here in detail about the finer and finer analysis made by Iskovskih in the study of the linear systems $|H'-2Z'|$ and $|H'-3Z'|$. (See [I2],5.4).

Therefore it suffices here to say that Iskovskih makes now *extra assumptions* which, roughly, amount to the hypothesis that the base points $z_i' \in Z'$ $(i=1,\dots,m+1)$ of $L = \mathrm{Tr}_{Z'}|H'-2Z'|$ are *resolved by blowing them up one time*. Let then $\mu: \bar{V} \to V'$ be the blow up of V' in the curves

Z_i^0 (i=1,...,m+1) , $\bar{Z}_i^0 = \mu^{-1}(Z_i^0)$ and $\bar{H}' = \mu^{-1}(H')$, $\bar{Z}' = \mu^{-1}(Z')$. Then the linear system $|\bar{H}'-2\bar{Z}'-\sum_i \bar{Z}_i^0|$ determines the map $\psi_{2Z} = \phi_{2Z'} : \bar{V} \dashrightarrow W$. Making the above mentioned extra assumptions, Iskovskih now shows (always with $g \geq 7$):

i) ψ_{2Z} is a *morphism*,

ii) if $g \geq 9$, then ψ_{2Z} is a *birational* morphism (with possibly one exception),

iii) if $g \geq 9$, the $\mathrm{Pic}(W) = \mathbb{Z}.\mathcal{O}_W(1)$,

iv) if $g \geq 9$, W is smooth $\Leftrightarrow$ $\psi_{2Z}|\bar{Z}'$ gives an isomorphism; moreover if this is the case then W is a Fano variety of index $r = 2$.

See ([I2],5.4) for the precise statements.

4.15. CLASSIFICATION THEOREM ([I2],§6)

Let V be a smooth Fano threefold with index $r(V) = 1$, Picard number $\rho(V) = 1$, with very ample $|-K_V|$ and embedded by means of $|-K_V|$ and $g = g(\Gamma)$ with Γ the general codim. 2 section.

With the above introduced notations and methods Iskovskih classifies now these Fano's by given for each g a description of the variety and conversely such varieties are Fano's with the corresponding g . The degree $d = 2g - 2$ always. Now the result is as follows:

$g = 3$; $d = 4$, $V = V(4) \subset \mathbb{P}_4$.

Unirational: general one *unknown*.

Non-rational: all [I-M].

$g = 4$; $d = 6$, $V = V(3,2) \subset \mathbb{P}_5$.

Unirational: all (Enriques).

Non-rational: [B2] for general one; [I4] all.

$g = 5$; $d = 8$, $V = V(2,2,2) \subset \mathbb{P}_6$.

Unirational: all.

Non-rational : all (see for instance [B2]).

$g = 6$; $d = 10$, $V = Gr(1,4).V(2).H.H' \subset \mathbb{P}_7$.

Unirational: all (Fano, [I4],[P]).

Rational: *unknown;* general one: non-rational [B2].

$g = 7$; $d = 12$, $\pi_{2Z}: V \dashrightarrow \mathbb{P}_1$, general fibre is a Del Pezzo surface of degree 5 with 8 blown up points. Also: $\pi_Z: V \dashrightarrow V^* = V(2,2,2) \subset \mathbb{P}_6$ with $V^* \supset R^*$ a smooth rational cubic scroll.

Rational: all (Fano).

$g = 8$; $d = 14$, $\pi_{2Z}: V \dashrightarrow \mathbb{P}_2$, general fibres are curves of genus 2.

EXAMPLE: $V = Gr(1,5).H_1H_2H_3H_4H_5 \subset \mathbb{P}_9$. For those we have:

V is birational with cubic threefold (Fano, [I4],[P]).

Hence: all unirational, all non-rational.

$g = 9$; $d = 16$, $\pi_{2Z}: V \dashrightarrow W = \mathbb{P}_3$ *birational.* Inverse transformation given by $|7\mathcal{O}_W(1)-Y|$ where Y is smooth curve of degree 7 and genus 3 (Y is obtained as the contraction of the surface $Q \subset V$ of conics meeting Z , see 4.10).

$g = 10$; $d = 18$, $\pi_{2Z}: V \dashrightarrow W \subset \mathbb{P}_4$ with W a *quadric.* Again π_{2Z} is *birational;* the inverse is given by $|5\mathcal{O}_W(1)-2Y|$ where Y a smooth curve of genus 2 and degree 7.

$g = 11$; does not exist.

$g = 12$; $d = 22$, $\pi_{2Z}: V \dashrightarrow W \subset \mathbb{P}_6$ birational, with W a Fano of index 2, degree 5, $Pic(W) = \mathbb{Z}$ and at most one singular point. The inverse is given by $|3\mathcal{O}_W(1)-2Y|$, Y a normal rational curve of degree 5.

$g > 12$; do not exist.

REMARK. The cases $g = 9,10,12$ are all rational. This is clear for $g = 9,10$ and follows for $g = 12$ since W is rational (see 2.3).

REFERENCES

[B1] BEAUVILLE, A., *Surfaces algébriques complexes*, Astérisque No. 54, Paris (1978).

[B2] BEAUVILLE, A., *Variétés de Prym et Jacobiennes intermédiares*, Ann. Sci. Ecole norm. sup. (4), 10 (1977).

[Bo] BONARDI, M.T., *Varietà di Fano sul corpo complesso*, Sem. dell' Istituto di Mat., No.13, Genova (1979).

[C.-G.] CLEMENS, C.H. and GRIFFITHS, P.A., *The intermediate jacobian of the cubic threefold*, Ann. of Math., 95 (1972).

[F1] FANO, G., *Sopra alcune varietà algebriche a tre dimensioni aventi tutti i generi nulli*, Atti Acc. Torino, 43 (1908).

[F2] FANO, G., *Sulle varietà algebriche a tre dimensioni aventi tutti i generi nulli*, Atti Congr. Internaz. Bologna, IV, (1931).

[F3] FANO, G., *Su alcune varietà algebriche a tre dimensioni aventi curve sezioni canoniche*, Scritti Mat. offerti a L. Berzolari, Pavia (1936).

[F4] FANO, G., *Sulle varietà algebriche a tre dimensioni a curve sezioni canoniche*, Mem. Acc. d'Italia, VIII, (1937).

[F5] FANO, G., *Nuove ricerche sulle varietà algebriche a tre dimensioni a curve sezioni canoniche*, Comm. Pont. Acc. Sc., 11 (1947).

[G] GALLARATI, D., *Les variétés algébriques a courbes sections canoniques*, Troisième Coll. Géom. Alg., Bruxelles, (1959).

[H] HARTSHORNE, R., *Algebraic Geometry*, Springer Verlag, New York (1977).

[I.-M.] ISKOVSKIH, V.A. and MANIN, J.I., *Three dimensional quartics and counter-examples to the Lüroth problem*, Mat. Sbornik, 86(1971), engl. transl. Math. U.S.S.R., Sb. 15(1971).

[I1] ISKOVSKIH, V.A., *Fano 3-folds I*, Izv. Akad. Nauk, 41(1977), engl. trans. Math. U.S.S.R., Izv. 11 (1977).

[I2] ISKOVSKIH, V.A., *Fano 3-folds II*, Izv. Akad. Nauk, 42(1978), engl. trans. Math. U.S.S.R., Izv. 12 (1978).

[I3] ISKOVSKIH, V.A., *Anticanonical models of three-dimensional algebraic varieties*, Itogi Nauki i Tekhniki, Sovr. Prob. Mat. 12(1979), engl. transl. J. Soviet Math., 13 (1980).

[I4] ISKOVSKIH, V.A., *Birational automorphisms of three-dimensional algebraic varieties*, Itogi Nauki i Tekhniki, Sovr. Prob. Mat. 12(1979), engl. transl. J. Soviet Math., 13(1980).

[Ma] MANIN, J.I., *Cubic Forms*, North Holland Publ. Co., Amsterdam (1974).

[Mo] MORI, S., *Threefolds whose canonical bundles are not numerically effective*, Proc. Nat. Acad. Sci. U.S.A. 77(1980).

[M.-M.] MORI, S. and MUKAI, S., *Classification of Fano 3-folds with the second Betti number* ≥ 2, manuscr.math.36(1981) 163-178.

[M] MUMFORD, D., *Varieties defined by quadratic equations*, C.I.M.E. lectures, Varenna (1969).

[N] NAGATA, M., *On rational surfaces, I, II*, Mem. Coll. Sc. Kyoto, 32 and 33 (1960).

[P] PUTS, P.J., Forthcoming Ph.D. Thesis, Univ. Leiden.

[Ra] RAMANUJAM, C.P., *Remarks on the Kodaira Vanishing Theorem*, J. of Ind. Math. Soc. 36(1972), reprinted in C.P. Ramanujam, A. Tribute, Springer-Verlag, Berlin (1978).

[R] REID, M., *Lines on Fano 3-folds according to Shokurov*, Mittag-Leffler Report No. 11 (1980).

[Ro] ROTH, L., *Algebraic Threefolds*, Springer-Verlag, Berlin (1955).

[S.D] SAINT-DONAT, B., *Projective models of K3-surfaces*, Am. J. of Math. 96(1974).

[S.-R.] SEMPLE, J.G. and ROTH, L., *Introduction to algebraic geometry*, Oxford (1949).

[Se] SERPICO, M.E., Forthcoming.

[S1] SHOKUROV, V.V., *Smoothness of the general anticanonical divisor on a Fano 3-fold*, Izv. Akad. Nauk. 43(1979), engl. transl. Math. U.S.S.R., Izv. 14(1980).

[S2] SHOKUROV, V.V., *The existence of a straight line on Fano 3-folds*, Izv. Akad. Nauk. 43(1979), engl. trans. Math. U.S.S.R., Izv. 15(1980).

[Z1] ZARISKI, O., *Algebraic Surfaces (2nd ed.)*, Springer-Verlag, Berlin (1971).

[Z2] ZARISKI, O., *An introduction to the theory of algebraic surfaces*, Springer Lect. Notes 83(1969).

Degeneration Techniques in the Study of Threefolds

C.H. CLEMENS

LECTURE #1
Mixed Hodge Theory-Topological Considerations

Our aim in this first lecture is to explain a construction carried out several years ago in "Degeneration of Kähler Manifolds" (Duke Math. J., Vol. 44, No. 2, 1977, pp. 215-90). The situation is

(1) $$p : V \longrightarrow \Delta_\varepsilon$$

a proper morphism of an analytic manifold V onto the complex disc of radius ε, where the fibre V_t of p over t smooth and reduced except possibly over $t = 0$. We want to construct a continuous action of the semi-group

(2) $$S = ([0,1] \times \mathbb{R})$$

on V which is equivariant, via (1), with the natural action

$$S \times \Delta_\varepsilon \longrightarrow \Delta_\varepsilon$$
$$((r,\theta)\ ,\ t) \longmapsto re^{2\pi i\theta}t \quad .$$

Using resolution of singularities in the analytic category and the fact that our action will be the trivial action on V_0 it will suffice to restrict our attention to the case in which

(3) $$p^{-1}(0) = \bigcup_{j \in I} D_j = V_0$$

is a normal crossing variety, since, if we construct an action on a resolution of (1) along its central fibre, this action will descend to an action on (1) itself. Even under the assumption (3) however, $p^{-1}(0)$ may not be reduced, that is, some of the components D_j of the central fibre may be acquired with multiplicity greater than one. This does not complicate our construction in any essential way but it does make it notationally more cumbersome and involves use of what are called V-manifolds. So, for purposes of exposition, we will make the following additional assumption:

(4) The divisor of the analytic function p on V is

$$\sum_{j \in I} m_j D_j \quad \text{with all } m_j = 1 \ .$$

Let's first consider the case in which the family (1) is topologically trivial. Already in this case we can say some things that will indicate the properties of the general construction. We proceed as follows. Since V_0 is smooth, we can cover V_0 by a finite collection $\mathcal{U}$ of open sets in V such that each $U \in \mathcal{U}$ has analytic coordinates

$$(z_U, w_U)$$

with $z_U = 0$ defining $(V_0 \cap U)$ and $w|_{(V_0 \cap U)}$ a global coordinate system for $(V_0 \cap U)$. Next we let

(5) $$\{\sigma_U\}_{\mathcal{U}}$$

be a C^∞ partition of unity on V_0 which is subordinate to the covering

$$\{V_0 \cap U\}_{\mathcal{U}}$$

of V_0. So on each set $U \times (V_0 \cap U)$ we can define

$$v_U(y,x) = w_U(y) - w_U(x) \quad ,$$

and $\{y \in U : v_U(y,x) = 0\}$ is a complex analytic disc meeting V_0 transversely at x. So locally we have defined a normal disc bundle to V_0 with analytic fibres.

Of course we want to fit these disc bundles together preserving as much of the complex structure as we can. We do this via the partition of unity (5). For each U and U' in $\mathcal{U}$ and $x \in (U \cap U')$, we define

$J_{U'U}(x)$ = (the Jacobian matrix associated to changing from the coordinate system $v_{U'}(y,x)$ to the coordinate system $v_U(y,x)$) .

Here x is fixed and y is allowed to vary over a small neighborhood of x in $(U \cap U' \cap V_0)$. We compute the Jacobian at the point $y = x$. Now we define for $(y,x) \in U \times (V_0 \cap U)$:

$$A_U(y,x) = \sum_{U' \in \mathcal{U}} \sigma_{U'}(x)\, v_{U'}(y,x) J_{U'U}(x) \quad . \tag{6}$$

Again $\{y \in U : A_U(y,x) = 0\}$ is a complex disc meeting V_0 transversely at x . However, since the function (6) is analytic only in the y variable but not in the x variable, the normal analytic disc changes only in a C^∞ way as x moves along V_0 .

However in losing a certain amount of analyticity we have gained compatibility of this normal fibration over the different open sets of our covering. Namely, the chain rule for Jacobian matrices at a point gives the identity

$$A_{U''}(y,x) = A_U(y,x) J_{UU''}(x) \quad .$$

So the normal disc to V_0 at x is well defined, independently of the choice of open set $U \in \mathcal{U}$ (as long as $x \in U$) . So we have a tubular neighborhood U_0 of V_0 in V and a C^∞ mapping

$$\mu : U_0 \longrightarrow V_0$$

such that, for each $x \in V_0$, $\mu^{-1}(x)$ is an *analytic* disc meeting V_0 transversely at x . p maps $\mu^{-1}(x)$ isomorphically to Δ_ε so we simply use this isomorphism to lift the action of S to $\mathcal{V}$.

Now if $V_0 = D_1 \cup D_2$ we can make the same sort of construction, in fact, in such a way that we have *two* dimensions of analyticity along the singular locus $D_{12} = D_1 \cap D_2$. More precisely there are C^∞ normal disc bundles

$$\mu_j : U_j \longrightarrow D_j \quad , \quad j = 1,2$$

which have the following additional property along

$$U_{12} = U_1 \cap U_2 :$$

(7) $$\mu_1 \circ \mu_2 = \mu_2 \circ \mu_1 \underset{\text{def.}}{=} \mu_{12} ,$$

and

$$\mu_{12} : U_{12} \longrightarrow D_{12}$$

is a C^∞ normal bundle with *analytic* polydiscs as fibres. Furthermore, μ_1 and μ_2 are both analytic projections when restricted to any fibre of μ_{12}.

(8)

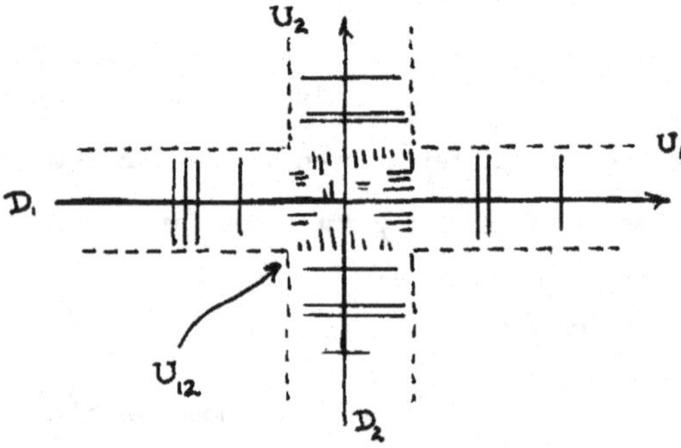

Let us explain the next step in detail in case $\dim D_{12} = 0$. In U_{12} we will have coordinates z_1 and z_2 whose zero sets define D_1 and D_2 respectively. We must make the additional assumption on the choices of the z_j that

$$z_1 \cdot z_2 = t$$

where t is the coordinate of Δ_ε pulled back to V by the mapping p, and $\mu_j|_{U_{12}}$ is given by z_j = constant.

To define the action of S in U_{12}, we will need to make one more auxiliary construction. Consider the following picture in a small square in $\mathbb{R}^2$:

(9) 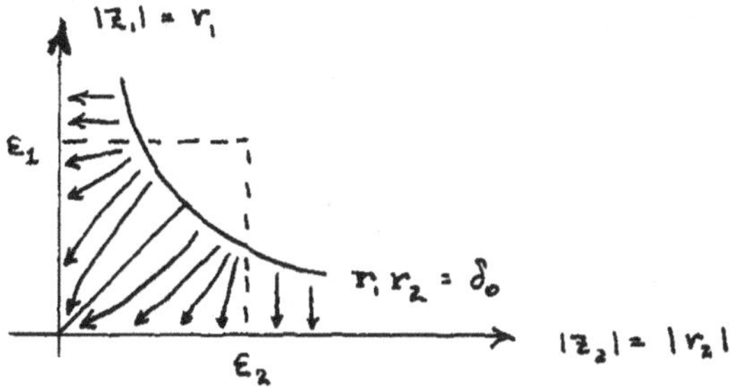

Define $C_\delta = \{(r_1,r_2) : r_1 \cdot r_2 = \delta\}$ and $D_{\delta_0} = \bigcup_{0\leq\delta\leq\delta_0} C_\delta$.

Define a smoothly varying family of arcs between C_{δ_0} and C_0 as in the above picture. The salient features of this family are:

i) if $r_2 \geq \varepsilon_2$ they are vertical lines, if $r_1 \geq \varepsilon_1$ they are horizontal;

ii) all arcs in a neighborhood of the diagonal end at $(0,0)$;

iii) by associating to each arc its endpoint we obtain a continuous mapping

$$D_{\delta_0} \longrightarrow C_0$$

which is C^∞ on $(D_{\delta_0} - C_0)$.

Then (9) determines a continuous mapping

$$R\colon [0,1] \times D_{\delta_0} \longrightarrow D_{\delta_0}$$

with the following properties:

i) $R|_{[0,1]\times(D_{\delta_0}-C_0)}$ is C^∞ ,

ii) $R(r,\ )(C_\delta) = C_{r\delta}$,

iii) $R(1,\ )$ is the identity map,

iv) $R(r,\ )|_{C_0}$ is the identity map for all r .

Using

$$R = (R_1,R_2)$$

we define an action of the semi-group $[0,1]$ on U_{12} in terms of our chosen coordinate system by

(10)
$$\begin{aligned} [0,1]\times U_{12} &\longrightarrow U_{12} \\ (r\,,\,(z_1,z_2)) &\longrightarrow (R_1(r;|z_1|,|z_2|)\cdot z_1/|z_1|\;, \\ &\qquad R_2(r;\;|z_1|,|z_2|)\cdot z_2/|z_2|)\;. \end{aligned}$$

This action is clearly compatible with the previously defined action of

$$[0,1] \leqslant S$$

in the sets

$$U_j' = U_j - U_{12} \quad .$$

Now we have an open covering of

(11)
$$(U_1 \cup U_2) \quad - \quad (D_1 \cup D_2)$$

by two open sets

$$\begin{aligned} A_j = \;&(\text{interior of the set of all } y \text{ such that} \\ &0\cdot y \in D_j \text{ under the action of } [0,1]) \end{aligned}$$

for $j = 1,2$. We choose a C^∞ partition of unity $\{\lambda_j\}$ subordinate to this covering of (11) and define an action of $\mathbb{R}$ on $(U_{12} - (U_{12} \cap V_0))$ by

(12)
$$\begin{aligned} \mathbb{R}\times U_{12} &\longrightarrow U_{12} \\ (\theta\,,\,(z_1,z_2)) &\longmapsto (e^{2\pi i\lambda_1\theta} z_1\;,\;e^{2\pi i\lambda_2\theta} z_2)\;. \end{aligned}$$

Again this action is compatible with the action of $\mathbb{R} \leqslant S$ which we previously defined on the U_j' . Our construction (9) insures that this action extends to an action on all of U_{12} by defining it to be trivial on $(V_0 \cap U_{12})$.

In fact we can insure that the actions (10) and (12) commute by simply requiring that our partition of unity $\{\lambda_j\}$ be chosen so that the λ_j's are functions only depending on $(|z_1|,|z_2|)$ and so that they are constant along arcs in (9). This then gives the required action of S on V, or at least on some neighborhood of V_0 in V, which is all we are interested in anyway.

In the general case, where the dimension of V is arbitrary and V_0 has many components the idea is exactly the same, it is simply much more complicated to carry out. This is done in detail in the Duke Journal article. However one added difficulty of the general case should probably be explained a bit here. Namely, the construction of the action of S on U_{12} depends heavily on the choice of local coordinates (z_1,z_2) there. Suppose now $\dim D_{12} > 0$. How do we know that we can make compatible choices along all of D_{12} ?

To do this, choose a covering $\mathcal{U}$ of D_{12} by coordinate charts of V. Choose a partition of unity $\{\rho_U\}_{\mathcal{U}}$ on D_{12} which is subordinate to the covering $\{U \cap D_{12}\}_{\mathcal{U}}$. In each $U \in \mathcal{U}$ we have

$$z_{1,U} \ , \ z_{2,U}$$

defining D_1 and D_2 respectively such that

$$z_{1,U} \cdot z_{2,U} = t \ .$$

Now define a one-form ω_j at $y \in U_{12}$ by

(13) $$\omega_j = \frac{1}{2\pi i} \Sigma_{\mathcal{U}_{12}} \rho_U(\mu_{12}(y)) d \log z_{j,U} \ .$$

Notice that, restricted to any fibre of $\mu_{12}: U_{12} \longrightarrow D_{12}$, the one-form ω_j is meromorphic with simple pole along the intersection of the fibre with D_j. Also

(14) $$\omega_1 + \omega_2 = \frac{1}{2\pi i} d\log t + \mu_{12}^*(\alpha)$$

for some one-form α on D_{12} . So if $x \in D_{12}$ the coordinates we want on $\mu_{12}^{-1}(x)$ are

$$(15) \qquad z_j = k_j \, e^{2\pi i \int_{y_0}^{y} \omega_j}$$

for some appropriately chosen base point

$$y_0 \in (\mu_{12}^{-1}(x) - (V_0 \cap \mu_{12}^{-1}(x))) \quad ,$$

and some non-zero constant k_j .

The important thing about the choice of k_j and the basepoint y_0 in (15) is that they be coordinated so that

$$(16) \qquad r_j = |z_j|$$

is a well-defined function on all of U_j . This can clearly be done since the normal bundle to D_j has an orthogonal structure.

Once we have μ_{12} and the coordinates (15) we define

$$\mu_1' \big|_{U_{12}}$$

to have fibres given by $z_2 = \text{constant}$ and redefine

$$\mu_1 : U_1 \longrightarrow D_1$$

to be given by μ_1' inside U_{12} . Notice that we have arranged things so that, as we leave U_{12} in the direction of U_1 , the differential ω_2 has the form

$$\mu_1^*(\beta)$$

for some one-form β on D_1 . So we can extend the differential ω_1 in (13) to a C^∞ form on all of $(U_1 - D_1)$ such that

$$\omega_1 = \begin{cases} \frac{1}{2\pi i}\,\mathrm{dlog}\, t + \mu_1^*(\alpha) & \text{on} \quad (U_1 - U_{12}) \\ \frac{1}{2\pi i}\,\mathrm{dlog}\, t - \omega_2 + \mu_{12}^*(\beta) & \text{on} \quad U_{12} \end{cases} .$$

Similarly we extend ω_2 after redefining μ_2 .

The outcome of the (rather laborious) pursuit of these constructions in the general case is the content of Theorems 5.7 and 6.9 of the Duke Journal article which we will reproduce here:

Theorem: For each $I \subseteq \{1,\ldots,N\}$, there exists a tubular neighborhood U_I of D_I in V and a C^∞ normal projection

$$\mu_I : U_I \longrightarrow D_I$$

such that

i) the fibres of μ_I are (open) holomorphic submanifolds of V ;

ii) for all $I,J \subseteq \{1,\ldots,N\}$,

$$(U_I \cap U_J) = U_{(I \cup J)} \quad ;$$

iii) if $I \supseteq J$, then on U_I ,

$$\mu_I \circ \mu_J = \mu_I \quad .$$

Furthermore, setting $U_j = U_{\{j\}}$, then for each $j = 1,\ldots,N$, there exists a C^∞ one-form

$$\omega_j$$

defined on $(U_j - D_j)$ such that if $x \in D_I$, $j \in I$, and $X = \mu_I^{-1}(x)$, then:

iv) $\omega_j|_X$ is a closed meromorphic one-form on X with simple pole (and residue 1) along $(D_j \cap X)$;

v) if $y_0 \in (X - (V_0 \cap X))$, then the functions

$$z_{j,x}(y) = \exp(2\pi i \int_{y_0}^{y} \omega_j) , \quad j \in I ,$$

give a system of holomorphic coordinates on X such that

$$\prod_{j\in I} z_{j,x}^{m_j} = (\text{constant})\cdot t\Big|_X$$

(the constant will depend on x and the choice of y_0) and such that on $U_{(j,k)}$, $j \neq k$, $z_{j,x}$ is constant on fibres of $\mu_{(k)}$.

<u>Theorem:</u> There exists an action

$$S \times V \longrightarrow V$$

of the semi-group $S = [0,1] \times R$ on V with the following properties:

i) the mapping $S \times V \longrightarrow V$ is continuous, and the restriction of the mapping to the subdomain $S \times V^*$ is C^∞ ;

ii) $(r_1,\theta_1)\cdot(r_2,\theta_2)\cdot y = (r_1 r_2, \theta_1 + \theta_2)\cdot y$ for all $(r_i,\theta_i) \in S, y \in V$;

iii) $(1,0)\cdot y = y$ for all $y \in V$;

iv) $(r,\theta)\cdot y = y$ for all $y \in V_0$ and all $(r,\theta) \in S$;

v) the diagram

$$\begin{array}{ccc} S \times V & \longrightarrow & V \\ \downarrow (\text{identity}) \times p & & \downarrow p \\ S \times \Delta_{\delta_0} & \longrightarrow & \Delta_{\delta_0} \\ ((r,\theta),t) & \longrightarrow & re^{2\pi i \theta}t \end{array}$$

is commutative;

vi) if $x \in (D_I - (D_I \cap \cup_{k \notin I} U_k))$, then S preserves $\mu_I^{-1}(x)$, that is, S respects the fibration μ_I on

$$(U_I - \bigcup_{k \notin I} U_{(I \cup \{k\})}) .$$

LECTURE #2
Mixed Hodge Theory-Analytic Considerations

To make the closest possible relation between the asymptotic topology of the family

(1) $$p: V \longrightarrow \Delta_\varepsilon$$

of Lecture #1 and the analysis, we will construct the smallest possible system of complexes in which to do our Hodge theoretic computations.

We begin by constructing a C^∞ retraction mapping

(2) $$f: V \longrightarrow V_0 \subseteq V \quad .$$

The mapping $y \longmapsto (0,0)\cdot y$ from our S-action is almost good enough except that it is not differentiable at the singular points of V_0. We can fix this up by moving things slightly - what we will lose is the property that $f|_{V_0} = \text{identity}$. We proceed as follows. We define a C^∞ homeomorphism

(3) $$V_0 \longrightarrow V_0$$

with the following properties:

i) in

$$U_I' = U_I - \bigcup_{J \supsetneq I} U_J$$

the mapping (3) commutes with $\mu_I: U_I \longrightarrow D_I$;

ii) at a point

$$x \in D_I \cap U_I' \quad ,$$

all partial derivatives of (3) of all orders in the directions of $\mu^{-1}(x)$ are zero;

iii) the mapping (3) is a topological homeomorphism which is homotopic to the identity map.

We then define f in (2) to be the composition of the mapping $y \longmapsto (0,0)\cdot y$ with the mapping (3). Then f is a C^∞ mapping when considered as a mapping from V to itself. Also f is a homotopy equivalence.

Given any differential α on V_0 (i.e. a compatible set of differentials on the various components D_j), then

$$(4) \qquad f^*(\alpha)$$

is a well-defined C^∞ differential on V and is invariant under the action of the semi-group S. We will call a differential of the form (4) a *horizontal differential*. The complex

$$(5) \qquad (H, d)$$

of horizontal differentials is quasi-isomorphic to the deRham complex of V under the natural inclusion.

Now the complex obtained by restricting the forms of H to $V^* = (V - V_0)$ has the same cohomology that H itself did. Furthermore we can increase this complex a bit without changing its cohomology. The differential algebra generated by the functions (from Lecture #1)

$$(6) \qquad \lambda_j(y)\ ,\ j\in I\ , \text{ and } \log r_j((r,0)\cdot y)\ ,\ j\in I \text{ and } r\in(0,1)\ ,$$

is everywhere locally acyclic. So we can adjoin these to $H\big|_{V^*}$ to generate a differential algebra

$$(7) \qquad (\hat{H}, d)$$

which is quasi-isomorphic to H under the natural inclusion. It is the complex (7) which we will use instead of the deRham complex to compute the cohomology of V.

Next we need a "small" complex with which to compute the cohomology of V^*. We let

(8) $$(L\ ,\ d)$$

be the subcomplex of the deRham complex of V^* consisting of differentials which can be written everywhere locally as a sum of terms of the form

(9) $$\alpha \wedge \omega_{j_1} \wedge \cdots \wedge \omega_{j_r} \quad ,$$

$j_1,\ldots,j_r \in I$, where α is the restriction of a form of $\hat{H}$ and the ω_j's are as in Lecture #1. L is acutally a complex because $d\,\omega_j \in H$. Also, since $d \log r_j \in \hat{H}$, the real and imaginary parts of ω_j are in L so that L is defined over the real numbers. The semi-group S acts on the complex L. In fact, if we define

$$\gamma_\theta : V^* \longrightarrow V^* \quad ,$$

$$y \longmapsto (1,\theta)\cdot y$$

then

(10) $$\gamma_\theta^*(\omega_j) = \omega_j + \theta d\lambda_j \quad .$$

This last formula follows immediately from the definition of the action of θ in (12) of Lecture #1. From (10) it is clear that

$$T = \log \gamma_1^*$$

is well-defined on L, and, since the log of the product is the sum of the logs,

(11) $$T(\alpha \wedge \omega_{j_1} \wedge \cdots \wedge \omega_{j_r}) = \textstyle\sum_{k=1}^{r} (\alpha \wedge \omega_{j_1} \wedge \cdots \wedge \omega_{j_{k-1}} \wedge d\lambda_{j_k} \wedge \omega_{j_{k+1}} \cdots)$$

A slightly more elegant way to express (11) is to let

(12) $$X = (\text{vector field associated to the action of } \mathbb{R}(\leq S) \text{ on } V^*) \quad .$$

Then

$$p_*(X) \quad = \quad 2\pi \frac{\partial}{\partial\theta}$$

where $\log t = \log|t| + i\theta$ on Δ_ε . In our distinguished coordinates on V^* we can write

(13) $$X \quad = \quad 2\pi \sum_{j\in I} \lambda_j \frac{\partial}{\partial\theta_j} \quad .$$

From (9) and the elementary properties of the Lie derivative, we have on L:

(14) $$L_X \quad = \quad T$$

where L_X is Lie differentiation with respect to the vector field X .

We will want to put "Hodge" and "weight" filtrations on L , and on certain quotient complexes of L , to obtain *mixed Hodge structures*. The naive Hodge filtration (from types of differential forms) is not the correct one, it must be 'twisted". To this end, define

(15) $$J(\phi) \quad = \quad \sum_{k=0}^{\infty} \frac{(-1)^k}{k!} \left(\frac{\log|t|}{2\pi i}\right)^k L_X^{(k)}(\phi) \quad .$$

J has some very nice properties:

(16) i) $dJ(\phi) \quad = \quad J(d\phi) - \frac{d\log|t|}{2\pi i} \wedge J(L_X(\phi))$

ii) $L_X \circ J \quad = \quad J \circ L_X$

iii) $J \circ \gamma_\theta^* \quad = \quad \gamma_\theta^* \circ J$.

Next form the differential ideal $I_{|t_0|}$ in L which is generated by the zero-form

$$(\log|t| - \log|t_0|)$$

and the one-form

$$\frac{d\log t}{2\pi i}$$

for t_0 fixed in $\Delta_\varepsilon^* = \Delta_\varepsilon - \{0\}$. Define

(17) $$M_{|t_0|} = L/I_{|t_0|} .$$

The natural restriction mapping

(18) $$M_{|t_0|} \longrightarrow A_{t_0} = (\text{deRham complex of } V_{t_0})$$

for fixed $t_0 \neq 0$ is a quasi-isomorphism. The action of $(r,0) \in S$ induces isomorphisms

(19) $$M_{|t_0|} \cong M_{r|t_0|} ,$$

and we have, for each $(1,\theta) \in S$, a commutative diagram,

$$\begin{array}{ccc} M_{|t_0|} & \longrightarrow & A_{e^{2\pi i\theta}t_0} \\ \gamma_\theta^* \downarrow & & \gamma_\theta^* \downarrow \\ M_{|t_0|} & \longrightarrow & A_{t_0} \end{array} .$$

The cohomology

(20) $$H^*$$

of (any complex of) the system (19) is what we will mean by the asymptotic cohomology of our original family $\{V_t\}$. It is clear from the construction that the logarithm of the monodromy transformation on $H^*(V_{t_0})$ corresponds under the quasi-isomorphisms (18) to the Lie differentiation L_X on any complex of the system (19).

At this point, if the operator J were well-defined on all of the deRham complex of V^* , we would have no more need for our small complex L . Unfortunately the series (15) does not reduce to a finite sum when ϕ is an arbitrary smooth form. We proceed as follows. We define a decreasing filtration

$$F^p\mathcal{H} = f^*(\text{usual Hodge filtration on forms on } V_0)\ .$$

Then

$$F^p(\hat{\mathcal{H}}) = \widehat{(F^p\mathcal{H})}\ ,$$

that is, the augmentation of $F^p\mathcal{H}$ obtained by adjoining forms of the differential algebra generated by the functions (6). Finally we let

(21) $$F^p\mathcal{L}$$

be the subcomplex of $\mathcal{L}$ consisting of differentials which can be written everywhere locally as a sum of terms of the form (9) with

$$\alpha \in F^q\hat{\mathcal{H}} \quad \text{and} \quad (q+r) \geq p\ .$$

The filtration (21) induces a filtration $\hat{F}^p\mathcal{M}_{|t_0|}$ on these complexes from (19). The problem with this filtration is that it does not induce a flat filtration on $H^*(V_{t_0})$ under the action of $\mathbb{R} \leq S$. We define

$$F^p\mathcal{M}_{|t_0|} = \mathcal{J}(\hat{F}^p\mathcal{M}_{|t_0|})$$

where $\mathcal{J}$ is the twisting morphism (15). Notice that, by (16)i), $\mathcal{J}$ is acutally a chain mapping. We then define the *Hodge filtration on the asymptotic cohomology* by

(22) $$F^pH^* = \lim_{|t_0|\to 0} F^pH^*(\mathcal{M}_{|t_0|})\ .$$

We would have gotten the same filtration by using the filtration $\hat{F}^pH^*(V_{t_0})$ induced by the usual Hodge filtration on V^* and defining

$$F^pH^* = \lim_{|t_0|\to 0} \sum_{k=0}^{\infty} \frac{(-1)^k}{k!}\left(\frac{\log t_0}{2\pi i}\right)^k T^{(k)}\ \hat{F}^p(H^*(V_{t_0}))$$

which is exactly the way things are done in Schmid's representation - theoretic construction of the asymptotic Hodge filtration.

The *weight filtration* on H^* is much easier since it comes from an S-invariant filtration on L. The basic recipe is this:

(23) i) ω_j has weight 2,

ii) $d\lambda_j$ has weight 0,

iii) everything else in L has weight equal to its degree.

A little more precisely:

(24)
$$W_r(L) = \{\text{forms which can be written everywhere locally as a sum of terms } \alpha \wedge d\lambda_{j_1} \wedge \cdots \wedge d\lambda_{j_{r'}} \wedge \omega_{k_1} \wedge \cdots \wedge \omega_{k_{r''}} \text{ with } (\deg\alpha + 2r'') \leqslant r\}.$$

The basic properties are:

(25) i) The weight filtration is defined over $\mathbb{R}$.

ii) J preserves the weight filtration.

iii) $L_X(W_r(L)) \subseteq W_{r-2}(L)$.

The property iii) follows directly from the definition of the weight filtration and formulas (11) and (14). We also need to know the action of L_X on the asymptotic Hodge filtration. From the definitions themselves ((11) and (14) again), it is clear that

$$L_X(F^p(L)) \subseteq F^{p-1}(L).$$

Now use that L_X commutes with the twisting map J and with passage to the limit to get

$$L_X(F^pH^*) \subseteq F^{p-1}H^* . \tag{26}$$

The weight and Hodge filtrations on H^* induce a *mixed Hodge structure* on the asymptotic cohomology, in such a way that the logarithm T of the monodromy transformation is an endormorphism of mixed Hodge structures of type $(-1,-1)$. This last assertion is exactly the content of (25)iii) and (26). Rather than explore further the formal implications of these assertions, we will pass on in the next lecture to a specific geometric situation in which the preceding formalism is very natural and has surprisingly strong corollaries.

LECTURE #3
Applications to threefolds

I would now like to apply the constructions of the first two lectures to a very specific situation. Suppose we have a generic linear pencil of hypersurfaces

(1) $$V_t \subseteq \mathbb{P}^4$$

with

(2) $$V_0 = D' \cup D''$$

where D' and D'' are smooth and reduced and intersect transversely along a smooth surface

(3) $$A = D' \cap D'' .$$

In this situation there is a base locus to the pencil which has the form

(4) $$B' \cup B''$$

and $B' \cap B'' = B' \cap A = B'' \cap A =$ a smooth irreducible curve which we call

(5) $$C .$$

To achieve the situation of the first two lectures, we let

(6) $$\mathcal{V} = \{(t,x) : x \in V_t\} \subseteq \Delta \times \mathbb{P}^4 .$$

An easy calculation shows that $\mathcal{V}$ is smooth in a neighborhood of $t = 0$ except that $\mathcal{V}$ contains

$$\{0\} \times C$$

as an ordinary double curve. We obtain

(7) $$p: \tilde{\mathcal{V}} \longrightarrow \Delta_\varepsilon$$

by blowing up the double curve of $\mathcal{V}$ and restricting our attention to a small neighborhood of the fibre over $t = 0$. We can apply the constructions of the first two lectures to the family (7).

Let's return to our original family (1) before the blowing up. The only non-trivial homology of V_t is in dimension three. Let's informally describe a homology basis for $H_3(V_t)$. Take a tubular neighborhood U of A in $\mathbb{P}^4$. The intersection of U with a nearby V_t has the homotopy type of a circle bundle over A except that the circles over points of C have been collapsed to points. We therefore have a Gysin map

(8) $$\phi\colon H_2(A,C) \longrightarrow H_3(V_t) \quad .$$

Furthermore all the third homology of D' and D'' can be supported off A and so "lifts" into a nearby V_t (non-canonically). In fact, using the standard tools of elementary algebraic topology, we have the following diagram which is both horizontally and vertically exact:

(9)
$$\begin{array}{ccccccccc}
 & & H_2(A) & \to & H_2(A,C) & \to & H_1(C) & \to & 0 \\
 & & & & \downarrow \phi & & & & \\
 & & & & H_3(V_t) & & & & \\
 & & & & \downarrow & \searrow^{\nu} & & & \\
0 & \to & H_3(D') + H_3(D'') & \to & H_3(V_0) & \to & H_2(A) & \xrightarrow{\chi} & H_2(D') + H_2(D'') \\
 & & & & \downarrow & & & & \\
 & & & & 0 & & & &
\end{array}$$

The kernel of χ is just $H_2(A)^0$, the primitive homology of A. Let $\nu\colon H_3(V_t) \longrightarrow H_2(A)$ be the composite map in (9). Then it is not hard to show that, in the notation of the first two lectures

$$\begin{array}{rcl}
\gamma_{1*}\colon H_3(V_t) & \longrightarrow & H_3(V_t) \\
\alpha & \longmapsto & \alpha + \phi(\nu(\alpha)) \quad .
\end{array}$$

Thus

$$T_* = \log \gamma_{1*} = \phi \circ \nu$$

and

$$T_*^2 = 0 .$$

The sequences (9) define an increasing filtration on $H_3(V_t)$ given by

$$W_2 = \phi(H_2(A))$$

$$W_3 = \ker \nu$$

$$W_4 = H_3(V_t)$$

which, via duality, defines an increasing filtration on $H^3(V_t)$. (Cohomology will always be with *complex* coefficients, homology with integral coefficients.) This filtration on $H^3(V_t)$ is exactly the weight filtration for the asymptotic mixed Hodge structure associated to the family (7). The only somewhat mysterious point here is why the (homology and) cohomology coming from $H_1(C)$ in (9) should have weight three instead of weight two or one, since a "rule of thumb" is that, the deeper into the singular locus of V_0 that a cycle is supported, the lower its weight. The explanation is that these cycles are *not* supported near the singular set of $\tilde{V}_0$, the central fibre of the resolved family (7).

A consequence of the study of the asymptotic cohomology introduced in the second lecture is the Local Invariant Cycle Theorem, which, in the case now under consideration translates to mean that the sequence

(10) $$0 \longrightarrow W_3H^3(V_t) \longrightarrow H^3(V_t) \xrightarrow{T} W_2H^3(V_t) \longrightarrow 0$$

is exact. In fact all the maps are defined over $\mathbb{Z}$ and the sequence is exact over $\mathbb{Q}$ but not, as we shall see, over $\mathbb{Z}$.

The exactness of (10) together with the fact that the homology of a hypersurface has no torsion implies that we have an injection

$$\phi: \frac{H_2(A)}{\{H_A\}} \longrightarrow H_3(V_t) \tag{11}$$

where H_A is a hyperplane section of A. It is easy to show that the image of ϕ in (11) is a $\mathbb{Z}$-module direct summand of $H_3(V_t)$ and is perpendicular, with respect to the intersection pairing to the kernel of ν. In fact, we have the formula

$$(\phi(\alpha) \cdot \beta)_{V_t} = (\alpha \cdot \nu(\beta))_A . \tag{12}$$

Furthermore the symplectic module $H_1(C)$ can be lifted into $H_2(A,C)$ in such a way that for the composition

$$\iota: H_1(C) \longrightarrow H_3(V_t) \tag{13}$$

induced by ϕ in (9) we have

$$(\gamma \cdot \delta)_C = -(\iota(\gamma) \cdot \iota(\delta))_{V_t} .$$

Similarly, the symplectic modules $H_3(D')$ and $H_3(D'')$ lift into $H_3(V_t)$ as direct summands.

We frame $F^2H^3(V_t)$, cocycles of type $(3,0) + (2,1)$ in the *usual* Hodge filtration, by choosing the dual basis in this space to the basis

(14) $\phi(\alpha_j)$, $j = 1,\ldots,r$, where the α_j's give a basis for $H_2(A)/\{H_A\}$

δ_k, $k = 1,\ldots,s$ running through bases for maximal totally isotropic subspaces of $(H_1(C)$, $H_3(D')$ and $H_3(D'')$ of $(F^2H^3(V_t))^*$.

Since the cycles in (14) are all invariant under the monodromy action γ_1, they give a well-defined framing of $F^2H^3(V_t)^*$ via which we can extend this vector bundle over Δ_ε^* to a (trivial) analytic vector bundle over all of Δ_ε. Let

(15) $$\omega_1(t) \ldots , \omega_r(t) , \eta_1(t) , \ldots , \eta_s(t)$$

be the vectors of $F^2H^3(V_t)$ which give the framing of this bundle dual to the framing (14) of $F^2H^3(V_t)^*$.

Next we recall from the second lecture that we have a commutative diagram

(16) $$\begin{array}{ccc} H^3 & \xrightarrow{1+\theta T} & H^3 \\ \downarrow{\tau_t} & & \downarrow{\tau_{t_0}} \\ H^3(V_t) & \xrightarrow{\gamma_\theta^*} & H^3(V_{t_0}) \end{array}$$

where $t = e^{2\pi i\theta}t_0$ and H^3 is the asymptotic third cohomology group. Using (12), we complete (14) to a symplectic basis of $H_3(V_{t_0})$ as follows:

(17) Γ_j, $j = 1,\ldots,r$, such that the two-cycles $\nu(\Gamma_j)$ in A give a basis for $H_2(A)^0$ which is dual to the basis $\{\alpha_j\}$ of $H_2(A)/\{H_A\}$;

ε_k, $k = 1,\ldots,s$ completing the δ_k's to symplectic bases of $H_1(C)$, $H_3(D')$ and $H_3(D'')$.

So, via Poincare duality, we have a symplectic basis

(18) $$\Gamma_j^* , \phi(\alpha_j)^* , \delta_k^* , \varepsilon_k^*$$

for $H^3(V_{t_0})$ and, via τ_{t_0} in (16) we can transfer this basis to a basis of H^3. To avoid an impending notational nightmare we shall use the same symbols (18) to denote the resulting basis elements in H^3. It is in terms of the basis (18) that we wish to compute the subspace

$$F^2H^3 \leq H^3 .$$

We write the following identities in H^3:

$$\tau_t^{-1}(\omega_i(t)) = \Gamma_i^* + \sum_j \hat{\Omega}(\Gamma,\omega)_{ij}\phi(\alpha_j)^* + \sum_k \Omega(\varepsilon,\omega)_{ik}\delta_k^*$$

$$\tau_t^{-1}(\eta_i(t)) = \sum_j \Omega(\Gamma,\eta)_{ij}\phi(\alpha_j)^* + \varepsilon_i^* + \sum_k \Omega(\varepsilon,\eta)_{ik}\delta_k^* .$$

Here $\hat{\Omega}(\Gamma,\omega)_{ij} = \int_{\Gamma_j(t)} \omega_i(t)$, where

$$\Gamma_j(t) = \text{image of } \Gamma_j \text{ under the action of } \left(\frac{|t|}{|t_0|}, \frac{\arg t - \arg t_0}{2\pi}\right) \in S$$

So, for example,

$$\hat{\Omega}(\Gamma,\omega)_{ij}(e^{2\pi i}t) = \hat{\Omega}(\Gamma,\omega)_{ij}(t) + \int_{T_*(\Gamma_j)} \omega_i(t) .$$

Let's compute this a bit more explicitly:

$$T_*(\Gamma_j) = \phi(\nu(\Gamma_j)) = \phi(\textstyle\sum_k (\nu(\Gamma_j)\cdot\nu(\Gamma_k))\cdot\alpha_k)$$

$$= \textstyle\sum_k (\nu(\Gamma_j)\cdot\nu(\Gamma_k))\phi(\alpha_k) .$$

Thus

$$\hat{\Omega}(\Gamma,\omega)_{ij}(e^{2\pi i}t) = \hat{\Omega}(\Gamma,\omega)_{ij}(t) + (\nu(\Gamma_i)\cdot\nu(\Gamma_j)) ,$$

so that

$$\Omega(\Gamma,\omega)_{ij}(t) = \hat{\Omega}(\Gamma,\omega)_{ij}(t) - (\nu(\Gamma_i)\cdot\nu(\Gamma_j)) \frac{\log t}{2\pi i}$$

is a well-defined analytic function on the punctured disc Δ_ε^* . The other integrals are simpler:

$$\Omega(\varepsilon,\omega)_{ik}(t) = \int_{\varepsilon_k(t)} \omega_i(t)$$

$$\Omega(\Gamma,\eta)_{ij}(t) = \int_{\Gamma_j(t)} \eta_i(t)$$

$$\Omega(\varepsilon,\eta)_{ik}(t) = \int_{\varepsilon_k(t)} \eta_i(t)$$

are all already well-defined analytic functions on Δ_ε^* .

Now let's see what happens when we apply Schmid's twisting operator:

$$\tau_t^{-1}(\sum_k \frac{(-1)^k}{k!} (\frac{\log t}{2\pi i})^k T^{(k)}(\omega_i(t))) =$$

$$\tau_t^{-1}(\omega_i(t)) - (\frac{\log t}{2\pi i})\, \tau_t^{-1}((\nu(\Gamma_i)\cdot\nu(\Gamma_j))\phi(\alpha_j)^* =$$

$$\Gamma_i^* + \sum_j \Omega(\Gamma,\omega)_{ij}\phi(\alpha_j)^* + \sum_k \Omega(\varepsilon,\omega)_{ik}\delta_k^*$$

whereas

$$\tau_t^{-1}(\sum_k \frac{(-1)^k}{k!} (\frac{\log t}{2\pi i})^k T^{(k)}(\eta_i(t))) =$$

$$\sum_j \Omega(\Gamma,\eta)_{ij}\phi(\alpha_j)^* + \varepsilon_i^* + \sum_k \Omega(\varepsilon,\eta)_{ik}\delta_k^* \ .$$

So, in terms of the basis (18) of the asymptotic cohomology group H^3 , the subspace F^2H^3 is given by the limit, as t approaches zero, of the subspace spanned by the rows of the matrix

(19) $$\begin{pmatrix} I & \Omega(\Gamma,\omega)(t) & 0 & \Omega(\varepsilon,\omega)(t) \\ 0 & \Omega(\Gamma,\eta)(t) & I & \Omega(\varepsilon,\eta)(t) \end{pmatrix} .$$

The Regularity Theorem of Griffiths, which was the starting point for the construction of Schmid, says that all entries in the matrix (19) are meromorphic functions of t , so that the limit which we are after is actually well-defined.

In order to apply these constructions, the essential point is to use the full power of the formalism of the asymptotic mixed Hodge structure to compute the orders of the entries of the matrix (19) at $t = 0$. The morphism

$$T: W_4H^3/W_3H^3 \longrightarrow W_2H^3 \tag{20}$$

is given by the rule

$$T(\Gamma_j^*) = \sum_k (\nu(\Gamma_j)\cdot\nu(\Gamma_k))\phi(\alpha_k)^* .$$

Now the content of Schmid's SL_2-orbit Theorem in this situation is that the mapping (20) is an isomorphism of Hodge structures. The Hodge structure on W_2H^3 is simply the Hodge structure of $H^2(A)/\{c_1(H_A)\}$ via the isomorphism

$$\begin{array}{ccc} W_2H^3 & \longrightarrow & H^2(A)/\{c_1(H_A)\} \\ \phi(\alpha_j)^* & \longmapsto & \alpha_j^* \end{array} .$$

The Hodge structure W_4H^3/W_3H^3 is just that of $H^2(A)^0$, the primitive cohomology, via the isomorphism

$$\begin{array}{ccc} W_4H^3/W_3H^3 & \longrightarrow & H^2(A)^0 \\ \Gamma_j^* & \longmapsto & \nu(\Gamma_j)^* \end{array} . \tag{21}$$

This way the mapping (20) is simply the natural inclusion

$$H^2(A)^0 \longrightarrow H^2(A)/\{c_1(H_A)\} .$$

Finally

$$W_3H^3/W_2H^3 \cong H^1(C)\oplus H^3(D')\oplus H^3(D'') .$$

These facts, together with the first Riemann relation on the matrix (19) will allow us to determine the orders of growth of some of the entries of (19) at $t = 0$.

We let

$$r_0 = h^{2,0}(A) \ .$$

There must exist an $(r_0 \times r)$ matrix $C^{2,0}(t)$ of maximal rank such that the forms

$$C^{2,0}(t)\begin{pmatrix}\omega_1(t)\\ \vdots \\ \omega_r(t)\end{pmatrix}$$

lie in $F^3H^3(V_t)$ and the limit of the span of the rows of

(22) $$C^{2,0}(t)\cdot(I \ \ \Omega(\Gamma,\omega) \ \ 0 \ \ \Omega(\varepsilon,\omega))$$

is precisely $F^3H^3/(F^3 \cap W_3)$. So the entries of the matrix (22) must stay bounded as $t \to 0$ and the rows of $C^{2,0}(0)$ must be the period matrix

$$\left(\int_{\alpha_j} \xi_\ell\right)_{\begin{pmatrix}\ell = 1,\dots,r_0\\ \alpha_j = 1,\dots,r\end{pmatrix}}$$

for some basis $\{\xi_\ell\}$ of $H^{2,0}(A)$. Next, suppose that, for some i ,

$$\nu(\Gamma_i) \text{ is an algebraic cycle in } H_2(A)^0 \ .$$

Then we claim that

(23)
$$\int_{\Gamma_i(t)} \omega_j(t) \text{ has at most logarithmic growth,}$$
$$\int_{\Gamma_i(t)} \eta_j(t) \text{ is bounded ,}$$

as $t \to 0$. To see this, we use the first Riemann relation to change the integrals (23) to

$$\int_{\Gamma_j(t)} \omega_i(t) \qquad \int_{\varepsilon_j(t)} \omega_i(t) \quad ,$$

and these integrals are entries in the i-th row of

$$(I \quad \hat{\Omega}(\Gamma,\omega) \quad 0 \quad \Omega(\varepsilon,\omega)) \quad ,$$

that is, to the element

$$\Gamma_i^* \;+\; \sum \hat{\Omega}(\Gamma,\omega)_{ij}\phi(\alpha_j)^* \;+\; \sum \Omega(\varepsilon,\omega)_{ik}\delta_k^*$$

of H^3. What we need to prove, therefore, is that the element

$$(24) \qquad \Gamma_i^* \;+\; \sum \Omega(\Gamma,\omega)_{ij}\phi(\alpha_j)^* \;+\; \sum \Omega(\varepsilon,\omega)_{ik}\delta_k^*$$

stays bounded as $t \to 0$. This is the same as showing that the limit of the one-dimensional subspace spanned by the vector (24) does not lie in W_3H^3. Via the isomorphism (21), this is exactly the statement that $\nu(\Gamma_i)^*$ lies in $F^1H^2(A)^0$, that is, that

$$(\textstyle\sum_j(\int_{\alpha_j} \xi_\ell)(\nu(\Gamma_j)^*)) \wedge (\nu(\Gamma_i)^*) \;=\; 0$$

for each $\xi_\ell \in H^{2,0}(A)$. But the left-hand-side above is

$$\int_{\sum(\nu(\Gamma_i)\cdot\nu(\Gamma_j))\alpha_j} \xi_\ell$$

and so is

$$\int_{\nu(\Gamma_i)} \xi_\ell \quad ,$$

and so is zero, since $\nu(\Gamma_i)$ is algebraic.

LECTURE #4
The Neron Model

At the end of the last lecture, we were treating the period matrix

(1) $$\begin{pmatrix} I & \frac{\log t}{2\pi i}M + \Omega(\Gamma,\omega)(t) & 0 & \Omega(\varepsilon,\omega)(t) \\ 0 & \Omega(\Gamma,\eta)(t) & 0 & \Omega(\varepsilon,\eta)(t) \end{pmatrix}$$

of our degenerating family of threefolds. Implicit in our final argument was the statement that there is a matrix of functions $C^{1,1}(t)$ such that

(2) $$\lim_{t\to 0} \begin{pmatrix} C^{2,0}(t) \\ C^{1,1}(t) \end{pmatrix} \begin{pmatrix} I & \Omega(\Gamma,\omega) & 0 & \Omega(\varepsilon,\omega) \end{pmatrix}$$

was a matrix whose rows give $F^3H^3/F^3 \cap W_3$. Here

$C^{2,0}(0)$ was of maximal rank r_0

$C^{1,1}(0)$ is of maximal rank $r - 2r_0$

where r_0 is the geometric genus of the double surface A and r is its primitive second Betti number. So, in particular all entries in the matrix of functions (2) stay bounded as $t \to 0$.

But we have arranged things so that

(3) $$C^{2,0}(t)\begin{pmatrix} I & \frac{\log t}{2\pi i}M + \Omega(\Gamma,\omega)(t) & 0 & \Omega(\varepsilon,\omega)(t) \end{pmatrix}$$

is actually a partial period matrix for $F^3H^3(V_t)$. So, by Griffiths' infinitesimal period relation, the *derivative* of this matrix with respect to t must have rows giving elements of $F^2H^3(V_t)$. In fact, if

(4) $$\left.\frac{\partial}{\partial t} C^{2,0}(t)\right|_{t=0}$$

has rows which span

$$H^2(A)^0 / F^1H^2(A)^0$$

under the isomorphism

$$W_4H^3 / W_3H^3 \longrightarrow H^2(A)^0 \quad ,$$

then we can frame $F^2H^3(V_t)$ with differentials

$$\begin{pmatrix} C^{(2,0)}(t) \\ C^{(1,1)}(t) \end{pmatrix} \begin{pmatrix} \omega_1(t) \\ \vdots \\ \omega_r(t) \end{pmatrix}$$

together with the derivatives of

$$C^{(2,0)}(t) \begin{pmatrix} \omega_1(t) \\ \vdots \\ \omega_r(t) \end{pmatrix}$$

with respect to the Gauss-Manin connection, instead of using the differentials $\omega_1(t),\ldots,\omega_r(t)$ themselves. I am sure that this condition on the matrix (4) is satisfied for a generic degeneration of hypersurfaces (although, to my knowledge, no proof has been written down yet). In my paper on Double Solids in Advances in Mathematics, the proof is given in the case of a generic degeneration of a double branched covering of $\mathbb{P}^3$ into a union of two components meeting transversely. Anyway we will assume the condition on the matrix (4) - in the final examples toward which we are heading the condition is vacuously satisfied.

The advantage of making this assumption is that, since the derivative of the matrix (3) has only first order poles and logarithmic terms, and since the rows of

$$\begin{pmatrix} \frac{\partial C^{2,0}(t)}{\partial t} & \frac{C^{2,0}M}{2\pi i t} + \frac{\partial (C^{2,0}\Omega(\Gamma,\omega))}{\partial t} & 0 & \frac{\partial C^{2,0}\Omega(\varepsilon,\omega)}{\partial t} \end{pmatrix}$$

must span a space which, as $t \to 0$, gives exactly $F^2W_2H^3$, we can conclude that, in fact, *all* entries of the matrix $\Omega(\varepsilon,\omega)$ in (1) must stay bounded as $t \to 0$, and all entries of $\Omega(\Gamma,\omega)$ had at worst first-order poles and

$$\lim_{t\to 0} t\Omega(\Gamma,\omega)(t)$$

has rank exactly r_0. So $\Omega(\Gamma,\eta)$ stays bounded by the first Riemann relation, and $\Omega(\varepsilon,\eta)$ converges to the period matrix of $H^1(C) + H^3(D') + H^3(D'')$. These facts, together with the fact that the matrix

$$M = ((\nu(\Gamma_i) \cdot \nu(\Gamma_j))_{ij})$$

in (1) is of maximal rank incorporate themselves into the following result:

(5) The family

$$J(V_t) = (F^2H^3(V_t))^* / H_3(V_t)$$

of intermediate Jacobians of our hypersurfaces V_t can be continued over $t = 0$ by inserting there a generalized complex torus J. J fits in an exact sequence of abelian complex Lie groups

$$1 \to (\mathbb{C}^*)^r \to J \to J(C) + J(D') + J(D'') \to 0 .$$

The reason for this conclusion is that $J(V_t)$ is formed by dividing $\mathbb{C}^{r+s}$ by the lattice generated by the columns of the matrix (1). As t approaches zero, this lattice "converges" to the partial lattice generated by the columns of

$$\begin{pmatrix} I & 0 & \Omega(\varepsilon,\omega)(0) \\ 0 & I & \Omega(\varepsilon,\eta)(0) \end{pmatrix} .$$

J is simply the quotient of $\mathbb{C}^{r+s}$ by the (lattice generated by the) columns of this last matrix.

However, the Lie group J is not "big enough" because there are sections of the "Jacobian bundle"

$$(6) \qquad \mathcal{J} \longrightarrow \Delta$$

which do not extend over $t = 0$ when we use J as the fibre of (6) over $t = 0$. To see the difficulty, let's pass to the simplest case in which it occurs. We start with the degeneration $\mathcal{E}$ of elliptic curves with fibre

$$E_t = \frac{\mathbb{C}}{\mathbb{Z} + \mathbb{Z}\left(\frac{\log t}{2\pi i}\right)} .$$

We "base-extend" by forming the fibred product

$$\begin{array}{ccc} \mathcal{D} & \dashrightarrow & \mathcal{E} \\ \downarrow & & \downarrow \\ \Delta & \longrightarrow & \Delta \\ s & \longmapsto & t = s^2 \end{array} .$$

The surface $\mathcal{D}$ as an ordinary node over the double point of E_0. We blow up this node in $\mathcal{D}$ to obtain a smooth manifold

$$\tilde{\mathcal{D}} \longrightarrow \Delta$$

with fibre D_s an elliptic curve isomorphic to E_{s^2} when $s \neq 0$ and

$$D_0 = D' \cup D'' ,$$

where D' and D'' are two rational curves meeting at two points:

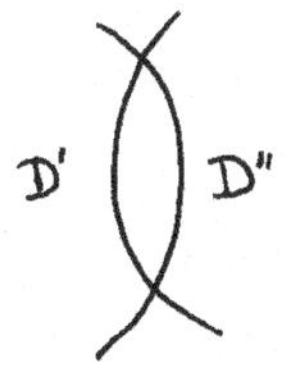

If D' is the proper transform of E_0, then the analogue of our family (6) of tori in this situation would be the family

$$(\tilde{\mathcal{D}} - D'') \longrightarrow \Delta \ . \tag{7}$$

But it is clear that a disc in $\tilde{\mathcal{D}}$ which meets D_0 at a smooth point lying in D'' gives a section of (7) which does not extend over $t = 0$.

This example is even more instructive when we look at the family of period matrices associated to $\tilde{\mathcal{D}}$:

$$D_s = \frac{\mathbb{C}}{\mathbb{Z} + \mathbb{Z}\left(\frac{2 \log s}{2\pi i}\right)} \ .$$

It is the factor of "2" in front of the $\log s$ which is causing the problem. Analogously we should expect problems because, in (1),

$$\det M = (\text{degree of } A \text{ in } \mathbb{P}^4) \tag{8}$$

is almost never equal to one.

So now, to build the Neron model, that is, the object that allows the extension of sections of

$$J \longrightarrow \Delta$$

over Δ^* to sections over all of Δ, we should first give a little more information about the sections which we are trying to extend. This is contained in the following result:

(9) Theorem: Let $\{Z_t\}_{t \in \Delta}$ be a family of algebraic one-cycles in $\{V_t\}$ such that:

i) $Z_t \subseteq U_t$ for each t, where $\{U_t\}$ is a holomorphic family of smooth ample divisors on $\mathbb{P}^4$ which meet V_0 transversely,

ii) each Z_t is homologous to zero in V_t via a multivalued family of three-chains $\Gamma_t \subseteq V_t$ with

$$\partial\Gamma_t = Z_t ,$$

iii) $Z_0 = Z' - Z''$ with $Z' \subseteq D'$ and $Z'' \subseteq D''$ and with an *algebraic* cycle $\alpha \in H_2(A)$ having the property that

$$\begin{aligned} Z' &\sim \alpha \quad \text{in} \quad H_2(D') \\ Z'' &\sim \alpha \quad \text{in} \quad H_2(D'') . \end{aligned}$$

Then the integrals

$$\int_{\Gamma_t} \omega_i$$

have at most logarithmic growth at $t = 0$ and the integrals

$$\int_{\Gamma_t} \eta_i$$

extend holomorphically over $t = 0$.

The proof of this theorem is very similar in concept to the proof at the end of Lecture #3 of the analogous statements for

$$\int_{\Gamma_i}$$

in case that $\gamma_i = \nu(\Gamma_i)$ is algebraic. The idea is to deform the singular varieties

$$V_t \cup U_t$$

in a new direction to obtain a two-parameter family

(10)
$$\{X_{(s,t)}\}_{(s,t) \in \Delta\times\Delta}$$

where $X_{(s,t)}$ is a smooth divisor in $\mathbb{P}^4$ unless $s = 0$ or $t = 0$. By taking chains

$$\Sigma_t \subseteq U_t$$

with

$$\partial\Sigma_t = Z_t$$

we build a cycle

$$\tilde{\Gamma}(0,t) = \Gamma_t - \Sigma_t$$

in $X_{(0,t)}$ and deform it to a cycle

$$\tilde{\Gamma}_{(s,t)}$$

in the family (10). Now work of Cattani and Kaplan, together with the several variables form of the Schmid Nilpotent Orbit Theorem allows us to put a mixed Hodge structure on the asymptotic cohomology

$$\tilde{H}^3$$

of the two-variable degeneration (10), such that there is a natural morphism of mixed Hodge structures

$$\begin{array}{ccc} \tilde{H}^3 & \longrightarrow & H^3(X_{(s,t)}) \\ \downarrow & & \downarrow \text{ projection onto a symplectic direct summand} \\ H^3 & \longrightarrow & H^3(V_t) \end{array} .$$

Under the hypothesis of Theorem 9, $\tilde{\Gamma}^*_{(s,t)}$ is formally the same in $\tilde{H}^3$ as Γ_i^* with $\nu(\Gamma_i)$ algebraic was in H^3. The proof of boundedness or logarithmic growth of

$$\int_{\tilde{\Gamma}(s,t)}$$

is completely analogous to the proof given at the end of the third lecture.

The degeneration, for *fixed* t,

$$X_{(s,t)} \longrightarrow (V_t \cup U_t)$$

is exactly the kind of degeneration studied in the third lecture. Here V_t plays the role of D' and U_t the role of D''. Here, as before, we frame

$$F^2H^3(X_{(s,t)})$$

by a basis $\tilde{\omega}_i$, $\tilde{\eta}_i$ such that, say, the first, $(r+s)$ $\tilde{\eta}_i$ restrict to

$$\begin{array}{ccl} \omega_1,\ldots,\omega_r,\eta_1,\ldots,\eta_s & \in & F^2H^3(V_t) \\ 0 & \in & F^1H(\text{"base curve"}) \\ 0 & \in & F^2H^3(U_t) \end{array} .$$

Then if $i = 1,\ldots,r+s$

$$\lim_{s\to 0} \int_{\tilde{\Gamma}(s,t)} \tilde{\eta}_i = \begin{cases} \int_{\Gamma_t} \omega_i & \text{if } i = 1,\ldots,r \\ \int_{\Gamma_t} \eta_{(i-r)} & \text{if } i = r+1,\ldots,s \end{cases} .$$

Moreover we have the explicit formula for differentials in $F^2H^3(V_t)$:

$$\int_{\Gamma_{e^{2\pi i}t}} = \int_{\Gamma_t} + \int_{\phi(\alpha)} \tag{11}$$

where $\alpha \in H_2(A)$ is the cycle appearing in Theorem 9. Notice that α does not necessarily lie in $H_2(A)^0$ so that there need not exist any element

$$\Gamma(t) \in H_3(V_t)$$

such that

$$\begin{aligned} \Gamma(e^{2\pi i}t) - \Gamma(t) &= \Gamma_{e^{2\pi i}t} - \Gamma_t \\ &= \phi(\alpha) . \end{aligned}$$

In fact, this last equation will hold for some $\Gamma(t) \in H_3(V_t)$ exactly when $\alpha \in H_2(A)^0$, that is, exactly when the section

$$\int_{\Gamma_t}$$

of the bundle (6) can be extended across $t = 0$. In essence, if $\alpha \notin H_2(A)^0$, there is no way to modify $\int_{\Gamma_t}$ by a period to get an integral which stays bounded as $t \to 0$.

We remedy this situation as follows. Let

(12) $$G = H_2(A)\big/(H_2(A)^0 + \mathbb{Z}\,(\text{hyperplane section}))\ .$$

Then G is a finite cyclic group whose order is equal to the degree of the "double surface A in $\mathbb{P}^4$". The transformation

$$T:\ W_4H_3(V_t)\big/W_3H_3(V_t) \longrightarrow W_2H_3(V_t)\ ,$$
$$\Gamma \longmapsto \phi(\nu(\Gamma))$$

which is the logarithm of the monodromy transformation, can be tensored with $\mathbb{Q}$. Let

$$\Lambda = T_{\mathbb{Q}}^{-1}(W_2H_3(V_t))\ .$$

Then $T_{\mathbb{Q}}$ induces an isomorphism

(13) $$\Lambda\big/W_4H_3(V_t) \longrightarrow G\ .$$

We identity G with the left-hand group of (13), and so, with the group of sections

(14) $$\int_\lambda \in J(V_t)\ ,\quad \lambda \in \Lambda\ ,$$

of the Jacobian bundle J over Δ^* .

Now take $|G|$ copies of the bundle J and index them by elements $\lambda \in G$. We identify a point x in the fibre of J_{λ_1} over $t \neq 0$ with a point y in the fibre of J_{λ_2} over the same t if and only if

$$(x-y) = \int_{\lambda_1-\lambda_2} \in J(V_t) .$$

The result is a smooth complex manifold

$$(15) \qquad \hat{J} \longrightarrow \Delta$$

which is isomorphic to J over Δ^* and whose fibre $\hat{J}$ over $t = 0$ fits into the exact sequence

$$(16) \qquad 0 \to J \to \hat{J} \to G \to 0 .$$

In fact, since the sections (14) of J over Δ^* (corresponding to the elements of G) extend to sections of $\hat{J}$ over all of Δ, the sequence (16) is split.

Now let $\int_{\Gamma_t}$ be a section of J over Δ^* where $\partial\Gamma_t = Z_t$ is a holomorphic family of curves satisfying the conditions of Theorem 9. Then suppose that, under the isomorphism (13), the element $\lambda \in \Lambda$ corresponds to $\alpha \in H_2(A)/(H_2(A)^0 + \mathbb{Z}\{H_A\})$. Here α is the algebraic curve on A which Theorem 9 associates to the family $\{Z_t\}$. Then

$$(17) \qquad \int_{\Gamma_t} - \int_{\lambda}$$

is a well defined element of $(F^2H^3(V_t))^*$ over Δ^* with at most logarithmic growth at $t = 0$. Thus the element (17) is in fact *bounded* at $t = 0$ and so gives a well-defined section of the Jacobian bundle J. So

$$(18) \qquad \int_{\Gamma_t}$$

gives a well-defined section of the Neron model $\hat{J}$ which, over $t = 0$, passes through the component of $\hat{J}$ indexed by λ. Furthermore, the extension of (18) over $t = 0$ varies holomorphically with respect to auxiliary parameters.

Finally there is the question of compactifying J. We will do this in a non-standard way which does not have many of the advantages of compactifications of families of abelian varieties by Mumford and others, but which is quite simple, and adequate for our purposes. We begin by dividing out $\hat{J} \to \Delta$ by the finite group G of sections to obtain a family of tori

$$T \longrightarrow \Delta \quad . \tag{19}$$

The point of all this is that the monodromy matrix of the family (19) with respect to the correctly chosen basis

$$\lambda_j \quad , \quad \phi(\alpha_j) \quad , \quad \delta_k \quad , \quad \varepsilon_k$$

of the lattice now has the form:

$$\begin{matrix} I & I & 0 & \} \; r \quad \text{rows} \\ 0 & I & 0 & \} \; r \quad \text{rows} \\ 0 & 0 & I & \} \; 2s \quad \text{rows} \; . \end{matrix}$$

Now we need that the period matrix for the family T have the form

$$\begin{matrix} I & \frac{\log t}{2\pi i} I + \Omega_{r,r}(t) & 0 & \Omega_{r,s}(t) \\ 0 & \Omega_{s,r}(t) & I & \Omega_{s,s}(t) \end{matrix}$$

where the matrix $\Omega_{r,r}(t)$, as well as the others, stays bounded as $t \to 0$. We can only achieve this under the assumption, which we will now make, that

$$r_0 = \operatorname{rank}(F^2H^3 \cap W_2H^3) = 0 ,$$

that is, that *all* of the cycles in the kernel of the natural map

$$H_2(A) \longrightarrow H_2(D') + H_2(D")$$

are algebraic.

Under this assumption, which lowers the degree of our family of hypersurfaces $\{V_t\}$ to four or less, it is easy to compactify the fibre of over zero - this degeneration is "essentially" the same as a product of degenerations

$$E_t \longrightarrow E_0$$

of an elliptic curve to a rational curve with one node and so the compactification of the fibre of $\mathcal{T}$ over 0 is "essentially" the same as a product of singular curves E_0 .

We call the compactified family $\overline{\mathcal{T}}$. We have a map

(20) $$\hat{\mathcal{J}} \longrightarrow \mathcal{T} \subseteq \overline{\mathcal{T}} .$$

The image of the fundamental group of $\hat{\mathcal{J}}$ in $\overline{\mathcal{T}}$ gives rise to the maximal covering space

$$\overline{\mathcal{J}} \longrightarrow \overline{\mathcal{T}}$$

through which the mapping (20) factors. In fact the induced mapping

$$\hat{\mathcal{J}} \hookrightarrow \overline{\mathcal{J}}$$

is an open immersion, fibred over Δ . $\overline{\mathcal{J}}$ is the desired compactification of $\hat{\mathcal{T}}$

It is instructive to pursue the analogue of the construction of $\hat{\mathcal{J}}$ and $\mathcal{J}$ in the case of the degenerating family

$$(\tilde{\mathcal{D}} - D'') \longrightarrow \Delta$$

of elliptic curves which we looked at earlier (see (7)). The analogue of $\hat{\mathcal{J}}$ is

$$\tilde{\mathcal{D}} - (\text{two singular points of fibre of } \tilde{\mathcal{D}} \text{ at } t = 0) .$$

The analogue of $\mathcal{J}$ is exactly $\tilde{\mathcal{D}}$.

LECTURE #5
The Quartic Double Solid

Our goal in this lecture is to give a compelling example of the techniques which we have been discussing.* We have made all our constructions only for a pencil $\{V_t\}$ of hypersurfaces in $\mathbb{P}^4$, but they hold more generally. In particular they hold for a pencil of hypersurfaces in a weighed projective space, and so, in the following situation.

Let

$$\pi_t : V_t \longrightarrow \mathbb{P}^3 \tag{1}$$

be a family of double covers of $\mathbb{P}^3$ with branch locus

$$B_t = (\text{fourth degree surface in } \mathbb{P}^3) . \tag{2}$$

We assume that the B_t move in a linear pencil so that

$$B_0 = 2A$$

for a smooth quadric $A \subseteq \mathbb{P}^3$ and that

$$C = B_t \cdot A , \quad t \neq 0 , \tag{3}$$

is smooth and reduced.

In this case,

$$D' \cong D'' \cong \mathbb{P}^3 ,$$

*In fact it was the suggestion of F. Oort that the constructions in this example "looked like the Neron model" which led me to work out these techniques in general.

and (genus C) = 9 . So

$$h^{3,0}(V_t) = h^{3,0}(D') + h^{3,0}(D'') + h^{2,0}(A) = 0$$

$$h^{2,1}(V_t) = h^{2,1}(D') + h^{2,1}(D'') + h^{1,0}(C) + h^{1,1}(A) - 1 = 10 .$$

We have one transverse cycle $\Gamma = \Gamma'-\Gamma''$

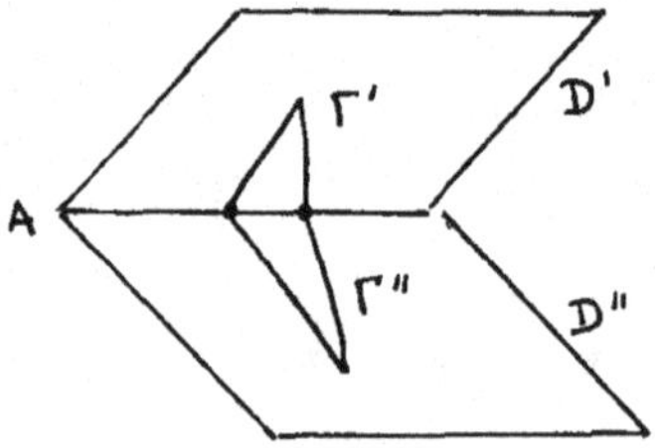

where $\partial\Gamma' = \gamma$, a generator of $H_2(A)^0$. γ is homologous to a cycle

$$L - M ,$$

where L is a line from one of the two rulings of A , and M is a line from the other.

As for the "vanishing cycles", they are generated by

$$\phi(L) = \phi(M) \in H_3(V_t) .$$

This is because

$$\{L\} \equiv \{-M\} \text{ mod hyperplane section of } A .$$

So if we again use T_* to denote the logarithm of the monodromy transformation on homology

$$T_*(\Gamma) = \phi(\gamma) = \phi(L) - \phi(M) = 2\phi(L) .$$

That is, with respect to our basis

$$\Gamma \ , \ \phi(L) \ , \ \delta_1,\ldots,\delta_9 \ , \ \varepsilon_1,\ldots,\varepsilon_9$$

of $H_3(V_t)$, the monodromy matrix has the form

(4)
$$\begin{pmatrix} \begin{matrix} 1 & 2 \\ 0 & 1 \end{matrix} & 0 \\ 0 & I \end{pmatrix} .$$

Now V_t is polarized by the inverse image of a hyperplane in $\mathbb{P}^3$. We will call an effective algebraic one-cycle on V_t a *line* if it is of degree one with respect to this polarization. It is easy to see that every line in V_t lies over a bitangent to the quartic surface B_t in $\mathbb{P}^3$, and that, if we have chosen our pencil

$$B_t \longrightarrow 2A$$

generically, every bitangent to B_t , $t \neq 0$, has over it two distinct lines in V_t . Thus:

(5) The variety S_t of lines on V_t is an unbranched double cover of the surface T_t of bitangents to the quartic B_t .

That S_t is connected can be seen by specializing B_t to a smooth quartic containing a line in which case the double covering of T_t becomes branched at that point.

The surface

$$T_t \subseteq Gr(2,4)$$

is easily seen to be smooth - for example, if $K \subseteq \mathbb{P}^3$ is a hyperplane such that $K \cdot B_t$ is non-singular, then the variety of lines in K intersects T_t in 28 distinct points, each of multiplicity 1 , and every line in T_t lies in some such K . We will also be very interested in the asymptotic behavior of T_t as $t \to 0$. We can make this analysis "one plane at a time." That is, fix K such that K meets both A and the base curve C transversely. Then, as t approaches zero,

$$(6) \qquad (K \cdot B_t) \longrightarrow 2(K \cdot A) \quad ,$$

that is, we have a family of plane quartics degenerating to a conic counted twice. The 28 bitangents of B_t specialize to the

$$\binom{8}{2} = 28$$

bisecants of $(K \cdot C)$. Thus

$$T_0 = (\text{variety of bisecants of } C),$$

and T_0 is acquired with multiplicity one.

In fact, T_0 is non-singular *except* along those points corresponding to lines which actually lie inside the surface A . Let L be a generic line in one of the rulings of A :

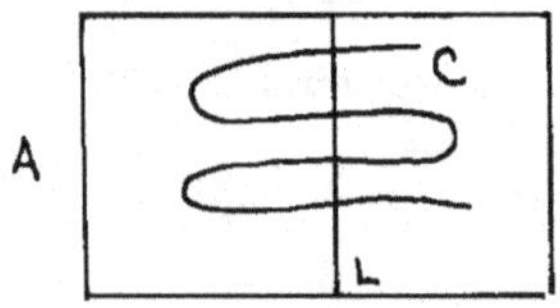

Such a line L never lies in a plane K which meets A transversely. To see what happens at such a line $L \subseteq A$, choose a generic plane K containing L . We obtain a family of plane quartic curves

$$(B_t \cdot K) \tag{7}$$

which degenerates to $2(A \cdot K) = 2L + 2M$. This degeneration is studied by base extension $(s^2 = t)$ and normalization. The family of curves parametrized by s degenerates to the union of the two elliptic curves obtained as double covers of L and M branched along their intersection with C. These two elliptic curves meet at the two points lying over the point $L \cap M$. Since the monodromy transformation for the degeneration in the variable s is the identity modulo 2, the 28 bitangents of the curve (7) are s-invariant. However as $t \to 0$, six of the twenty-eight bitangents become coincident with L, six others become coincident with M, and the

$$4 \times 4 = 16$$

others stay separate from all the others, each joining a point of $(L \cap C)$ with a point of $(M \cap C)$. As t goes once around zero, each of the 6 lines associated to L goes to a partner - that is, the monodromy breaks the set of 6 lines up into three sets of two lines each. These same considerations continue to hold if $\{K_t\}$ is a variable family of planes instead of just a constant plane. The only thing that matters is whether K_0 is transverse or tangent to the quadric A.

Now let

$$\mathcal{T} = \bigcup_{t \in \Delta} \{t\} \times T_t' \subseteq \Delta \times Gr(2,4) .$$

Then, if $\tilde{\mathcal{T}}$ = normalization of $\mathcal{T}$ and

$$\tilde{T}_t = \text{fibre of } \tilde{\mathcal{T}} \longrightarrow \Delta , \tag{8}$$

the above discussion shows that $\tilde{T}_0$ has two disjoint curves E_1 and E_2 which are each three-sheeted branched covers of one of the $\mathbb{P}^1$'s which comprise the singular locus of T_0. In fact, we can describe the curves E_1 and E_2 more precisely. They occur in a construction of S. Recillas for realizing the Jacobian of a curve with a g^1_4 as the Prym variety of a curve with a g^1_3. Namely, the normalization of the surface T_0 is $C^{(2)}$, the second symmetric product of our "base curve" C. $C^{(2)}$ has two disjoint distinguished curves

(9)
$$D_1 = \{(p,q) \in C^{(2)} : \text{the line } \overline{pq} \text{ lies in the first ruling of } A\}$$
$$D_2 = \{\ \ldots\ldots\ \text{second ruling of } A\}.$$

Then there is a natural mapping

(10)
$$D_j \longrightarrow E_j .$$

Again by studying the monodromy of the family of curves (7), one concludes that this map (10) sends (p,q) and (r,s) to the same point of E_j exactly if

$$(p+q+r+s) \in g^1_4(j)$$

where $g^1_4(j)$ is the g^1_4 on C cut out by the j-th ruling of A. Thus the double cover (10) is unbranched.

Now go back to the situation in which we have a family of planes

$$K_t \subseteq \mathbb{P}^3$$

such that K_0 meets A transversely. Then the family of surfaces

$$\pi_t^{-1}(K_t) \subseteq V_t$$

meets all the conditions laid down in Theorem 9 of the last lecture, so that, if

I make two continuous choices of lines

(11) $$L_t \, , \, M_t \subseteq \pi_t^{-1}(K_t) \quad ,$$

then the "normal functions"

(12) $$\int_{M_t}^{L_t} \in J(V_t) \, , \, t \neq 0 \quad ,$$

extend to sections of the Neron model

(13) $$\hat{J} \longrightarrow \Delta \quad .$$

The monodromy matrix (4) tells us that the fibre $\hat{J}$ of (13) over 0 has two components.

So if we let

(14) $$S = \bigcup_{t \in \Delta} (\{t\} \times S_t) \quad ,$$

where S_t is, as before, the family of lines on V_t, and if we choose a section $\{M_t\}$ of

$$S \longrightarrow \Delta \, ,$$

the formula (12), now for arbitrary $\{L_t\} \in S_t$ allows us to define a morphism

(15) $$\begin{array}{ccc} \hat{S} & \xrightarrow{\quad \Phi \quad} & \hat{J} \\ & \searrow \quad \swarrow & \\ & \Delta & \end{array} \quad ,$$

where $\hat{S} = S$-(singular set of S_0) . Also by (17) and (18) of the last lecture, a section $\{L_t\}$ will map to the component of 0 in $\hat{J}$ if and only if L_0 and M_0 are in the same component of V_0 .

Furthermore the mapping Φ in (15) can be computed over $t = 0$. The component J of zero in $\hat{J}$ is given by an exact sequence

(16) $$1 \to \mathbb{C}^* \to J \to J(C) \to 0$$

with obstruction to splitting given by

(17) $$g_4^1(1) - g_4^1(2) \in \mathrm{Pic}^0 C = J(C)$$

where, as before, these two g_4^1's are those cut out on C by the two rulings of A . If we let

$$\hat{S} = (\text{fibre of } \hat{S} \to \Delta \text{ over } t = 0) ,$$

then $\hat{S}$ has two disjoint components S' and S'' each of which is isomorphic to

(18) $$C^{(2)} - (D_1 \cup D_2) .$$

The mapping Φ on S' is the natural embedding of (18) into J which sends D_1 to zero and D_2 to infinity. The basepoint, that is, the point which goes to the identity of J is $(M_0 \cap C)$.

Now the family of threefolds

$$\pi_t : V_t \longrightarrow \mathbb{P}^3$$

carries a natural involution given by "sheet interchange" of the double coverings of projective space:

(19) $$\iota : V_t \longrightarrow V_t \quad .$$

This involution induces one on S which interchanges the two components of $\hat{S}$. ι also induces the map (- identity) on cohomology and so also on $J(V_t)$. So the mapping of S'' unto the other component of $\hat{J}$ must fit to make a commutative diagram

$$\begin{array}{ccccc} & \nearrow & S' & \longrightarrow & J \\ C^{(2)} - (D_1 \cup D_2) & & & & \downarrow \chi \\ & \searrow & S'' & \longrightarrow & (\text{other comp. of } \hat{J}) \end{array}$$

where χ is the composition of the group-theoretic inverse mapping on J and a translation.

We next let

$$\tilde{S} \quad = \quad (\text{normalization of } S)$$

and claim there is an extension of (15) to a morphism

(20) $$\Phi : \tilde{S} \longrightarrow \bar{J}$$

where $\bar{J}$ is the compactification of the Neron model. To see this, first go back to the degeneration (7). Corresponding to it is a degeneration

$$\pi_t^{-1}(K) \longrightarrow \pi_0^{-1}(K)$$

of rational surfaces. An elementary study of the monodromy of the family of lines on $\pi_t^{-1}(K)$ shows that locally S is an unbranched double cover of T and so:

(21) The natural mapping

$$\begin{array}{ccc} \tilde{S} & \longrightarrow & \tilde{T} \\ & \searrow \quad \swarrow & \\ & \Delta & \end{array}$$

is an *unbranched* double covering.

Therefore the fibre of $\tilde{S} \to \Delta$ over $t = 0$, which we will call $\tilde{S}_0$, is an unbranched double cover of $\tilde{T}_C$. $\tilde{S}_0$ consists of two copies of $C^{(2)}$ with D_j in one copy identified to D_j in the other copy via the natural fix-point-free involution on D_j.

So given $\ell_0 \in \tilde{S}_0$ such that ℓ_0 corresponds to a line $L_0 \subseteq A$, let $\{L_s\}_{s \subset \Delta}$, $s^2 = t$, be a generic "double section" of

$$\tilde{S} \longrightarrow \Delta$$

which specializes to ℓ_0 at $s = 0$. We can then "base extend" to obtain a family of varieties

$$W_s = V_{s^2}$$

parametrized by the s-disc, a family of intermediate Jacobians $\{J(W_s)\}$ parametrized by the punctured s-disc and a normal function

$$\int_{M_{s^2}}^{L_s} \in J(W_s) .$$

Here $M_{s^2} = M_t$ is as in (12). The Neron model over the s-disc has four components, arranged cyclicly. The two "new" components can be obtained by forming the fibred product

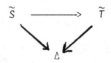

$$\begin{array}{ccc} H & \dashrightarrow & \bar{J} \\ \downarrow & & \downarrow \\ \Delta & \longrightarrow & \Delta \\ s & \longmapsto & s^2 = t \end{array}$$

and then blowing up the two components of the singular set of H .

We let

$$\omega \, , \, \eta_1 \, , \, \ldots \, , \, \eta_9$$

be the framing of $F^2H^3(W_s) = F^2H^3(V_{s^2})$ used in previous lectures. Then, just as before:

$$\text{(21)i)} \quad \int_{M_{s^2}}^{L_s} \omega = \frac{\log s}{2\pi i} + \quad \text{(bdd. hol. fn. of } s)$$

$$\text{ii)} \quad \int_{M_{s^2}}^{L_s} \eta_k = \text{(bdd. hol. fn. of } s) \quad .$$

A corollary of (21)i) is that all adherence values of Φ in (15) at ℓ_0 lie in $\bar{J} - \hat{J}$. Also, as before, the value of the functions (21)ii) at $s = 0$ can be computed and seen to give a point

$$(pq) \in D_1 \cup D_2 \subseteq C^{(2)} \subseteq J(C)$$

such that p and q lie on the line L_0 . Thus the set of adherence values is finite.

In case the line $L_0 \subseteq A$ is tangent to the curve C , the situation is a bit more complicated but the outcome is the same - the set of adherence values is finite. So, by Zariski's Main Theorem, we obtain a morphism

(22) $$\Phi : \tilde{S} \longrightarrow \mathcal{J}$$

over Δ . Since Φ is a closed embedding on the fibre over $t = 0$, it must in fact be an embedding in a neighborhood of the fibre over zero.

Many consequences flow easily from the embedding (22). One is the isomorphism

(23) $$\mathrm{Alb}(S_t) \longrightarrow J(V_t)$$

for generic t as well as the fact that $H_1(S_t)$ is torsion-free. Also the cohomological invariants of the surface T_t of bitangents to a quartic can be computed from those of the fibre of $\tilde{T} \to \Delta$ over $t = 0$.

To me, the deepest use of this asymptotic construction has been made by Welters in his Utrecht thesis. I would like to end this lecture by describing some of his results. We start with the *incidence curve* with respect to $s \in S_t$:

$$D_{s'} = \{s \in S_t : L_s \cap L_{s'} \neq \phi\} ,$$

where $L_s \subseteq V_t$ is the line parametrized by $s \in S_t$. Now let s' specialize to a point in $S' \cong C^{(2)} - (D_1 \cup D_2)$. Let L_0 denote the corresponding bisecant of C in D' , and let p and q denote the two points of intersection of C and L_0 . Then $D_{s'}$ specializes to a curve in $\tilde{S}_0$ with three components:

$$\alpha' \quad = \quad \{(x,y) \in C^{(2)} : \ p,q,x,y \text{ are coplanar}\} \subseteq S'$$

$$\beta''_p \quad = \quad \{(p,x) \in C^{(2)}\} \subseteq S''$$

$$\beta''_q \quad = \quad \{(q,x) \in C^{(2)}\} \subseteq S'' \quad .$$

α' meets D_1 in three points lying over $\{L\}$ where $p \in L \subseteq A$ and three more over $\{L\}$ with $q \in L \subseteq A$. Similarly α' meets D_2 in two sets of three points each. β''_p meets D_1 in 3 points corresponding under the involution on D_1 to the first three points in which α' meets D_1, etc.

These six points pair to $\alpha' \cdot D_1$ under the involution on D_1

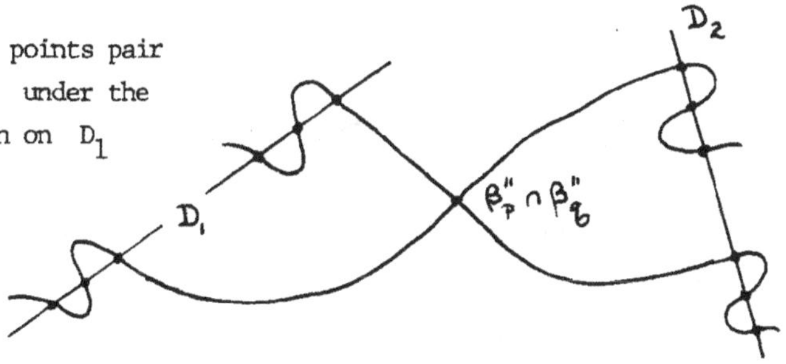

these six points pair to $\alpha' \cdot D_2$ under the involution on D_2

Welters shows that, if $\Theta_t \subseteq J(V_t)$ is the theta-divisor, then

$$(24) \qquad (\Theta_t \cdot S_t)_{S_t} \sim D_{s_1} + D_{s_2} + D_{s_3}$$

($\sim$ denotes algebraic equivalence). He shows that, by specializing s_1, s_2 and s_3 each to a different line in $\tilde{S}_0$, we obtain a divisor

$$(25) \qquad (\alpha'_1 + \alpha'_2 + \beta'_3) \quad + \quad (\beta''_1 + \beta''_2 + \alpha''_3)$$

in $\tilde{S}_0$ such that

(26) $$\dim H^0(\widetilde{S}_0\ ,\ \mathcal{O}\ (\text{divisor } (25))) \quad = \quad 1\ .$$

This is done via a Mayer-Vietoris argument to reduce the result to computations on $C^{(2)}$, and these in turn are accomplished by looking at restrictions to various curves. The result (26), together with semicontinuity, imply that, for generic t and generic choice of $s_1, s_2, s_3 \in S_t$,

(27) $$\dim H^0(S_t;\ \mathcal{O}(D_{s_1} + D_{s_2} + D_{s_3})) \quad = \quad 1\ .$$

Next, there is a commutative diagram of isomorphisms

$$\begin{array}{cccc} s & \mathrm{Alb}(S_t) & \xrightarrow{\ \lambda\ } & \\ \big\downarrow & \big\downarrow & & J(V_t) \\ \{D_s - D_{s_0}\} & \mathrm{Pic}^0(S_t) & \xleftarrow{\ \lambda^*\ } & \end{array}$$

so that the map

$$\begin{array}{ccc} J(V_t) & \longrightarrow & \mathrm{Pic}^0(S_t) \\ a & \longmapsto & \{(a+\Theta_t)\cdot S_t \ - \ 3D_{s_0}\} \end{array}$$

is generically surjective. So by (27), for generic $a \in J(V_t)$

(28) $$\dim H^0(S_t;\ \mathcal{O}((a+\Theta_t)\cdot S_t)) \quad = \quad 1\ .$$

Also the closure of the set of divisors

$$\{(a+\Theta_t)\cdot S_t\ :\ a \in J(V)\}$$

is irreducible and must contain a divisor from every linear equivalence class which is algebraically equivalent to $(\Theta_t \cdot S_t)$. So, by (27), this closure must contain the divisor

(29) $$D_{s_1} + D_{s_2} + D_{s_3}$$

for generic, and therefore for *all* $s_1, s_2, s_3 \in S_t$. Therefore, if two divisors of the form (29) are linearly equivalent, the only possibility is that the translate of Θ_t corresponding to their linear equivalence actually *contains* S_t .

To see that there must in fact be a large number of $a \in J(V_t)$ such that

$$(a + \Theta_t) \supseteq S_t \quad ,$$

notice that, if for example, L_{s_1} and L_{s_2} are incident, then $(L_{s_1} + L_{s_2})$ is a member of a rational family of conics containing several other degenerate conics, and so several other pairs of lines. One might hope that there is a seven-dimensional family of translates $(S_t - a)$ lying inside Θ_t and that

$$\dim H^0(S_t; \mathcal{O}_{S_t}(\Theta_t + a)) > 1$$

if and only if $S_t \subseteq (\Theta_t + a)$.

LECTURE #6

The Abel-Jacobi Mapping

The previous lecture was devoted to a very detailed analysis of one example. It was meant to illustrate the uses of very precise information about the Neron model in obtaining very precise information about the intermediate Jacobian at the generic point. In this lecture we will suggest a more general approach to the solution to the problem of determining if an Abel-Jacobi map

(1) $$\mathrm{Alb}(S) \longrightarrow J(V)$$

is an isomorphism. This program will not require such detailed knowledge of a very specific degeneration, but rather a lesser amount of knowledge about the generic degeneration.

We will again illustrate the method with a specific family of Fano threefolds, namely the family of quartic hypersurfaces in $\mathbb{P}^4$. Let S be the surface of conics on a quartic threefold V . For generic V , S is smooth and irreducible. Choosing a basepoint $s_0 \in S$, we map

(2) $$\begin{array}{ccc} S & \longrightarrow & J(V) \\ s & \longmapsto & \displaystyle\int_{C_{s_0}}^{C_s} \end{array}$$

where C_s is the conic parametrized by s . Our first task is to show the surjectivity of the mapping

(3) $$H_1(S) \longrightarrow H_1(J(V)) = H_3(V)$$

induced by (2).

To accomplish this, let W be a generic quartic *fourfold*, and $\{V_t\}_{t \in \mathbb{P}^1}$ a Lefschetz pencil of hyperplane sections. Associated to this pencil we have an *intermediate Jacobian bundle*

(4) $$\bar{J} \longrightarrow \mathbb{P}^1$$

whose fibre over t_0 when V_{t_0} has a node is described as follows. Let $\tilde{V}_{t_0}$ be the desingularization of V_{t_0} and let A denote the exceptional quadric surface in $\tilde{V}_{t_0}$ lying over the node in V_{t_0}. If J denotes the fibre of the Neron model over t_0, then we have an exact sequence

(5) $$1 \to \mathbb{C}^* \to J \to J(\tilde{V}_{t_0}) \to 0 \ .$$

The fibre of (4) over t_0 is obtained from J by putting in a copy of $J(\tilde{V}_{t_0})$ at 0 and at ∞, then by glueing these two copies together via a translation. To compute which translation, the extension (5) is given by a certain line bundle of degree zero on $J(\tilde{V}_{t_0})$, and so by a point $a \in J(\tilde{V}_{t_0})$. The section of $\bar{J}$ at 0 is identified to the section at ∞ via translation by $(-a)$.

To compute the value of a, we let $\Gamma_t \subseteq V_t$ be a three-cycle which has intersection number one with the vanishing cycle of our degeneration

(6) $$V_t \longrightarrow V_{t_0} \ .$$

Then, by our previous analysis of the topology of a degeneration, it is not hard to see that, if

$$\tilde{\Gamma}_0 = (\text{"proper transform" of } \Gamma_0 \text{ in } \tilde{V}_{t_0}) \ ,$$

then

$$\partial\tilde{\Gamma}_0 \quad = \quad \text{generator of } H_2(A)^0 \quad .$$

So, as before, we have

$$(7) \qquad a \quad = \quad \int_{\tilde{\Gamma}_0} \quad \in \quad J(\tilde{V}_{t_0}) \quad .$$

To insure the non-triviality of the Abel-Jacobi map, it is enough to realize the cycle Γ_t as a one-cycle on S_t, that is, to have

$$\Gamma_t \quad = \quad \bigcup_{s \in \gamma_t} C_s \subseteq V_t$$

for some one-cycle γ_t in S_t. To achieve this, we claim without proof that:

(8) i) S_{t_0} is irreducible;

ii) there is a point $s_0 \in S_{t_0}$ such that C_{s_0} has the singular point of V_{t_0} as a simple point;

iii) there are two mappings

$$f_j : \Delta \longrightarrow S_{t_0} \ , \ j = 1,2 \ ,$$

such that, if $u \in \Delta^*$, $C_{f_j(u)}$ does not pass through the singular point of V_{t_0}, and in $\tilde{V}_{t_0}$:

$$\lim_{u \to 0} C_{f_j(u)} \quad = \quad (\text{proper transform of } C_{s_0}) \ + \\ + \ (\text{member of the } j\text{-th ruling of } A)$$

Then we can construct $\tilde{\Gamma}_0$ by taking a path $\tilde{\gamma}_0$ in $\tilde{S}_{t_0}$, the normalization of S_{t_0}, such that a neighborhood of its boundary points lie in the proper transforms of the $f_j(\Delta)$, and letting

(9) $$\tilde{\Gamma}_0 \quad = \quad \bigcup_{s \in \tilde{\gamma}_0} C_s \quad \subseteq \quad \tilde{V}_{t_0} \quad .$$

Then γ_0, the image of $\tilde{\gamma}_0$ in S_{t_0} is a closed one-cycle. We need to know that γ_0 is the specialization at t_0 of a one-cycle on S_t for t near t_0, but this is very easy to conclude. S_{t_0} has an ordinary double curve consisting of those conics C_s in V_{t_0} which pass through the singular point, and $\tilde{S}_0$ has two disjoint components lying over this double curve, one for each of the two rulings of A .

So once we have succeeded in making the above construction, we know that the mapping

(10) $$H_1(S_t) \longrightarrow H_3(V_t)$$

has in its image a transverse, and therefore also the vanishing cycle of the degeneration (6). Now by classical Lefschetz theory, all the vanishing cycles are conjugate and that they generate $H_3(V_t)$. Since the mapping (10) is equivariant with respect to the monodromy representation, we conclude that the mapping (10) is a surjection and that

(11) $$K_t \quad = \quad \text{kernel of (10)}$$

is a subspace of $H_1(S_t)$ invariant under the monodromy representation.

If we had chosen for our example the family of conics on the intersection of three quadrics in $\mathbb{P}^6$ we would already be finished at this point. This is because of a result of Welters, based on work of Beauville and Tjurin, that the Abel-Jacobi mapping (1) is an isogeny in that case. Surjectivity of the homology map (3) implies that the isogeny must in fact be an isomorphism.

In our example, however, one *might* proceed as follows:

I. Show that if T is the variety of conics on the generic quartic fourfold W has the property that

$$H_1(T) = 0 .$$

II. The set T' of conics twice incident to the base locus of the pencil $\{V_t\}$ form an ample divisor in T, so $H_1(T') = 0$. But

$$T' = \bigcup_{t\in\mathbb{P}^1} S_t .$$

III. Show that, for each degeneration (6) of the pencil, the degeneration-induced map

$$H_1(S_t) \longrightarrow H_1(S_{t_0})$$

has only a one-dimensional kernel.

If this line of reasoning can be made rigorous, then the elements of K_t in (11) must be individually invariant for the monodromy representation and so the mapping

$$H_1(S_t) \longrightarrow H_1(T')$$

must restrict to an injection on K_t. Thus $K_t = 0$ and the mapping (10) is an isomorphism.

My very strong feeling is that all this can be made precise in this case, and in the case of other Fano threefolds whose anti-canonical divisor generates the Picard group. It seems to me worth the effort to try an approach like this, which would work in general, rather than relying too heavily on the particular geometry of each special class of Fano threefold.

Threefolds whose canonical bundles are not numerically effective

by

SHIGEFUMI MORI

Nagoya University, Japan and Harvard University, U.S.A *

Introduction.

Let X be a non-singular projective 3-fold over an algebraically closed field k of charactertistic 0 such that the canonical bundle K_X is not numerically effective. These X form an important class of 3-folds from the viewpoint of classification of 3-folds with Kodaira dimension $-\infty$ [17] and the study of minimal models (1.6).

It is shown (see §2 for precise statements) that one has (i) X contains an exceptional divisor of several types, which we classify explicitly in (2.3) and (2.4), (ii) X has a morphism to a projective non-singular surface whose fibers are conics (smooth conics, reducible conics, or double lines) (2.5.1), (iii) X has a morphism to a projective non-singular curve whose fibers are irreducible reduced surfaces D such that ω_D^{-1} is ample (2.5.2), or (iv) X is a Fano 3-fold (i.e. $-K_X$ is ample) and the Picard number is 1 (2.5.3). The 3-folds X in case (iv) were classified by Iskovskih [8] and [9]. The Fano 3-folds with Picard number ≥ 2 are classified up to deformation by S. Mukai and myself using our theory of extremal rays [14].

In this note, we give a simpler argument than [13] about the main results, though [13] is sometimes quoted. The results of §2 is proved (or sometimes explained) in §§3 - 6. After the preliminary in §3, the result of §2 is treated in 3 cases (§§4 - 6).

As for notation, by irreducible divisors on a variety is meant divisors which are irreducible and reduced unless otherwise mentioned. For a sheaf F on a variety X over k, $h^i(X, F)$ denotes the dimension of $H^i(X, F)$ over k.

We thank Ms. Barbara Moody for nice typing.

Partially supported by NSF Grant MCS 77-15524 and Educational project for Japanese Mathematical Scientists.

§1. Cone of 1-cycles.

Let X be an n-dimensional irreducible reduced projective variety over an algebraically closed field k of characteristic $p \geq 0$. By a 1-cycle, we understand an element of the free abelian group generated by all the irreducible reduced subvarieties of dimension 1 (i.e. irreducible curves) of X. A 1-cycle $Z = \Sigma\, n_C\, C$ ($n_C \in \mathbb{Z}$) is called effective if $n_C \geq 0$ for all C. Cartier divisors (resp. effective Cartier divisors) are called (n-1)-cycles (resp. effective (n-1)-cycles). Let $(\cdot)$ be the intersection pairing of 1-cycles and (n-1)-cycles. For $r = 1$ or $n-1$, we say that an r-cycle C is numerically trivial (resp. numerically effective) and write $C \approx 0$ if $(C\cdot D) = 0$ (resp. $(C\cdot D) \geq 0$) for all effective (n-r)-cycles D. Then the real vector spaces

$$N_r(X) = (\{\text{r-cycles}\}/\approx) \otimes_{\mathbb{Z}} \mathbb{R}$$

for $r = 1$ and $n-1$ are dual to each other via $(\cdot)$ and of finite dimension $\rho(X)$, the Picard number of X [11]. We will give $N_r(X)$ a norm $\|\ \|$ and usual topology. Let $NE(X) \subset N_1(X)$ be the cone generated by effective r-cycles which is closed under multiplication by $\mathbb{R}_+ = \{a \in \mathbb{R} \mid a \geq 0\}$. Let $\overline{NE}(X)$ be the closure of $NE(X)$. Let $\overline{NE}(X)^*$ be the dual of $\overline{NE}(X)$:

$$\overline{NE}(X)^* = \{y \in N_{n-1}(X) \mid (x\cdot y) \geq 0 \text{ for all } x \in \overline{NE}(X)\}.$$

Now Kleiman's criterion for ampleness says:

(1.1) A Cartier divisor D on X is ample iff. $(D.Z) > 0$ for all $Z \in \overline{NE}(X) \cap \{Z \in N(X) \mid \|Z\| = 1\}$, i.e. iff. D lies in the interior of $\overline{NE}(X)$ [11].

In particular $\overline{NE}_1(X)$ does not contain any non-zero real vector subspaces.

From now we will use the notation $N(X)$, $N(X)^*$ instead of $N_1(X)$, $N_{n-1}(X)$, respectively, since we are interested in 1-cycles. We also assume that X is a non-singular projective variety of dimension n over k from now on. Let K_X be the canonical bundle.

The cone $\overline{NE}(X)$ just introduced is rational polyhedral if $-K_X$ is ample.

<u>Theorem</u> (1.2). If $-K_X$ is ample, then X contains finitely many rational curves $C_1, \ldots, C_s$ such that $(-K_X \cdot C_i) \leq n+1$ for all i and $\overline{NE}(X) = \mathbb{R}_+[C_1] + \ldots + \mathbb{R}_+[C_s]$, where $[Z]$ denotes the class of cycle Z.

We remark that a rational curve means an irreducible reduced curve whose normalization is $\mathbb{P}^1$. Threefolds X with ample $-K_X$ are called Fano 3-folds and $\overline{NE}(X)$ is an important invariant of X when $\rho(X) \geq 2$ [14].

Let

$$\overline{NE}_-(X) = \{Z \in \overline{NE}(X) \mid (-K_X\ Z) \leq 0\}.$$

A half line $R = \mathbb{R}_+Z \subset \overline{NE}(X)$ is an <u>extremal ray</u> if (i) $(-K_X \cdot Z) > 0$ and (ii) Z_1 and Z_2 in $\overline{NE}(X)$ satisfy $Z_1 \in R$ and $Z_2 \in R$ if $Z_1 + Z_2 \in R$. R is <u>numerically effective</u> if so is Z. An irreducible reduced curve C in X is called an <u>extremal rational curve</u> if (i) normalization of C is isomorphic to $\mathbb{P}^1$, (ii) $\mathbb{R}_+[C]$ is an extremal ray and (iii) $(-K_X\ C) \leq n+1$.

Now $\overline{NE}(X)$ for general X is "rational polyhedral modulo" $\overline{NE}_-(X)$:

Theorem (1.3). $\overline{NE}(X)$ is the smallest closed convex cone containi $\overline{NE}_-(X)$ and all the extremal rays. For an arbitrary open convex cone U in $N(X)$ containing $\overline{NE}_-(X) - \{0\}$, there exist only a finite number of extremal rays which do not lie in $U \cup \{0\}$. Every extremal ray is spanned by an extremal rational curve.

We remark that (1.3) implies (1.2). Indeed if $-K_X$ is ample, then $\overline{NE}_-(X) = 0$ (1.1) and there are only finite number of extremal rational curves C modulo algebraic equivalence because $(-K_X\ C) \leq n+1$

The following property of extremal rays is used later.

Corollary (1.4). If R is an extremal ray, then there is a closed convex cone B of $N(X)$ such that $R \not\subset B$ and $\overline{NE}(X) = R + B$.

Proof. Let L be an ample divisor of X. Since R is generatec by a curve C such that $(-K_X\ C) > 0$, one can find a small enough real number $\varepsilon > 0$ such that $R \not\subset \bar{U}$, where

$$\bar{U} = \{Z \in \overline{NE}(X) \mid (-K_X\ Z) \leq \varepsilon(L\ Z)\}.$$

Let $R = R_1, \ldots, R_s$ be all the extremal rays not lying in $\bar{U}$ (1.3). Since $\overline{NE}(X)$ does not contain a straight line (1.1), it is easy to see that $V = R_1 + \ldots + R_s + \bar{U}$ and $B = R_2 + \ldots + R_s + \bar{U}$ are closec convex cones. Then $V = \overline{NE}(X)$ (1.3) and $R \not\subset B$ because R is an extremal ray. q.e.d.

One has an obvious

Corollary (1.5). X has an extremal ray iff. K_X is not numerically effective (i.e. iff. $\overline{NE}(X) \neq \overline{NE}_-(X)$).

Let us describe the extremal rational curves in the case of $\dim X = 2$.

Theorem (1.6). An irreducible curve C on X is an extremal rational curve iff. (i) C is an exceptional curve of the first kind, (ii) X has a structure of a $\mathbb{P}^1$-bundle over a curve such that C is one of its fibers, or (iii) $X \simeq \mathbb{P}^2$ and C is a straight line.

The idea of the proof for the only-if part when $\rho(X) \geq 2$ is as follows: Let C be an extremal rational curve. Since C is on the boundary of $\overline{NE}(X)$ by $\rho(X) \geq 2$, one must have $(C^2) \leq 0$. Indeed if $(C^2) > 0$, then every divisor D whose class [D] is close enough to $\mathbb{R}_+[C]$ satisfies $(D^2) > 0$ and hence some (positive or negative) multiple νD of D satisfies $h^0(\nu D) > 0$ by Grothendieck [5], whence D or $-D \in \overline{NE}(X)$. Since [D] is close to $\mathbb{R}_+[C]$, one has $D \in \overline{NE}(X)$ and thus C is in the interior. Now from $(C^2) \leq 0$ and $(C\cdot K_X) < 0$ follows (a) $(C\cdot K_X) = (C^2) = -1$, or (b) $(C\cdot K_X) = -2$ and $(C^2) = 0$, by the arithmetic genus formula: $(C^2) + (C\cdot K_X) = 2\,p_a(C) - 2 \geq -2$. (a) implies (i) and it is easy to see that (b) implies (ii). The case $\rho(X) = 1$ is essentially Castelnuovo's criterion for rationality.

(1.6) is generalized to 3-folds by the use of the contraction

$\phi : X \longrightarrow Y$ to a normal projective variety (2.1) to make the statement less ambiguous (see the remark after (2.3)), where ϕ for a surface X is the contraction of C in case (i), the $\mathbb{P}^1$-bundle morphism in case (ii), and the structure morphism $X \longrightarrow \mathrm{Spec}\ k$ in case (iii). To be more explicit, an exceptional curve of the first kind is generalized to five types of exceptional divisors explicitly classified in (2.3) and (2.4); a $\mathbb{P}^1$-bundle is generalized in two ways, to conic bundles (2.5.1) and to del Pezzo fiber spaces (2.5.2), and $\mathbb{P}^2$ is generalized to Fano 3-folds with Picard number 1 (2.5.3).

§2. 3-folds with extremal rays (characteristic 0).

From now on, we add the assumption that the characteristic of k is 0. We remark that X has an extremal ray iff. K_X is not numerically effective (1.5).

Theorem (2.1). To each extremal ray R, there exists a morphism $\phi : X \longrightarrow Y$ to a projective variety Y such that (i) $\phi_* \mathcal{O}_X = \mathcal{O}_Y$ and (ii) for an arbitrary irreducible curve C in X, $[C] \in R$ iff. $\dim \phi(C) = 0$.

We remark that such a ϕ is unique up to an isomorphism. Indeed, for two points x_1 and x_2 of X, one has: $\phi(x_1) = \phi(x_2)$ iff. there exists a connected reduced curve C joining x_1 and x_2 such that $[C] \in R$. Such a ϕ is called the contraction of R and is denoted by $\mathrm{cont}_R : X \longrightarrow Y = \mathrm{cont}_R(X)$.

From (2.1), easily follows

Corollary (2.2). $-K_X$ is ϕ-ample and $R^i\phi_*\mathcal{O}_X = 0$ for all $i > 0$.

Proof. For relative ampleness, it is enough to show that $\mathcal{O}_V(-K_X)$ is ample for every irreducible reduced subvariety V of X such that $\dim \phi(V) = 0$ [3, III, (4.7.1)]. Let C be an irreducible curve generating R and L an ample divisor on X. By (ii) of (2.1), the image of the natural homomorphism $N(V) \longrightarrow N(X)$ is generated by [C]. Thus the image of $N(X)^* \longrightarrow N(V)^*$ is 1-dimensional and one has

$$\mathcal{O}_V(-K_X)^{\otimes(L\cdot C)} \approx \mathcal{O}_V(L)^{\otimes(-K_X\cdot C)}.$$

Thus $\mathcal{O}_V(-K_X)$ is ample because $(-K_X\cdot C) > 0$. Let M be an ample divisor on Y. Since $-K_X$ is ϕ-ample, $\nu M - K_X$ is ample on X and $H^i(\mathcal{O}_Y(\nu M)\otimes R^j\phi_*\mathcal{O}(-K_X)) = 0$ for $i > 0$, $j \geq 0$, and $\nu >> 0$. By Kodaira vanishing and Leray spectral sequence $H^0(\mathcal{O}_Y(\nu M)\otimes R^j\phi_*\mathcal{O}(-K_X)) = 0$ for $j > 0$ and $\nu >> 0$. This implies $R^i\phi_*\mathcal{O}_X = 0$ for $i > 0$. q.e.d.

The structure of cont_R is given by:

<u>Theorem</u> (2.3). If the extremal ray R is not numerically effective, then there exists a unique irreducible divisor D on X such that $(D\cdot C) < 0$ for irreducible curves C such that $[C] \in R$. The contraction $\phi : X \longrightarrow Y$ induces an isomorphism on X - D and $\dim \phi(D) \leq 1$, and one has:

(2.3.1) $\phi(D)$ is a non-singular curve and Y is non-singular, $\phi|_D : D \longrightarrow \phi(D)$ is a $\mathbb{P}^1$-bundle;

(2.3.2) $Q = \phi(D)$ is a point and Y is non-singular, $D \simeq \mathbb{P}^2$ and $\mathcal{O}_D(D) \simeq \mathcal{O}_{\mathbb{P}}(-1)$;

(2.3.3) $Q = \phi(D)$ is a point, $D \simeq \mathbb{P}^1\times \mathbb{P}^1$, $\mathcal{O}_D(D)$ is of bidegree (-1, -1) and $s\times\mathbb{P}^1 \not\approx \mathbb{P}^1\times t$ on X $(s, t \in \mathbb{P}^1)$;

(2.3.4) $Q = \phi(D)$ is a point, $D \simeq$ an irreducible reduced singular quadric surface in $\mathbb{P}^3$, $\mathcal{O}_D(D) \simeq \mathcal{O}_D\otimes\mathcal{O}_{\mathbb{P}}(-1)$; or

(2.3.5) $Q = \phi(D)$ is a point, $D \simeq \mathbb{P}^2$ and $\mathcal{O}_D(D) \simeq \mathcal{O}_{\mathbb{P}}(-2)$.

This uniquely determined D is called the <u>exceptional divisor associated to</u> R which is not numerically effective.

If the associated exceptional divisor D is $\mathbb{P}^1\times\mathbb{P}^1$ with $\mathcal{O}_D(D)$ of bidegree (-1, -1), we are in case (2.3.1) or (2.3.3) according as $s\times\mathbb{P}^1 \approx \mathbb{P}^1\times t$ or $s\times\mathbb{P}^1 \not\approx \mathbb{P}^1\times t$ on X $(s, t \in \mathbb{P}^1)$. Thus the description of D and $\mathcal{O}_D(D)$ is not enough to study extremal rays. This is the main reason why we introduced cont_R to study R.

Corollary (2.4). Under the notation and assumptions of (2.3), we have:

(2.4.1) X is the blowing-up of Y by the ideal defining the reduced closed subscheme $\phi(D)$,

(2.4.2) the divisor class group of the local ring $\mathcal{O}_{Y,Q}$ of Y at Q is 0 in cases (2.3.3) and (2.3.4), and $\mathbb{Z}/2\mathbb{Z}$ in case (2.3.5), and

(2.4.3) the completion $\mathcal{O}_{Y,Q}\hat{}$ of $\mathcal{O}_{Y,Q}$ is given by

$$\mathcal{O}_{Y,Q}\hat{} \simeq \begin{cases} k[[x,y,z,u]]/(x^2+y^2+z^2+u^2) & (2.3.3) \\ k[[x,y,z,u]]/(x^2+y^2+z^2+u^3) & (2.3.4) \\ k[[x,y,z]]^{(2)} & (2.3.5), \end{cases}$$

where $k[[x,y,z]]^{(2)}$ is the subring of invariants of $k[[x,y,z]]$ for the involution $(x,y,z) \longmapsto (-x,-y,-z)$.

For proof of (2.4), we refer the reader to [13]. The idea is to show that $R^i\phi_*\mathcal{O}(-jD) = 0$ for all $i > 0$ and $j \geq 0$ (cf. (2.2)) and $\phi_*\mathcal{O}(-jD) = I_Q^j$ for all $j \geq 0$, where I_Q is the maximal ideal at Q.

Theorem (2.5). If the extremal ray R is numerically effective, then $Y = \mathrm{cont}_R(X)$ is non-singular, $\rho(X) = \rho(Y)+1$, and we have:

(2.5.1) dim Y = 2, and for an arbitrary closed point y of Y, the fiber X_y is isomorphic to a conic of $\mathbb{P}^2$ as a scheme (i.e. X_y is $\mathbb{P}^1$, a reducible conic, or a double line);

(2.5.2) dim Y = 1, and for an arbitrary closed point y of Y,

$F = X_y$ is an irreducible reduced surface such that ω_F^{-1} is ample; or

(2.5.3) dim $Y = 0$, and $-K_X$ is ample.

Supplement (2.6). In case (2.5.1), the discriminant locus

$$\Delta = \{y \in Y \mid X_y \text{ is singular}\}$$

is a divisor with only ordinary double points as singularities, and X_y is a double line iff. y is a singular point of Δ. $\phi : X \longrightarrow Y$ is usually called a conic bundle.

In case (2.5.2), one has $1 \le (\omega_F{}^2) \le 6$ or $(\omega_F{}^2) = 8, 9$; X is a $\mathbb{P}^2$-bundle over Y if $(\omega_F{}^2) = 9$; X is embedded in a $\mathbb{P}^3$-bundle P over Y such that X_y is an irreducible reduced quadric of $\mathbb{P}^3 \simeq P_y$ for all $y \in Y$ if $(\omega_F{}^2) = 8$. We call $\phi : X \longrightarrow Y$ a del Pezzo fibering or a del Pezzo fiber space.

In case (2.5.3), such Fano 3-folds X with $\rho = 1$ are classified by Iskovskih [8] and [9].

The first part of the supplement is done by Beauville [1], and we refer the reader to [13] for the second part.

§3. Preliminaries.

In this section, we introduce some notation and prove a few results which is used later.

Let M be a finite dimensional real vector space and M^* its dual. For an element or a subset F or M, let $F^{\perp}$ be the vector subspace of M^* given by:

$$F^{\perp} = \{v \in M^* \mid (v, m) = 0 \text{ for all } m \in F\} .$$

For $v \in M^*$ and $R = \mathbb{R}_+ m \subset M$ with $m \in M$, we write $(v \cdot R) > 0$ (resp. $= 0$, < 0) if $(v \cdot m) > 0$ (resp. $= 0$, < 0).

PROPOSITION 3.1. Let R be an extremal ray on X. Let R^* be the subset of $R^{\perp} \subset N(X)^*$ defined by

$$R^* = \{H \in \overline{NE}(X)^* \mid H^{\perp} \cap \overline{NE}(X) = R\} .$$

Then R^* is a non-empty open convex cone of $R^{\perp}$. Furthermore if H and E are divisors on X such that $H \in R^*$ and $(E \cdot R) > 0$, then $nH + E$ is ample for $n >> 0$.

Proof. By (1, 4), there is a closed convex cone B of $N(X)$ such that $R \not\subset B$ and $\overline{NE}(X) = R + B$. Then an element $H \in R^{\perp}$ belongs to R^* if $(H \cdot Z) > 0$ for all $Z \in B - \{0\}$. There exists such an H and R^* is open, because $R \not\subset B$ and B is a closed convex cone. As for the second assertion, $nH + E$ is ample iff $nH + E$ is positive on $\overline{NE}(X) - \{0\}$ as a real-valued linear function. Since $(H \cdot R) = 0$ and $(E \cdot R) > 0$, this is the same as asserting $nH + E$ is positive on $B - \{0\}$. Since $B \cap \{\|Z\| = 1\}$ is compact and the linear function H is positive on $B - \{0\}$, one sees that $nH + E$ is positive on $B - \{0\}$ for $n >> 0$.

q. e. d.

The basic idea of proving the results in §2 is to consider divisors H in R^*.

The first merit of such an H is

LEMMA 3.2. Let R be an extremal ray and H a divisor in R^*. Then one has $H^i(\mathcal{O}(nH)) = 0$ for $i = 1, 2, 3$ and $n >> 0$.

Proof. Since $(K_X \cdot R) < 0$, $nH - K_X$ is ample for $n >> 0$ by (3.1). Thus one can apply Kodaira vanishing: $H^i(\mathcal{O}(nH)) = 0$ for $i = 1, 2, 3$ and $n >> 0$.

q.e.d.

The numerical effectiveness of R is interpretted as:

LEMMA 3.3. Under the notation of (3.2), R is not numerically effective iff $(H^3) > 0$.

Proof. If R is not numerically effective, then there is an irreducible divisor D such that $(D \cdot R) < 0$. Then $aH - D$ is ample for $a >> 0$ (3.1). Thus $h^0(\mathcal{O}(maH - mD))$ is a cubic polynomial in m if $m >> 0$. Now $h^0(\mathcal{O}(maH)) \geq$ $\geq h^0(\mathcal{O}(maH - mD))$ shows that $(H^3) > 0$ because

$$h^0(\mathcal{O}(maH)) = \chi(\mathcal{O}(maH)) = \frac{1}{6} a^3 m^3 (H^3) + \cdots \quad . \tag{3.2}$$

Conversely if $(H^3) > 0$, then $h^0(\mathcal{O}(mH)) = \chi(\mathcal{O}(mH))$ (3.2) is a cubic polynomial in m by the Riemann-Roch formula. Thus D-dimension of H is 3 [7], (or see Ueno's text in this volume). Let L be an ample divisor on X. Since H has D-dimension 3, there is a positive number a and an effective divisor D on X such that $aH \sim L + D$ [7]. Then $(D \cdot R) < 0$.

q.e.d.

LEMMA 3.4. Let R be an extremal ray on X. If H is a divisor and if B is an irreducible divisor on X such that $H \in R^*$ and $H \cdot B \approx 0$ on X, then the image of the natural map $N(X)^* \to N(B)^*$ (or equivalently $N(B) \to N(X)$) is of dimension 1.

Proof. For this, it is sufficient to prove that every irreducible curve C in B satisfies $[C] \in R$. Now let L be an irreducible divisor on X such that $L \neq B$ and $L \supset C$. Then $L \cdot B - C$ is an effective 1-cycle and one has

$$0 \leq (C \cdot H) \leq (L \cdot B \cdot H) = (L \cdot (B \cdot H)) = 0$$

because H is numerically effective and $B \cdot H \approx 0$. Thus $(C \cdot H) = 0$ and

$$[C] \in H^{\perp} \cap \overline{NE}(X) = R \quad \text{since} \quad H \in R^{*} \quad .$$

q. e. d.

§4. The case where R is not n.e.

In this section, assuming that R is not numerically effective (not n.e., in short) we will prove (2.1) and (2.3).

Let H be a divisor in R^* (3.1) and let D be an irreducible divisor on X such that $(D \cdot R) < 0$. We will show that $\mathcal{O}(mH)$ is generated by global sections and induces a morphism φ which satisfies the properties in (2.3) for $m >> 0$.

The linear system $|mH|$ is ample outside D and its trace $\mathrm{Tr}_D |mH|$ on D is a complete linear system by:

LEMMA 4.1. Let a be an arbitrary natural number. If $m >> 0$, then one has

$$H^i(\mathcal{O}_{aD}(mH)) = H^i(\mathcal{O}_{aD}(mH + K_X)) = 0 , \qquad i > 1 ,$$

$$H^0(\mathcal{O}_X(mH)) \longrightarrow H^0(\mathcal{O}_{aD}(mH)) \qquad \text{is a surjection} ,$$

and there is a natural number b for each m such that $mH - bD$ is very ample.

q.e.d.

Now the intersection $D \cdot H$ is considered to be an element of $N(X)$, and D and $\mathrm{Tr}_D |mH|$ are studied by the following two lemmas.

LEMMA 4.2. Assume that $H \cdot D \neq 0$. Then D has a structure of a $\mathbb{P}'$-bundle $\psi : D \to Y$ over a non-singular curve Y, and H is the pull back of an ample divisor on Y by ψ. Furthermore if F is the fiber of ψ, then $(F \cdot D) = = (F \cdot K_X) = -1$.

LEMMA 4.3. Assume that $H \cdot D = 0$. Then D is isomorphic to $\mathbb{P}^2$ or an irreducible reduced quadric surface in $\mathbb{P}^3$ and $\mathcal{O}_D(H) \simeq \mathcal{O}_D$. Furthermore $\mathcal{O}_D(-D)$ and $\mathcal{O}_D(-K_X)$ are ample.

Let us see that these 3 lemmas imply (2.1) and (2.3). By (4.1), $|mH|$ does not have base points outside D and cuts out complete linear system $\mathrm{Tr}_D|mH|$ which is free from base points if $m >> 0$, (4.2) and (4.3). Thus for $m >> 0$, $|mH|$ induces a morphism $\varphi : X \to Y$ such that $\varphi_* \mathcal{O}_X = \mathcal{O}_Y$. The property that $H \in R^*$ (3.1) implies the condition (ii) of (2.1), and φ is now the contraction of R. Since $mH - bD$ is very ample $(b > 0)$ by (4.1), φ is an isomorphism on $X - D$. If $H \cdot D \neq 0$, then $\varphi|_D$ is a $\mathbb{P}^1$-bundle morphism. Since the fiber F has the property that $(F \cdot D) = -1$, it follows that Y is smooth. This is (2.3.1). If $H \cdot D = 0$, then $\varphi(D)$ is a point because $\mathcal{O}_D(H) \simeq \mathcal{O}_D$. By (4.3), D is isomorphic to $\mathbb{P}^2$, a singular quadric surface of $\mathbb{P}^3$, or $\mathbb{P}^1 \times \mathbb{P}^1$. In the first 2 cases, one has $\mathrm{Pic}\, D \simeq \mathbb{Z}$. Since $\mathcal{O}_D(-D)$ and $\mathcal{O}_D(-K_X)$ are ample divisors whose sum is equal to ω_D^{-1}, it follows that one has (2.3.2), (2.3.4), or (2.3.5). If $D \simeq \mathbb{P}^1 \times \mathbb{P}^1$, then the same argument shows that $\mathcal{O}_D(D)$ is of bidegree $(-1, -1)$. Since the image of the natural map $N(D) \to N(X)$ is of dimension 1 (3.4), this shows that $s \times \mathbb{P}^1 \approx \mathbb{P}^1 \times s$ on X. This is (2.3.3). Thus it is enough to prove (4.1), (4.2), and (4.3).

Before proving these, we quote the Kawamata-Viehweg vanishing theorem [10] (or see Viehweg's paper in this volume), which is the generalization of Kodaira vanishing and Ramanujam vanishing [16]:

THEOREM 4.4 (Kawamata-Viehweg). If V is an n-dimensional non-singular projective variety over an algebraically closed field of characteristic D and L a numerically effective divisor such that $(L^n) > 0$, then $H^i(\mathcal{O}(L + K_X)) = 0$ for $i > 0$.

Using this vanishing, we will prove (4.1).

Proof of (4.1). Consider the standard exact sequences

$$0 \longrightarrow \mathcal{O}(mH - aD) \longrightarrow \mathcal{O}(mH) \longrightarrow \mathcal{O}_{aD}(mH) \longrightarrow 0 ,$$

$$0 \longrightarrow \mathcal{O}(mH + K_X - aD) \longrightarrow \mathcal{O}(mH + K_X) \longrightarrow \mathcal{O}_{aD}(mH + K_X) \longrightarrow 0 .$$

For $m >> 0$, $mH - aD - K_X$, $mH - K_X$ and $mH - aD$ are all ample (3.1) and hence

$$H^i(\mathcal{O}(mH - aD)) = H^i(\mathcal{O}(mH)) = H^i(\mathcal{O}(mH + K_X - aD)) = 0 , \quad i > 0$$

by Kodaira vanishing. By Kawamata-Viehweg vanishing (4.4), one has $H^i(\mathcal{O}(mH + K_X)) = 0$, $i > 0$ because $(H^3) > 0$ (3.3). Thus one has proved (4.1) except for the very ampleness of $mH - bD$ for some $b > 0$. But this assertion follows immediately from the ampleness of $mH - D$ for $m >> 0$.

q. e. d.

The idea to prove (4.2) is to study each curve C such that $[C] \in R$. We need 2 lemmas:

LEMMA 4.5. If $H \cdot D \neq 0$, then $h^0(\mathcal{O}_D(mH)) \to \infty$ as $m \to \infty$.

<u>Proof.</u> One sees first that $H \cdot D \in \overline{NE}(X) - \{0\}$. Indeed if L is a very ample divisor, then $D \cdot L \in \overline{NE}(X)$ and hence $(H \cdot D \cdot L) \geq 0$. Thus $H \cdot D \in \overline{NE}(X)^{**} = \overline{NE}(X)$ (cf. §1). From (4.1) and the Riemann-Roch formula, one has

$$h^0(\mathcal{O}_D(mH)) = \chi(\mathcal{O}_D(mH)) = \frac{1}{2} m(D \cdot H \cdot mH - K_X - D) + \chi(\mathcal{O}_D)$$

for $m >> 0$. Since $(K_X + D \cdot R) < 0$, $mH - K_X - D$ is ample if $m >> 0$. Hence $(D \cdot H \cdot mH - K_X - D) \geqq 1$ (1.1) because $H \cdot D \in \overline{NE}(X) - \{0\}$. Thus $h^0(\mathcal{O}_D(mH)) \geq \frac{1}{2} m + \chi(\mathcal{O}_D) \to \infty$.

q. e. d.

LEMMA 4.6. Assume that $H \cdot D \neq 0$. If C is a reduced connnected curve such that $[C] \in R$, then $C \simeq \mathbb{P}^1$, $N_{C/X} \simeq \mathcal{O}_C \oplus \mathcal{O}_C(-1)$, and $(-K_X \cdot C) = 1$.

<u>Proof.</u> Let $C_1, \ldots, C_r$ be the irreducible components of C. Since $[C_1] + \cdots + [C_r] \in R$ and since R is an extremal ray, one has $[C_i] \in R$ for all i. Hence $C_i \subset D$ because $(C_i \cdot D) < 0$. By (4.1) and (4.5), there exists M_i in $|m_i H|$ such that $M_i \not\supset D$ and $M_i \cap C_i \neq \emptyset$ for $m_i >> 0$. This implies that $M_i \cap D$ is 1-dimensional and $M_i \cap D \supset C_i$ because $(H \cdot C_i) = 0$. Let $M = M_1 + \cdots + M_r$ and $m = m_1 + \cdots + m_r$. Then one has $M \in |mH|$, $M \cap D$ is 1-dimensional, and $M \cap D \supset C$. Let C' be an arbitrary non-empty closed subscheme of X such that $C'_{red} \subseteq C$. Then I claim that

(4.6.1) $$\chi(\mathcal{O}_{C'}(K_X)) \geq 0 \quad .$$

Since $M \cap D \supset C$ and $C'_{red} \subset C$, there exists a natural number a such that $(aM) \cap (aD) \supset C'$. From the standard exact sequence

$$0 \longrightarrow \mathcal{O}_{aD}(nH - amH + K_X) \longrightarrow \mathcal{O}_{aD}(nH + K_X) \longrightarrow \mathcal{O}_{aM \cap aD}(nH + K_X) \longrightarrow 0 \quad ,$$

one has $H^1(\mathcal{O}_{aM \cap aD}(nH + K_X)) = 0$ by (4.1) for $n >> 0$. Let us define a coherent sheaf $\mathfrak{J}$ via the exact sequence:

$$0 \longrightarrow \mathfrak{J} \longrightarrow \mathcal{O}_{aM \cap aD}(nH + K_X) \longrightarrow \mathcal{O}_{C'}(nH + K_X) \longrightarrow 0 \quad .$$

Since $\mathfrak{J}$ is supported on a closed set of dimension ≤ 1, one has $H^2(\mathfrak{J}) = 0$ and hence $H^1(\mathcal{O}_{C'}(nH + K_X)) = 0$. Thus $\chi(\mathcal{O}_{C'}(nH + K_X)) \geq 0$. Since $(H \cdot C_i) = 0$, one has $\mathcal{O}_{C'}(H) \approx 0$ and $\chi(\mathcal{O}_{C'}(K_X)) = \chi(\mathcal{O}_{C'}(nH + K_X)) \geq 0$. Thus (4.6.1) is proved. If we choose as C' an irreducible component of C, then

$$\chi(\mathcal{O}_{C'}(K_X)) = \chi(\mathcal{O}_{C'}) + (K_X \cdot C') \geq 0 \quad .$$

Since $(K_X \cdot C') < 0$, one has $\chi(\mathcal{O}_{C'}) > 0$ and hence $p_a(C') = 0$. Thus $C' \simeq \mathbb{P}^1$ and also $(K_X \cdot C') = -1$, because $\chi(\mathcal{O}_{C'}) = 1$. If two distinct irreducible components A and B of C are note disjoint, then one sees $\chi(\mathcal{O}_{C'}) \leq 1$ for $C' = A \cup B$ from the exact sequence

$$0 \longrightarrow \mathcal{O}_{C'} \longrightarrow \mathcal{O}_A \oplus \mathcal{O}_B \longrightarrow \mathcal{O}_{A \cap B} \longrightarrow 0 \quad .$$

This contradicts (4.6.1) for $C' = A \cup B$ because

$$\chi(\mathcal{O}_{C'}(K_X)) = \chi(\mathcal{O}_{C'}) + (-K_X \cdot C') \leq 1 - 2 < 0 \quad .$$

Thus we have now $C \simeq \mathbb{P}^1$ and $(C \cdot K_X) = -1$. Let I be the defining ideal of C, and let

$$I/I^2 \simeq N_{C/X}{}^* \simeq \mathcal{O}(a) \oplus \mathcal{O}(b)$$

with $a+b=1$ and $a \geq b$. If $b \geq 0$, then $a = 1$ and $b = 0$ which is what we want to show. If $b < 0$, then let J be an ideal such that $I^2 \subset J \subset I$ and $J/I^2 = \mathcal{O}(a)$ in $I/I^2 = \mathcal{O}(a) \oplus \mathcal{O}(b)$. Then J defines a closed subscheme C' such that $C'_{red} = C$ and one has the exact sequence of $\mathcal{O}_X$-modules:

$$0 \longrightarrow \mathcal{O}_C(b) \longrightarrow \mathcal{O}_{C'} \longrightarrow \mathcal{O}_C \longrightarrow 0 \quad .$$

Thus $\chi(\mathcal{O}_{C'}) = 2 + b < 1$, while one has

$$\chi(\mathcal{O}_{C'}) + 2(K_X \cdot C) = \chi(\mathcal{O}_{C'}) - 2 \geq 0$$

from (4.6.1). This is a contradiction and one has $b \geq 0$.

q. e. d.

Now we can prove (4.2).

(4.7). <u>Proof of</u> (4.2): Let C be an irreducible reduced curve such that $[C] \in R$. Then $C \simeq \mathbb{P}^1$, $h^0(N_{C/X}) = 1$, and $h^1(N_{C/X}) = 0$ (4.6). From the obstruction theory, one has $T_{[C],H} \simeq H^0(N_{C/X})$ where H is the Hilbert scheme $\mathrm{Hilb}_{X/k}$ [4, n° 221] and $T_{[C],H}$ is the Zariski tangent space of H at $[C]$; and the obstruction to smoothness of H at $[C]$ lies in $H^1(N_{C/X})$. Thus H is smooth of dimension 1 at $[C]$. Let Y be the connected component of H containing $[C]$; $\psi : Z \to Y$ the universal closed subscheme on Y, and $\pi : Z \to H$ the natural map. We claim that

4.7.1. Y is a non-singular curve, and ψ is a $\mathbb{P}^1$-bundle.

Let F be an arbitrary fiber of ψ, let G be the largest coherent subsheaf of $\mathcal{O}_F$ such that the support of G has dimension ≤ 0, and let F' be the closed subscheme of F defined by G. Thus F' has no embedded components. Since $F' \approx C$ as 1-cycles, one has $(-K_X \cdot F') = (-K_X \cdot C) = 1$. Hence by applying (4.6) to each component of F', one sees that F' is irreducible reduced and hence $F' \simeq \mathbb{P}^1$. From the exact sequence

$$0 \longrightarrow G \longrightarrow \mathcal{O}_F \longrightarrow \mathcal{O}_{F'} \longrightarrow 0 ,$$

one has $\chi(G) = \chi(\mathcal{O}_F) - \chi(\mathcal{O}_{F'}) = 0$ because $\chi(\mathcal{O}_F) = \chi(\mathcal{O}_C) = 1$ by flatness of ψ. Since the support of G has dimension ≤ 0, one has $G = 0$, $F = F' \simeq \mathbb{P}^1$, and $N_{F/X} \simeq \mathcal{O}_F \oplus \mathcal{O}_F(-1)$ by 4.6. Again by obstruction theory, H is smooth of dimension 1 at $[F]$ and thus Y is a non-singular curve and ψ is a $\mathbb{P}^1$-bundle by Tsen's theorem because $F \simeq \mathbb{P}^1$. Thus 4.7.1 is proved. We will now show that $\pi : Z \to X$ is an embedding and $\pi(Z) = D$. Since every fiber F of ψ has the property that $(\pi(F) \cdot D) < 0$ by $[\pi(F)] \in R$, one sees that $\pi(F) \subset D$ and $\pi(Z) \subset D$. Since one also knows 4.6 that

$$T_{[F],Y} \simeq H^0(N_{F/X}) = H^0(\mathcal{O}_F \oplus \mathcal{O}_F(-1)) = k ,$$

it is easy to see that $\pi : Z \to X$ is unramified. Thus it is enough to show that π is injective. By definition of Z, $\pi_{|F}$ is an injection for all fibers F of ψ. If F_1 and F_2 are distinct fibers, then $\pi(F_1) \neq \pi(F_2)$ by definition of Y. Since $[\pi(F_1)], [\pi(F_2)] \in R$, one has $\pi(F_1) \cap \pi(F_2) = \emptyset$ by 4.6, whence π is an injection. Now we identify Z and D by π. Since $(H \cdot F) = 0$, $\pi^* H$ is of the form $\psi^* L$ for some invertible sheaf L on Y. Then L must be amply by 4.5. From $\omega_D \simeq \mathcal{O}_D(D + K_X)$, one has

$$-2 = (\omega_D \cdot F) = (\pi(F) \cdot D) + (\pi(F) \cdot K_X)$$

for all fibers F because ψ is a $\mathbb{P}^1$-bundle. Thus one has $(\pi(F) \cdot D) = (\pi(F) \cdot K_X) = -1$.

q. e. d.

4.8. Now we will prove 4.3. For the rest of this section, we will assume $H \cdot D = 0$.

By 3.4, there are an ample Cartier divisor L on D and positive integers a and b such that $-aL \approx \mathcal{O}_D(K_X)$ and $-bL \approx \mathcal{O}_D(D)$, because $(K_X \cdot R) < 0$ and $(D \cdot R) < 0$. Then $-(a+b)L \approx K_D$. One also has $\mathcal{O}_D(H) \approx \mathcal{O}_D$. We claim

(4.8.1) $$\chi(\mathcal{O}_D) = 1 .$$

By 4.1, one has $\chi(\mathcal{O}_D(nH)) \geq 0$ for $n >> 0$. Since $\mathcal{O}_D(H) \approx \mathcal{O}_D$, one has $\chi(\mathcal{O}_D) \geq 0$. Let $P(n)$ be a polynomial in n defined by $P(n) = \chi(\mathcal{O}_D(nL))$. Then P is a polynomial of degree 2 such that $P(0) = P(-a-b) = \chi(\mathcal{O}_D)$ (Serre duality) and $P(-a) = \chi(\mathcal{O}_D(K_X))$. By ampleness of $\mathcal{O}_D(-nH - K_X)$, one has $h^0(\mathcal{O}_D(nH + K_X)) = 0$ for all n, and hence one has $\chi(\mathcal{O}_D(nH + K_X)) = 0$ for $n >> 0$ by 4.1. Thus $P(-a) = \chi(\mathcal{O}_D(nH + K_X)) = 0$ because $\mathcal{O}_D(H) \approx \mathcal{O}_D$. Since P is of degree 2 and $P(0) = P(-a-b)$, one has $P(0) \neq 0$. Thus $\chi(\mathcal{O}_D) > 0$. Since $-K_D$ is ample, $h^2(\mathcal{O}_D) = 0$ and $\chi(\mathcal{O}_D) = 1 - h^1(\mathcal{O}_D) > 0$. Hence $\chi(\mathcal{O}_D) = 1$. This proves (4.8.1). About this $P(n)$, we can show

(4.8.2) $$P(n) = \frac{1}{ab}(n+a)(n+b) .$$

Indeed this follows from $P(-a) = 0$, $P(0) = 1$, and Serre duality $P(-a-b-n) = P(n)$.

The leading coefficient of $P(n)$ is known to be $(L^2)/2$, whence $ab(L^2) = 2$. Thus $(L^2) = 1$ or 2 and one has

(i) $(L^2) = 1$ and $h^0(L) \geq 3$, or

(ii) $(L^2) = 2$ and $h^0(L) \geq 4$.

Indeed $h^0(L) \geq \chi(L) = P(1) = (1+a)(1+b)/ab$ from $h^2(L) = h^0(\mathcal{O}_D(K_D - L)) = 0$. So the assertion above follows from $ab = 2/(L^2)$. Now it is easy to see that $|L|$ is very ample in cases (i) and (ii), by direct calculation. (Or see [2, §2].) Thus D is isomorphic to $\mathbb{P}^2$ or an irreducible quadric of $\mathbb{P}^3$. Since $\mathrm{Pic}\, D \simeq \mathbb{Z}$ or $\mathbb{Z} \oplus \mathbb{Z}$, one has $\mathcal{O}_D(H) \simeq \mathcal{O}_D$. Hence 4.3 is proved.

§5. The case where R is n.e. and $(H^2 \cdot -K_X) > 0$.

In this section, assuming that R is numerically effective (n.e., in short) and $(H^2 \cdot -K_X) > 0$ for some divisor H in R^*, we will prove 2.1 and (2.5.1).

Let H be a divisor in R^* 3.1 such that $(H^2 \cdot -K_X) > 0$. In §4, we proved that some multiple of H was base point free and induced the contraction of R under different situation. In our case, we cannot directly show that some multiple of H is base point free.

We first construct an open piece of contraction using $|aH|$ for some $a > 0$:

LEMMA 5.1. There exists a proper smooth morphism φ from an open dense subset U of X to a smooth surface V such that all fibers f have the properties $f \simeq \mathbb{P}^1$, $[f] \in R$, and $(f \cdot -K_X) = 2$.

This φ defines a morphism $V \to \mathrm{Hilb}_{X/k}$ by $v \mapsto [\varphi^{-1}(v)]$. Let Y be the irreducible reduced closed subscheme of $\mathrm{Hilb}_{X/k}$ dominated by V under this map, and let $Z \subset X \times Y$ be the induced family of closed subschemes parametrized by Y with natural morphisms $\psi : Z \subset X \times Y \to Y$ and $\pi : Z \subset X \times Y \to X$. Then,

LEMMA 5.2. There is a commutative diagram

$$\begin{array}{ccccc} Y & \longleftarrow & Z & \longrightarrow & X \\ \uparrow & & \uparrow & \nearrow & \\ V & \longleftarrow & U & & \end{array}$$

of natural morphisms. The vertical morphisms are open immersions, ψ is proper flat, and Z is an irreducible reduced projective 3-fold birational to X under π.

The following is similar to 4.6 and is proved by a similar method.

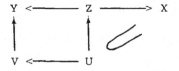

LEMMA 5.3. All fibers F of $\psi : Z \to Y$ are isomorphic to conics of $\mathbb{P}^2$ as schemes (smooth conics, reducible conics, or double lines).

Using 5.3, we finally show

LEMMA 5.4. $\pi : Z \to X$ is an isomorphism.

Now, under this identification of Z and X, every irreducible curve F of X which is sent to a point by $\psi : X \to Y$ has the property $[F] \in R$. This shows 2.1 and that $\rho(X) = \rho(Y) + 1$. Flatness of ψ implies Y is smooth. Thus 2.1 and (2.5.1) are proved.

We will prove the lemmas above now.

Proof of 5.1. Let H be a divisor in R^* such that $(H^2 \cdot -K_X) > 0$. By 3.3, one sees that $(H^3) = 0$. By 3.2 and the Riemann-Roch formula, one has

$$h^0(\mathcal{O}(nH)) = \frac{1}{4}(H^2 \cdot -K_X)n^2 + \cdots \qquad \text{for} \qquad n >> 0 \quad .$$

Thus H has D-dimension 2 [7] and there is a natural number a such that $|aH|$ induces a rational map onto a surface. Let M and F be the movable part and the fixed part of $|aH|$. Then $aH \sim M + F$. Let M_1, M_2, and M_3 be three general members of $|M|$ so that M_1 and M_2 are irreducible reduced by Bertini's theorem, F and M_1 intersect properly, and $M_1 \cap M_2 \not\subset F \cup M_3$. Let $M_1 \cdot M_2 = A + B$, where B is the largest sub 1-cycle of $M_1 \cdot M_2$ whose support is contained in $F \cup M_3$. (Hence A and B are effective, and $A \neq 0$.) Since H is numerically effective and $F \cdot M_1$ is effective, one has:

$$(H \cdot M_1 \cdot M_2) - (H \cdot A) = (H \cdot B) \geq 0 \quad ,$$

$$a(H^2 \cdot M_1) - (H \cdot M_1 \cdot M_2) = (H \cdot F \cdot M_1) \geq 0, \quad \text{and}$$

$$a^2(H^3) - a(H^2 \cdot M_1) = a(H^2 \cdot F) \geq 0 \quad .$$

Thus one has $0 \leq (H \cdot A) \leq a^2(H^3) = 0$, and $(H \cdot A) = (H \cdot M^2) = 0$. This implies $[A] \in R$ since $H \in R^*$. Since R is numerically effective, one sees $(A \cdot M) = (A \cdot F) = 0$. Since $A \cap (F \cup M_3)$ is of dimension ≤ 0 by definition of A, one sees $A \cap (F \cup M_3) = \emptyset$. Let $\alpha : X' \to X$ be the elimination of indeterminacy of $|M|$ so that $|M|$ induces a morphism onto a surface $\beta : X' \to W$ and α is an isomorphism outside the base locus $E \subset X$ of $|M|$. Let $\gamma : X' \to V'$ be the Stein factorization of β. Since $A \cap E = \emptyset$, $M_1 \cdot M_2 = A + B$ and $A \cap B = \emptyset$, one sees that $\alpha^{-1}(A)$ is a sum of fibers of γ, whence $\gamma^{-1}\gamma\alpha^{-1}(A) = \alpha^{-1}(A)$. Thus from $\alpha^{-1}(A) \cap \alpha^{-1}(E) = \emptyset$, one sees that $\gamma\alpha^{-1}(E) \cap \gamma\alpha^{-1}(A) = \emptyset$. In particular $\gamma\alpha^{-1}(E)$ is a proper closed set of V'. Let V be a non-empty smooth open set of V' such that $V' - V \supset \gamma\alpha^{-1}(E)$ and γ is smooth on $\gamma^{-1}V$. Let $U = \gamma^{-1}V$ and $\varphi = \gamma_{|U}$. Then $\alpha_{|U} : U \to X$ is an embedding because $X' - U \supset \alpha^{-1}(E)$. Since $(H \cdot M^2) = 0$, it is easy to see that there is a fiber f of φ such that $(f \cdot H) = 0$. Thus $[f] \in R$ for (one and hence) all fibers f of φ. Adjunction formula shows $\omega_f = \mathcal{O}_f(K_X + M + M)$, and this has negative degree because $(K_X \cdot R) < 0$ and $(M \cdot R) = 0$. Thus $f \simeq \mathbb{P}^1$ and $(f \cdot - K_X) = 2$.

q. e. d.

<u>Proof of</u> 5.2. Z is flat over an irreducible reduced surface Y with generic fiber $\mathbb{P}^1$ 5.1, whence Z is irreducible reduced of dimension 3. The existence of the commutative diagram follows from the property of $\mathrm{Hilb}_{X/k}$. Since $U \to Z$ is a section of $Z \to X$, $U \to Z$ is an open immersion. Hence $V \to Y$ is also an open immersion because $U = Z \times_Y V$ and ψ is faithfully flat. Clearly π is birational because it has a rational section $U \to Z$.

q. e. d.

For 5.3, we first need a sublemma which corresponds to (4.6.1) in the proof of 4.6.

LEMMA 5.5. Let f be an arbitrary fiber of ψ, and let $C = \pi(f)_{red}$. Let C' be an arbitrary closed subscheme of X such that $C'_{red} \subset C$. Then $\chi(\mathcal{O}_{C'}) \geq 0$.

Proof. Let D_1 and D_2 be two irreducible curves of Y such that $D_1, D_2 \not\subset Y - V$ and $D_1 \cap D_2 \ni \psi(f)$. Then it is easy to see that the irreducible divisors E_1 and E_2 on X whose supports are $\pi\psi^{-1}(D_1)$ and $\pi\psi^{-1}(D_2)$ have the properties $E_1 \neq E_2$, $(R \cdot E_1) = (R \cdot E_2) = 0$, and $C \subset E_1 \cdot E_2$. Now the rest is quite similar to that of (4.6.1) and left to the reader. We just use $H^i(\mathcal{O}(nH - a_1E_1 - a_2E_2)) = 0$ for $i > 0$ if $n >> a_1, a_2$ which follows from Kodaira vanishing by 3.1.

q.e.d.

The following lemma corresponds to the second part of proof of 4.6.

LEMMA 5.6. Let C be a connected reduced curve in X such that $\mathcal{O}_C(-K_X)$ is ample. Assume that $\chi(\mathcal{O}_{C'}) \geq 0$ for all closed subschemes C' of X such that $C'_{red} \subset C$. Then every irreducible component A of C is isomorphic to $\mathbb{P}^1$ and one has

$$N_{A/X} \simeq \mathcal{O}(p) \oplus \mathcal{O}(q) \qquad (p = 0 \text{ or } 1, \quad q \leq 0)$$

and $(K_X \cdot A) = -2-p-q$. Furthermore, if C_1 and C_2 are distinct irreducible components of C such that $C_1 \cap C_2 \neq \phi$, then C_1 and C_2 intersect at only one point and transversally.

Remark 5.6.1. In the description of $N_{A/X}$ above, one can actually show that $(p,q) = (0,0)$, $(0,-1)$, or $(1,-2)$ [13, (3.12)]. But we do not need this here.

Proof. Let A be an irreducible component of C. By the assertion, one has $\chi(\mathcal{O}_A) \geq 0$. If $\chi(\mathcal{O}_A) > 0$, then $\chi(\mathcal{O}_A) = 1$ and $A \simeq \mathbb{P}^1$. So assuming that

$\chi(\mathcal{O}_A) = 0$, we will derive a contradiction, which proves $A \simeq \mathbb{P}^1$. Since we assume $\chi(\mathcal{O}_A) = 0$, A is isomorphic to a plane cubic $E \subset \mathbb{P}^2$. From the exact sequence

$$0 \longrightarrow \mathcal{O}_E(-3) \longrightarrow \Omega^1_{\mathbb{P}} \otimes \mathcal{O}_E \longrightarrow \Omega^1_E \longrightarrow 0 ,$$

one easily sees $\chi(\Omega^1_E) = 0$. Since $A \simeq E$, one sees that A is locally of complete intersection in X. Thus $I/_{I^2}$ is locally free of rank 2 on A, where I is the sheaf of defining ideals of A in X. Thus one has an exact sequence

$$(5.6.2) \qquad 0 \longrightarrow I/_{I^2} \longrightarrow \Omega^1_X \otimes \mathcal{O}_A \longrightarrow \Omega^1_A \longrightarrow 0 .$$

Since $\chi(\mathcal{O}_A) = 0$, one has $\chi(\Omega^1_X \otimes \mathcal{O}_A) = (K_A \cdot A)$. Thus

$$\chi(I/_{I^2}) = \chi(\Omega^1_X \otimes \mathcal{O}_A) - \chi(\Omega^1_A) = (K_X \cdot A) ,$$

and one has

$$\chi(\mathcal{O}_X/I^2) = \chi(\mathcal{O}_A) + \chi(I/_{I^2}) = (K_X \cdot A) < 0 ,$$

which contradicts the assumption that $\chi(\mathcal{O}_{C'}) \geq 0$ for C', the subscheme defined by I^2. Thus one has $A \simeq \mathbb{P}^1$. The same method applied to $A = C_1 + C_2$ shows the last assertion on C_1 and C_2 in 5.6. Let us now find $I/_{I^2}$ for $A \simeq \mathbb{P}^1$.

Let

$$I/_{I^2} = \mathcal{O}_A(-p) \oplus \mathcal{O}_A(-q), \qquad p \geq q .$$

One sees $-p-q = \deg(I/_{I^2}) = (K_X \cdot A) + 2 \leq 1$ 5.6.2. Thus $(K_X \cdot A) =$

$= -2-p-q$ and $p+q \geq -1$. Now one has $p \geq 0$. If $p = 0$, then $q \leq 0$ and we are done. If $p = 1$, then $q \leq 1$. If $p = q = 1$, then

$$\chi(\mathcal{O}/_{I^3}) = \chi(\mathcal{O}_A) + \chi(I/_{I^2}) + \chi(I^2/_{I^3}) = -2 ,$$

because $I^2/_{I^3} \simeq S^2(I/_{I^2})$, the second symmetric power of $I/_{I^2}$. This is a contradiction and hence $q \leq 0$ and we are done if $p = 1$. We will derive a contradiction from $p \geq 2$, which finishes the proof of 5.6. Let J be the ideal of $\mathcal{O}_X$ such that $I^2 \subset J \subset I$ and $J/_{I^2} = \mathcal{O}(-q)$ in $I/_{I^2} = \mathcal{O}(-p) \oplus \mathcal{O}(-q)$. Then one has $J^2 \supset I^4$ and

$$\chi(\mathcal{O}/_{J^2}) = \chi(\mathcal{O}_A) + \chi(I/_{I^2}) + \chi(I^2/(I^3 + J^2)) + \chi(I^3/(J^2 \cap I^3))$$

$$= 6 - 7p - 2q$$

because

$$I^2/(I^3 + J^2) \simeq \mathcal{O}(-p-q) \oplus \mathcal{O}(-2p) \quad ,$$

$$I^3/(J^2 \cap I^3) \simeq \mathcal{O}(-3p) \quad ,$$

which is shown by easy calculation [13, (3.37)]. This is a contradiction because one has

$$\chi(\mathcal{O}/_{J^2}) \leq 6 - 7p - 2(-1-p) = 8 - 5p < 0$$

from $q \geq -1-p$.

q. e. d.

Proof of 5.3. Let F be an arbitrary fiber of $\psi : Z \to Y$. Let F' be the closed subscheme of F such that F' is free from embedded components and $F = F'$ except at finite number of points. One has $\chi(\mathcal{O}_{F'}) \leq \chi(\mathcal{O}_F) = 1$ because the general fiber of ψ is $\mathbb{P}^1$ 5.1. Now the proof is similar to that of (4.7.1): If we show that F' is a conic, then $\chi(\mathcal{O}_{F'}) = 1$ and $F = F'$. So it is enough to show that F' is a conic. First we know that $(F' \cdot -K_X) = 2$ and $(A \cdot -K_X) > 0$ for all irreducible components A of F' 5.1. Thus one has (i) F' is irreducible reduced, (ii) F' is reduced and has 2 components, or (iii) F'_{red} is irreducible and $F' = 2F'_{red}$ as 1-cycles. By 5.5, we can apply 5.6 to $C = F'$. So every irreducible component is isomorphic to $\mathbb{P}^1$, and two intersecting irreducible components intersect at only one point and transversally. Hence F' is a conic in cases (i) and (ii), and $F'_{red} = \mathbb{P}^1$ in case (iii). We consider case (iii). Let $A = F'_{red}$, and let I and J be the defining ideals of A and F' in X, respectively. Since $F' = 2A$ as 1-cycles, one sees that $I \supset J \supset I^2$ and I/J is an invertible sheaf on A. From $\chi(\mathcal{O}_{F'}) \leq 1$, one sees that $\chi(I/J) = \chi(\mathcal{O}_{F'}) - \chi(\mathcal{O}_A) \leq 0$ and hence $\deg(I/J) \leq -1$. Since one has the surjection $I/I^2 \twoheadrightarrow I/J$, one must have $I/I^2 = \mathcal{O}(-1) \oplus \mathcal{O}(-q)$ $(q \leq 0)$ and $\deg I/J = -1$ 5.6. Then $\chi(\mathcal{O}_{F'}) = 1$ and it is easy to see that F' is a conic.

q. e. d.

Proof of 5.4. Since π is a birational morphism to a smooth 3-fold, the exceptional set of π $\mathrm{Ex}(\pi) \subset Z$ is of pure codimension 1 if $\mathrm{Ex}(\pi) \neq 0$ [15, (3.20)]. Assuming that $\mathrm{Ex}(\pi)$ contains an irreducible divisor D, we will derive a contradiction, which finishes the proof. By 5.2, $D \subset Z - U$ and $S = \psi(D)$ is a curve. Hence D is an irreducible component of $Z_S = \psi^{-1}(S)$, and $\pi(D)$ is also an irreducible curve. We need to treat 3 cases depending on the type of the generic fiber of $Z_S \to S$. We treat here the cases where the genetic fiber is a smooth conic or a double line, and refer the reader to [13, (3.26.2)] for the reducible conic case. In the 2 cases we are going to consider, one has $D = (Z_S)_{red}$ and $Z_S \to S$ is a family of closed subschemes F of X such that $F_{red} = \pi(D)$ (constant). We are going to show that $Z_S \to S$ is actually a constant family, which will contradict the assumption that $S \subset \mathrm{Hilb}_{X/k}$. If the generic fiber of $Z_S \to S$ is $\mathbb{P}^1$, then S generally parametrizes the reduced closed subscheme $\pi(D)$ and $Z_S \to S$ is constant. Assume that the generic fiber is a double line. Let $C = \pi(D) \simeq \mathbb{P}^1$ 5.3. We will show that there is only one fiber F such that $F_{red} = C$. Let I and J be defining ideals of C and F in X , respectively. Since F is a double line, one has $I \supset J \supset I^2$ and $I/_J \simeq \mathcal{O}_C(-1)$. Now 5.6 shows that there is only one surjection $I/_{I^2} \longrightarrow\!\!\!\!\!\!\rightarrow \mathcal{O}_C(-1)$ up to multiplication by constants. Hence there is only one J and we are done.

q. e. d.

§6. The case where R is n.e. and $(H^2 \cdot -K_X) \leq 0$.

We assume in this section that R is numerically effective and that there exists a divisor H on X such that $H \in R^*$ and $(H^2 \cdot -K_X) \leq 0$.

If $\rho(X) = 1$, then $(K_X \cdot R) < 0$ shows that $-K_X$ is ample and $R = \overline{NE}(X)$. Thus the structure morphism $X \to \operatorname{Spec} k$ is the contraction and we have 2.1 and (2.5.3).

Assuming further that $\rho(X) \geq 2$, we will prove 2.1 and (2.5.2). Since H is numerically effective, one sees $H^2 \in \overline{NE}(X)$ [6] and $(H^3) = 0$ 3.3. Thus $(H^2 \cdot nH - K_X) \leq 0$ for ample divisors $nH - K_X$ if $n >> 0$ 3.1. Hence one sees by 1.1 that

6.1 $$H^2 = 0 \text{ in } N(X) \quad .$$

From this, easily follows the next lemma.

LEMMA 6.2. (i) $M \cdot N = 0$ for all $M, N \in R^{\perp}$ (see §3), and

(ii) $\rho(X) = 2$.

The essential part is to show

LEMMA 6.3. $h^0(\mathcal{O}_X(nH)) \to \infty$ as $n \to \infty$.

Let us finish the proof of 2.1 and (2.5.3) using these. By 6.3, there is a natural number a such that $\dim |aH| > 0$. Let B be the fixed component of $|aH|$ and $A = aH - B$. Since R is numerically effective, one sees $(A \cdot R) = 0$ from $A + B = aH$. It follows that $A \in R^*$ from $(A \cdot R) = 0$ because $\rho(X) = 2$ and $R^* \simeq \mathbb{R}_+$. By 6.2, one has $A^2 = 0$ and hence $|A|$ is free from base points. Let $\varphi : X \to Y$ be the Stein factorization of the morphism induced by $|A|$. Then Y is a non-singular curve. Every curve C in X sent to a point of Y by φ belongs to

R because $A \in R^*$. Thus one has 2.1, i.e. φ is the contraction of R, and hence $-K_X$ is φ-ample 2.2. It remains to prove that every fiber F of φ is irreducible reduced. If F is the generic fiber, then F is smooth and $-K_F \simeq \mathcal{O}_F(-F - K_X) \simeq \mathcal{O}_F(-K_X)$ is ample. Thus F is a del Pezzo surface and, in particular, one has $\chi(\mathcal{O}_F) = 1$, (see Murre's course text in this volume). Let F be an arbitrary fiber and G an irreducible component of F. 2.1 shows $(G \cdot R) = 0$ and $(G \cdot C) = 0$ for every irreducible curve C in F. This means $G = F_{red}$. Let a be a natural number such that $F = aG$. Since $G \cdot H = 0$, one has $\mathcal{O}_G(G) \approx \mathcal{O}_G$ by 3.4. Then

$$\begin{aligned}\chi(\mathcal{O}_F) &= \chi(\mathcal{O}_{aG}) \\ &= \chi(\mathcal{O}_G) + \chi(\mathcal{O}_G(-G)) + \cdots + \chi(\mathcal{O}_G(-a+1)G)) \\ &= a\chi(\mathcal{O}_G) \quad .\end{aligned}$$

Since we have $\chi(\mathcal{O}_F) = 1$ by flatness of φ, this means $a = 1$ and we are done.

For the rest of this section, we will consider 6.2 and 6.3.

Proof. of 6.2. (i) Let a be an arbitrary number. Then $aM + N \in R^\perp$ and $H \in R^*$. One sees that $nH \pm (aM + N) \in R^*$ if $n >> 0$ because R^* is open in $R^\perp$ 3.1. Let $D(\varepsilon) = nH + \varepsilon(aM + N)$ for $\varepsilon = \pm 1$. From $D(1) + D(-1) = 2nH$ and 6.1, one sees

$$D(1)^2 + 2D(1) \cdot D(-1) + D(-1)^2 = 0 \quad .$$

Since $D(1)$ and $D(-1)$ are numerically effective, one sees that $D(1)^2$, $D(1) \cdot D(-1)$, and $D(-1)^2 \in \overline{NE}(X)$ [6]. Thus $D(1)^2 = 0$ because $\overline{NE}(X)$ does not contain a straight line 1.1. One has $(nH + aM + N)^2 = 0$ for $n >> 0$, whence $(aM + N)^2 = 0$. Since a was arbitrary, one gets $M \cdot N = 0$.

(ii) Let

$$V = \{v \in N(X)^* \mid (v \cdot R) = 0,\ (v \cdot K_X^2) = 0\} \quad .$$

It is enough to show that $V = 0$ because $\dim V \geq \rho(X) - 2$. Let a be a positive number such that $aH - K_X$ is ample 3.1. Let $M = aH - K_X$. To show that $V = 0$, it is enough to show every divisor L in V is zero, because V is clearly generated by divisors in V. If L is a divisor such that $L \in R^{\perp}$ and $(R \cdot K_X^2) = 0$, then one has $(L \cdot M^2) = (L^2 \cdot M) = 0$ by (i). Whence $L = 0$ by [11, P306, Proposition 3].

q. e. d.

Now 6.3 is left. The proof is easy if $h^1(\mathcal{O}_X) > 0$.

6.3.1. Assume that $h^1(\mathcal{O}_X) > 0$. Let $\alpha : X \to A$ be the albanese mapping, where $\dim A = h^1(\mathcal{O}_X) > 0$. Let L be a very ample divisor on A. We claim $\alpha^* L \in R^* \simeq \mathbb{R}_+$. Indeed any morphism from an extremal rational curve in R to an abelian variety A is constant and $(\alpha^* L \cdot R) = 0$, i.e. $\alpha^* L \in R^{\perp} \simeq \mathbb{R}$. One has $\alpha^* L \in R^*$ because $|\alpha^* L| \neq \phi$ and $R^* \simeq \mathbb{R}_+$. Now the claim is proved. Thus there are natural numbers a and b such that $aH \approx b\alpha^* L$ because $\alpha^* L, H \in R^* \simeq \mathbb{R}_+$. Thus by 3.2, one has $h^0(\mathcal{O}_X(anH)) = h^0(\mathcal{O}_X(bn\alpha^* L)) \geq h^0(\mathcal{O}_A(bnL))$ for $n >> 0$. Since $h^0(\mathcal{O}_A(bnL)) \to \infty$ as $n \to \infty$, one sees $h^0(\mathcal{O}_A(nH)) \to \infty$ as $n \to \infty$ because $h^0(\mathcal{O}_A(nH))$ is a polynomial in n for $n >> 0$ by 3.2. Thus the proof is done if $h^1(\mathcal{O}_X) > 0$.

6.3.2. Assume that $h^1(\mathcal{O}_X) = 0$ in view of 6.3.1. Since $H^2 = 0$ 6.1, one sees that $\chi(\mathcal{O}(nH))$ is a polynomial in n of degree ≤ 1 by the Riemann-Roch formula. Thus there is a non-negative integer a such that

$$h^0(\mathcal{O}(nH)) = an + \chi(\mathcal{O}_X) \quad \text{for} \quad n >> 0$$

by 3.2. If $a > 0$, then we are of course done. So assuming that $a = 0$, we will derive a contradiction using the lemma 3.9 in [13]. This way, proof will be finished.

We first note that $h^3(\mathcal{O}_X) = h^0(K_X) = 0$ because $(K_X \cdot R) < 0$ and R is numerically effective. Thus by $a = 0$,

$$h^0(\mathcal{O}(nH)) = \chi(\mathcal{O}_X) = 1 + h^2(\mathcal{O}_X) \geq 1 \quad \text{for} \quad n >> 0 \quad .$$

Since $h^0(\mathcal{O}(nH))$ does not go to ∞ as $n \to \infty$, one must have $h^0(\mathcal{O}(nH)) = 1$ for $n >> 0$. We claim that the unique member H_n of $|nH|$ is a multiple of an irreducible divisor for $n >> 0$. Indeed if H_n contains two distinct irreducible (reduced) divisors Y_1 and Y_2, then Y_1 and Y_2 are proportional in $N(X)$ because $(Y_1 \cdot R) = (Y_2 \cdot R) = 0$ and $\rho(X) = 2$. Since $h^1(\mathcal{O}_X) = 0$, one has $N(X) = (\mathrm{Pic}\ X) \otimes_{\mathbb{Z}} \mathbb{R}$. So there are natural numbers a and b such that $\mathcal{O}(aY_1) \simeq \mathcal{O}(bY_2)$. Thus $aY_1, bY_2 \in |aY_1|$ and $1 \leqq \dim |aY_1| \leqq \dim |anH|$, which is a contradiction to $h^0(\mathcal{O}(anH)) = 1$. Hence it is clear that there is an irreducible (reduced) divisor Y of X such that H_n is a multiple of Y for all $n >> 0$. Thus from $Y \cdot H = 0$, one sees that ω_Y^{-1} is ample and $\mathcal{O}_Y(Y) \approxeq \mathcal{O}_Y$ 3.4. Since $\chi(\mathcal{O}_X(nY)) = 1$ for all n, one has $\chi(\mathcal{O}_Y) = \chi(\mathcal{O}_X) - \chi(\mathcal{O}_X(-Y)) = 0$. Now this contradicts the following.

LEMMA 6.4. ((3.9) in [13]). Assume that Y is an irreducible reduced projective surface (over k) which is locally Gorenstein, such that ω_Y^{-1} is ample. Then $\chi(\mathcal{O}_Y) \neq 0$.

Thus the proof is finished, but as for 6.4, one can say more:

THEOREM 6.5. (Goto and Mori). Under the assumption of 6.4, one has

(i) $\chi(\mathcal{O}_Y) = 1$ and furthermore $H^1(\mathcal{O}_Y(-L)) = 0$ if L is a numerically effective Cartier divisor such that $(L^2) > 0$, and

(ii) if the base field k is of characteristic > 0, then there are examples Y satisfying the assumptions of 6.4 with arbitrarily large $h^1(\mathcal{O}_Y)$.

Proof of this theorem is done together with the classification of non-normal surfaces Y (in arbitrary characteristics) satisfying the assumption of 6.4. This will be published elsewhere.

References

[1] A. Beauville, Variete de Prym et jacobiennes intermediaire, Ann. Scient. de l'Ecole Norm. Sup. 10, 1977, 304-392.

[2] T. Fujita, On the structure of polarized varieties with Δ-genera zero, J. Fac. Sci., U. of TokyO, Sec. IA, Vol. 22, No. 1, 1975, 103-115.

[3] A. Grothendieck, Elements de geometric algebrique, Publ. Math. IHES, No. 11, 1961.

[4] ______, Fondements de la geometrie algebrique, Secretariat Math. 11 Rue Pierre Curie, Paris 5^e, 1962.

[5] ______, Sur une note de Mattuck-Tate, Crelle, 20, 1958, 208-215.

[6] R. Hartshorne, Ample subvarieties of algebraic varieties, Lecture Notes in Math. 156, Springer-Verlag, 1970.

[7] S. Iitaka, On D-dimensions of algebraic varieties, J. Math. Soc. Japan, 23, 1971, 356-373.

[8] V. A. Iskovskih, Fano 3-folds I, Math. USSR, Izv. 11, 1977, 485-527.

[9] ______, Fano 3-folds II, Math. USSR, Izv. 12, 1978, 469-506.

[10] Y. Kawamata, A generalization of Kodaira-Ramanujam's vanishing theorem, to appear.

[11] S. Kleiman, Toward a numerical theory of ampleness, Ann. Math. 84, 1966, 293-344.

[12] S. Mori, Projective manifolds with ample tangent bundles, Ann. Math., 110, 1979, 593-606.

[13] ______, Threefolds whose canonical bundles are not numerically effective, to appear in Ann. Math.

[14] ______, and S. Mukai, Classification of Fano 3-folds with $B_2 \geq 2$, manuscr. math. 36 (1981) 147-162

[15] D. Mumford, Algebraic Geometry I, Complex Projective Varieties, Grundl. der Math. Wiss. 221, Springer-Verlag, New York, 1976.

[16] C. P. Ramanujam, Supplement to the article "Remarks on the Kodaira vanishing theorem", J. of the Indian Math. Soc. 38, 1974, 121-124.

[17] K. Ueno, Introduction to classification theory of algebraic varieties and compact complex spaces, Lecture Notes in Math. 412, Springer-Verlag, 1974, 288-332.

LES SINGULARITES DU DIVISEUR Θ DE LA JACOBIENNE INTERMEDIAIRE DE L'HYPERSURFACE CUBIQUE DANS $\mathbb{P}^4$

Arnaud BEAUVILLE

Centre de Mathématiques
Ecole Polytechnique
91128 Palaiseau

Introduction

La géométrie de l'hypersurface cubique de dimension 3 a été élucidée par Clemens et Griffiths [C-G] ; citons, parmi les résultats les plus frappants, le théorème de Torelli et le fait que la cubique n'est pas rationnelle (une partie des résultats de [C-G] avait été obtenue indépendamment par Tjurin [T2]). Clemens et Griffiths utilisaient des arguments délicats de dégénérescence ; Mumford a observé qu'on peut déduire l'irrationalité de la cubique de la théorie des variétés de Prym.

Nous donnons ici une approche différente des résultats de [C-G], basée également sur les variétés de Prym. Cette approche est moins générale, mais peut-être plus simple ; elle est entièrement algébrique et valable en toute caractéristique différente de deux. De manière précise, le résultat essentiel de cet exposé est le théorème suivant :

Théorème

Soient X une hypersurface cubique lisse dans $\mathbb{P}^4$ et Θ un diviseur thêta de sa jacobienne intermédiaire. Alors Θ a un seul point singulier, qui est un point triple ; le cône tangent à Θ en ce point est isomorphe au cône affine de base X.

De ce théorème résultent immédiatement le théorème de Torelli et (moins immédiatement) l'irrationalité de X. Cet énoncé est certainement bien connu des experts ; cependant il ne figure pas, à ma connaissance, dans la littérature (voir [T2] pour une approximation).

La démonstration comporte deux parties. On déduit d'abord l'unicité de la singularité (§3) de la description (due à Mumford) des singularités du diviseur thêta d'une variété de Prym. La structure de cette singularité résulte assez aisément d'un des points clés de [C-G], la paramétrisation de Θ par les différences de droites (§5) ; celle-ci est établie en interprétant en termes de variétés de Prym la "surface de Fano", c'est-à-dire la variété des droites contenues dans X (§4).

Les deux premiers paragraphes contiennent les préliminaires nécessaires sur les variétés de Prym et l'hypersurface cubique.

Toutes les variétés considérées sont définies sur un corps k algébriquement clos de caractéristique $\neq 2$.

§ 1 VARIETES DE PRYM

On rappelle dans ce paragraphe les résultats sur les variétés de Prym qui seront utilisés dans la suite, renvoyant à [M] pour les démonstrations.

Si Γ est une courbe lisse, on note $J^d\,\Gamma$, pour $d \in \mathbb{Z}$, la variété des classes de diviseurs de degré d sur Γ, modulo équivalence linéaire ; elle est isomorphe (non canoniquement) à la jacobienne $J\Gamma = J^o\Gamma$. Si D est un diviseur de degré d sur Γ, on note [D] la classe de D dans $J^d\Gamma$.

Soient C une courbe lisse de genre g, $\pi : \widetilde{C} \longrightarrow C$ un revêtement double étale connexe. La courbe $\tilde{C}$ est de genre $\tilde{g} = 2g - 1$. Soit σ l'involution de $\tilde{C}$ qui échange les deux feuillets du revêtement π ; on notera encore σ l'endomorphisme induit sur $\mathrm{Pic}(\tilde{C})$. L'application d'image directe $\pi_* : \mathrm{Div}(\tilde{C}) \longrightarrow \mathrm{Div}(C)$ induit un homomorphisme $N : \mathrm{Pic}(\tilde{C}) \longrightarrow \mathrm{Pic}(C)$.

(P.1) Posons $P = (1-\sigma)(J\tilde{C}) = (\mathrm{Ker}\, N)_o$; c'est une variété abélienne de dimension g-1, isomorphe à $J\tilde{C}/\pi^* JC$. On dit que P est la <u>variété de Prym</u> associée à $(\tilde{C}, C)$.

(P.2) Le groupe Ker(N) a deux composantes connexes, P_o et P_1, contenues dans $J\tilde{C}$, avec $P_o = P$; on a $P_i = (1-\sigma)(J^i\tilde{C})$ pour $i = 0,1$.

(P.3) Soit D un diviseur de degré d sur C ; il résulte de (P.2) que $N^{-1}([D])$ est réunion de deux sous-variétés P_o^D et P_1^D de $J^D\tilde{C}$ (isomorphes à P). Si $[\tilde{D}] \in P_o^D$ et si A est un diviseur sur $\tilde{C}$, la classe de $\tilde{D} + A - \sigma A$ appartient à P_i^D, avec $i \equiv \deg(A) \pmod 2$.

(P.4) Prenons D = K. On peut alors numéroter les deux composantes P_i^K de façon que, pour $[D] \in N^{-1}(K)$, on ait

$$[D] \in P_i^K \Longleftrightarrow h^o(D) \equiv i \pmod 2$$

Il est commode de considérer au lieu de P la variété P_o^K, qui lui est isomorphe, et qu'on notera P^*. On a donc ensemblistement

$$P^* = \{[D] \in J^{\tilde{g}-1}\tilde{C} \mid \pi_* D \equiv K \text{ et } h^o(D) \text{ pair}\}$$

(P.5) Soit Θ^* le diviseur de P^* formé des classes de diviseurs <u>effectifs</u>, c'est-à-dire $\Theta^* = \{[D] \in J^{\tilde{g}-1}\tilde{C} \mid \pi_* D \equiv K \text{ et } h^o(D) \text{ pair} \geq 2\}$. Lorsqu'on identifie P^* à P par translation, le diviseur Θ^* définit une <u>polarisation principale</u> sur P. Dans la suite on munira toujours P de cette polarisation.

Soit $\tilde{\Theta}$ le diviseur de $J^{\tilde{g}-1}\tilde{C}$ formé des classes de diviseurs effectifs (diviseur thêta de Riemann) ; on a l'égalité entre diviseurs $\widetilde{\Theta}_{|P^*} = 2\ \Theta^*$.

(P.6) L'égalité précédente permet de calculer les singularités de Θ^*. Un point $[D]$ de Θ^* est singulier dans l'un des deux cas suivants :

a) L'espace tangent à P^* en $[D]$ est contenu dans le cône tangent en $[D]$ à $\widetilde{\Theta}$;

b) $[D]$ est un point singulier de multiplicité ≥ 4 de $\widetilde{\Theta}$.

A l'aide du théorème de Riemann-Kempf, ces situations se traduisent comme suit :

a) On a $D \equiv \pi^* M + E$, où $|M|$ est un système linéaire mobile sur C et E un diviseur effectif sur $\widetilde{C}$, tel que $\pi_* E \in |K-2M|$;

b) On a $h^0(D) \geq 4$.

Les singularités du type a) sont dites spéciales ; elles n'apparaissent pas lorsque la courbe C est générique. Les singularités du type b), qui apparaissent sur toutes les variétés de Prym de dimension ≥ 9, seront dites (suivant Tjurin) stables.

§ 2 LA VARIETE DE PRYM ASSOCIEE A L'HYPERSURFACE CUBIQUE

On fixe désormais une hypersurface cubique lisse $X \subset \mathbb{P}^4$, ainsi qu'une droite $\ell \subset X$. Nous ferons sur ℓ l'hypothèse de position générale suivante:

Pour toute droite $f \subset X$ rencontrant ℓ, le plan $< \ell, f >$ découpe sur X trois droites distinctes.

Il n'est pas difficile de montrer qu'une droite générique de X vérifie cette condition (cf.[C-G] ou [Mu1]).

Soit X_ℓ la variété obtenue en éclatant ℓ dans X. La projection de centre ℓ définit un morphisme $p : X_\ell \longrightarrow \mathbb{P}^2$, dont les fibres sont des coniques ; l'hypothèse de position générale entraîne que ces coniques sont lisses ou réunion de deux droites distinctes. On en déduit facilement (cf.[B1],ch.1) que l'ensemble des points t de $\mathbb{P}^2$ tels que la conique $p^{-1}(t)$ soit dégénérée est une courbe lisse C. Soit $\tilde{C}$ la variété des droites contenues dans X et incidentes à ℓ ; la projection p définit un morphisme $\pi : \tilde{C} \longrightarrow C$, qui est un revêtement double étale (nous verrons plus loin que $\tilde{C}$ est connexe). L'involution σ associe à une droite f de $\tilde{C}$ la droite résiduelle de l'intersection $< \ell, f > \cap X$.

Proposition 1 (Mumford) :

La jacobienne intermédiaire[1] JX est isomorphe (comme variété abélienne principalement polarisée) à la variété de Prym P associée à $(\tilde{C}, C)$.

Nous nous contenterons d'indiquer la définition de l'isomorphisme $u : P \longrightarrow JX$, renvoyant à [T3] ou [B1] pour une démonstration complète.

Puisque $\tilde{C}$ paramètre une famille de courbes sur X, il existe une application d'Abel-Jacobi $\alpha : \tilde{C} \longrightarrow JX$ (définie à translation près), d'où un homomorphisme $a : J\tilde{C} \longrightarrow JX$. Comme la famille des fibres de p est paramétrée par $\mathbb{P}^2$, l'application d'Abel-Jacobi qui lui correspond est constante ; ceci entraîne que a est nul sur $\pi^* JC$. On en déduit que a se factorise en

$$a : J\tilde{C} \xrightarrow{1-\sigma} P \xrightarrow{u} JX ;$$

u est l'homomorphisme cherché.

---

[1] Si $k \neq \mathbb{C}$, JX sera par définition le groupe des cycles de dimension 1 sur X algébriquement équivalents à zéro modulo équivalence rationnelle, muni de la structure de variété abélienne principalement polarisée définie dans [Mu3] ou [B1],ch.3.

Dans le reste de ce paragraphe, on va établir quelques propriétés des courbes C et $\tilde{C}$.

(i) <u>La courbe</u> C <u>est une courbe plane lisse de degré</u> 5. On a vu que C est lisse ; son degré est le nombre d'intersection C.d, où d est une droite de $\mathbb{P}^2$. Lorsque d est assez générale, la surface cubique $p^{-1}(d)$ est lisse ; dans une telle surface, il y a dix droites rencontrant une droite donnée, réparties en cinq paires de droites concourantes (représenter la surface comme un plan projectif éclaté en 6 points), donc C est de degré 5.

(ii) <u>La courbe</u> $\tilde{C}$ <u>est connexe ; elle admet un pinceau</u> $|L|$ <u>sans point base, de degré</u> 5, <u>tel que</u> $\pi_* L \equiv H$.

Considérons en effet l'application $g : \tilde{C} \longrightarrow \ell$ qui associe à une droite de $\tilde{C}$ son point d'intersection avec ℓ. L'hypothèse de position générale entraîne que g est un morphisme fini de degré 5 (par un point x de ℓ, il passe 5 droites de X distinctes de ℓ). Si la courbe $\tilde{C}$ n'est pas connexe, elle est réunion de deux courbes isomorphes à C, et la restriction de g à l'une de ces courbes est de degré ≤ 2 : ceci entraîne que C est rationnelle ou hyperelliptique, d'où une contradiction.

Soit $x \in \ell$; posons $L = g^*[x]$. Le système $|L|$ est un pinceau sans point base ; le diviseur $\pi_* L$ est découpé sur C par la projection de l'hyperplan tangent en x à X, donc $\pi_* L \equiv H$.

Notons que (i) et (ii) impliquent $\dim(JX) = g(C)-1 = 5$.

Nous noterons H le diviseur (de degré 5) découpé sur C par une droite de $\mathbb{P}^2$.

(iii) <u>Soit</u> $|D|$ <u>un système linéaire mobile sur</u> C, <u>de degré</u> ≤ 5. <u>On a alors</u> $D \equiv H - p$ <u>ou</u> $D \equiv H - p + q$, <u>avec</u> p,q <u>dans</u> C.

Supposons $\deg(D) = 5$; par Riemann-Roch, les coniques passant par D doivent former un pinceau, ce qui impose que 4 des points de D sont alignés, c'est-à-dire $D \equiv H - p + q$. Il en résulte aussitôt que tout g_4^1 sur C est de la forme $|H-p|$ pour $p \in C$, et que C n'admet pas de g_3^1.

Notons η l'élément d'ordre deux de JC associé au revêtement π, de sorte qu'on a $\pi_* \mathcal{O}_{\tilde{C}} \cong \mathcal{O}_C \oplus \mathcal{O}_C(\eta)$.

(iv) <u>On a</u> $h^0(C,H+\eta) = 1$ <u>et</u> $h^0(\tilde{C},\pi^*H) = 4$. <u>En particulier, la classe de</u> π^*H <u>dans</u> $J^{10}\tilde{C}$ <u>appartient à</u> P^*.

On a

$$\pi_* \pi^* \mathcal{O}_C(H) \cong \mathcal{O}_C(H) \otimes \pi_* \mathcal{O}_{\tilde{C}} \cong \mathcal{O}_C(H) \oplus \mathcal{O}_C(H+\eta)$$

d'où
$$h^0(\tilde{C},\pi^*H) = h^0(C,H) + h^0(C,H+\eta) ;$$

les deux égalités énoncées sont donc équivalentes.

On a $h^0(C,H+\eta) \leq 1$: en effet dans le cas contraire, on aurait d'après (iii) $\eta \equiv p - q$ $(p,q \in C)$, d'où $2p \equiv 2q$, et C serait hyperelliptique. Il suffit donc de prouver l'inégalité $h^0(\pi^*H) \geq 4$. Puisque $H \equiv \pi_*L$ (cf.(ii)), on a $\pi^* H \equiv L + \sigma L$. Soient s,t deux sections de $\mathcal{O}_{\tilde{C}}(L)$ dont les diviseurs n'ont pas de point commun ; on a une suite exacte

$$0 \longrightarrow \mathcal{O}_{\tilde{C}}(\sigma L - L) \xrightarrow{(t,-s)} \mathcal{O}_{\tilde{C}}(\sigma L)^2 \xrightarrow{(s,t)} \mathcal{O}_{\tilde{C}}(\sigma L + L) \longrightarrow 0$$

d'où l'on déduit

$$h^0(\pi^*H) = h^0(L+\sigma L) \geq 2h^0(\sigma L) = 4,$$

ce qui achève la démonstration.

(v) Soit D un diviseur sur $\tilde{C}$ tel que $h^o(D) \geq 2$ et $\pi_* D \equiv H$. On a alors $D \equiv L$ ou $D \equiv \sigma L$.

Le système linéaire $|\pi^* H|$, qui est sans point base, définit un morphisme $j : \tilde{C} \longrightarrow \mathbb{P}^3$, génériquement injectif. La relation $\pi^* H \equiv D + \sigma D$ signifie alors que $j(\tilde{C})$ est contenue dans une quadrique $Q_D \subset \mathbb{P}^3$, les deux systèmes de génératrices (éventuellement confondus) de Q_D découpant sur $j(\tilde{C})$ la partie mobile des systèmes $|D|$ et $|\sigma D|$. Puisque la courbe $j(\tilde{C})$ est de degré 10, elle est contenue dans une quadrique au plus ; on a donc $Q_D = Q_L$ et, puisque $|L|$ est sans point base, $|D| = |L|$ ou $|\sigma L|$.

§ 3 SINGULARITES DU DIVISEUR Θ

Proposition 2 :

Le diviseur Θ de JX a un seul point singulier.

En particulier, Θ est normal et irréductible.

En vertu de la prop. 1, il revient au même de considérer un diviseur thêta de P, ou encore le diviseur Θ^* de P^*(P.5) ; nous allons montrer que celui-ci a un seul point singulier, à savoir $[\pi^* H]$. D'après (P.6), il y a deux types de singularités à étudier.

a) Singularités spéciales : Compte tenu du §2,(iii), les singularités spéciales de Θ^* correspondent aux diviseurs $\pi^*(H-p) + E$, avec $p \in C$ et $\pi_* E = 2p$, tels que $h^o(\pi^*(H-p)+E)$ soit pair. Si $\pi^{-1}(p) = \{p_1, p_2\}$, les diviseurs possibles sont $\pi^* H$ et $\pi^* H + p_i - \sigma p_i$ $(i = 1,2)$. Or $[\pi^* H]$ appartient à P^*(§2,(iv)), donc $[\pi^* H + p_i - \sigma p_i]$ n'appartient pas à P^*(P.3) : ainsi la seule singularité de ce type correspond au point $[\pi^* H]$.

b) Singularités stables : Soit $[D]$ un élément de P^* tel que $h^o(D) \geq 4$; il s'agit de prouver $D \equiv \pi^* H$. Soient s,t deux sections de $\mathcal{O}_{\tilde{C}}(L)$ dont les diviseurs n'ont pas de point commun (cf.§2,(ii)) ; on a une suite exacte

$$0 \longrightarrow \mathcal{O}_{\tilde{C}}(D-L) \xrightarrow{(t,-s)} \mathcal{O}_{\tilde{C}}(D)^2 \xrightarrow{(s,t)} \mathcal{O}_{\tilde{C}}(D+L) \longrightarrow 0,$$

d'où l'on déduit :

$$h^o(D-L) + h^o(D+L) \geq 2\ h^o(D) \geq 8.$$

D'autre part, on obtient par Riemann-Roch

$$h^o(D+L) = h^o(K-D-L) + 5 = h^o(\sigma D-L) + 5 = h^o(D-\sigma L) + 5,$$

d'où finalement $h^o(D-L) + h^o(D-\sigma L) \geq 3$.

Ainsi l'un des systèmes $|D-L|$ ou $|D-\sigma L|$ est un pinceau se projetant dans $|H|$, donc est égal à $|L|$ ou à $|\sigma L|$ (§2,(v)). Par conséquent D est linéairement équivalent à π^*H, 2L ou $2\sigma L$. Mais $2L \equiv \pi^*H + (L-\sigma L)$ et $2\sigma L \equiv \pi^*H + (\sigma L-L)$ n'appartiennent pas à P^* (P.3), donc $D \equiv \pi^*H$.

<u>Corollaire</u> : <u>La variété</u> X <u>n'est pas rationnelle</u>.

En effet la jacobienne intermédiaire d'une variété rationnelle est isomorphe à la jacobienne d'une courbe ou à un produit de telles jacobiennes ([C-G],§3), ce qui entraîne que le lieu singulier du diviseur thêta est de codimension ≤ 4. Puisque dim(JX) = 5, la prop. 2 montre que X n'est pas rationnelle.

§ 4 LA SURFACE DE FANO

Un rôle fondamental dans l'étude de JX est joué par la surface de Fano F, c'est-à-dire la variété des droites contenues dans X. C'est une surface <u>lisse et connexe</u> : la lissité et l'assertion sur la dimension résultent, via la théorie des déformations, d'un calcul facile de fibré normal, cf.[C-G]. On déduit la connexité de F de celle de $\tilde{C}$ (§2,(ii)) et du fait que $\tilde{C}$ rencontre toute courbe tracée sur F (puisque ℓ rencontre tout diviseur de X).

Notre but dans ce paragraphe est d'interpréter la surface F en termes du revêtement $\pi : \tilde{C} \longrightarrow C$.

La projection de centre ℓ détermine un morphisme $q' : F - \{\ell\} \longrightarrow \check{\mathbb{P}}^2$. Notons $\varepsilon : F_\ell \longrightarrow F$ l'éclatement du point ℓ de F ; on vérifie sans peine que q' se prolonge en un morphisme $q : F_\ell \longrightarrow \check{\mathbb{P}}^2$.

Soit f une droite assez générale de F, de sorte que l'hyperplan engendré par ℓ et f coupe X suivant une surface lisse Σ_f. Considérons l'ensemble D(f) des droites de $\tilde{C}$ qui sont incidentes à f. La surface cubique Σ_f contient 5 paires de droites de $\tilde{C}$ coplanaires, correspondant aux 5 coniques dégénérées au-dessus des 5 points de $p(f) \cap C$; dans chacune de ces paires, une et une seule des deux droites rencontre f. Autrement dit, l'image directe par π du diviseur $D(f) \subset \tilde{C}$ est le diviseur découpé sur C par la droite q(f).

Soit S la sous-variété de $\tilde{C}^{(5)}$ formée des diviseurs effectifs D tels que $\pi_* D \equiv H$. Il résulte de ce qui précède que l'application $f \longrightarrow D(f)$ définit une application rationnelle $D : F_\ell - - - \rightarrow S$, rendant commutatif le diagramme

$$\begin{array}{ccc} F_\ell & \overset{D}{- - - \rightarrow} & S \\ & {\scriptstyle q}\searrow \quad \swarrow{\scriptstyle \pi_*} & \\ & \check{\mathbb{P}}^2 = |H| & \end{array}$$

Il résulte de (P.3) que S est réunion disjointe de deux sous-variétés S_o et S_1, échangées par σ (avec les notations de (P.3), un diviseur effectif D sur $\tilde{C}$ appartient à S_i si et seulement si $[D] \in P_i^H$). Nous noterons par convention S_o celle de ces deux sous-variétés qui contient $D(F_\ell)$. Si $D_o \in S_o$, l'application $D \longrightarrow [D - D_o]$ de S dans $J\tilde{C}$ induit une application $\tilde{\varphi} : S_o \longrightarrow P$, qui ne dépend du choix de D_o que par une translation.

Nous noterons $\alpha : F \longrightarrow JX$ l'application d'Abel-Jacobi de F (définie à translation près), et $u : P \longrightarrow JX$ l'isomorphisme de la prop. 1.

<u>Proposition 3</u> :

<u>L'application D : $F_\ell \longrightarrow S_o$ est un isomorphisme. Le diagramme</u>

$$\begin{array}{ccc} F_\ell & \xrightarrow{\ D\ } & S_o \\ {\scriptstyle \alpha o \varepsilon}\downarrow & & \downarrow{\scriptstyle -\tilde{\varphi}} \\ JX & \xleftarrow[\ u\ ]{} & P \end{array}$$

<u>est commutatif à translation près.</u>

Rappelons ici la démonstration de [B2] : la surface S est normale (cela résulte d'une étude locale due à Welters, cf.[B2], §2) ; le morphisme π_* est fini, de degré $\frac{1}{2}\, 2^5 = 16$; le morphisme q est fini (à cause de l'hypothèse de position générale sur ℓ), et son degré est égal au nombre de droites de Σ_f ne rencontrant pas ℓ, c'est-à-dire $27 - 10 - 1 = 16$.

Pour prouver la première assertion, il suffit donc de montrer que D est génériquement injectif, c'est-à-dire que les seules droites incidentes aux 5 droites de D(f) (pour f assez générale) sont ℓ et f. Posons $D(f) = \sum_{i=1}^{5} f_i$; on peut trouver un morphisme $\varepsilon : \Sigma_f \longrightarrow \mathbb{P}^2$, éclatement de 6 points $p_1, \ldots, p_6$ de $\mathbb{P}^2$ en position générale, qui contracte f_i sur p_i pour $1 \le i \le 4$ et transforme f_5 en la droite $< p_4, p_5 >$ (on obtient ε en projetant Σ_f sur $\ell \times f$, puis $\ell \times f$ sur $\mathbb{P}^2$ par projection stéréographique). Il est clair sur cette représentation qu'il n'y a que deux droites de Σ_f incidentes à $f_1, \ldots, f_5$, à savoir les transformés des coniques passant par $p_1, \ldots, p_4$ et l'un des points p_5 ou p_6.

Il reste à vérifier que, pour une droite générique f de F, on a à une constante près

$$\alpha(f) = -\, u \circ \tilde{\varphi}(D(f))$$

ou encore

$$2\alpha(f) + 2\, u \circ \tilde{\varphi}(D(f)) = C^{te} ;$$

vu la définition de u (prop.1), il s'agit de prouver que l'élément $2\alpha(f) + \sum_i \alpha(f_i)$ de JX est indépendant de f. Or calculons dans $\mathrm{Pic}(\Sigma_f)$ l'élément $2f + \sum_i f_i$; utilisons le morphisme $\varepsilon : \Sigma_f \longrightarrow \mathbb{P}^2$ considéré plus haut, en posant $\varepsilon^{-1}(p_i) = e_i$ $(1 \leq i \leq 6)$, en notant h la classe dans $\mathrm{Pic}(\Sigma_f)$ d'une section hyperplane et d celle de $\varepsilon^* \mathcal{O}_{\mathbb{P}^2}(1)$. On a $f_i = e_i$ pour $1 \leq i \leq 4$; $f_5 = d - e_5 - e_6$; et(par exemple) $f = 2d - \sum_{i \neq 5} e_i$ et $\ell = 2d - \sum_{i \neq 6} e_i$. On a donc

$$2\ell + 2f + \sum_i f_i = 9d - 3\sum_j e_j = 3h.$$

On en déduit que le cycle $2f + \sum_i f_i$ est linéairement équivalent à $3h - 2\ell$ dans $\mathrm{Pic}(\Sigma_f)$, donc aussi dans JX, ce qui achève la démonstration de la proposition.

Corollaire :

L'application d'Abel-Jacobi $\alpha : F \longrightarrow JX$ est un plongement.

Démonstration :

Si $D \in S_0$, la fibre $\tilde{\varphi}^{-1}(\tilde{\varphi}(D))$ s'identifie au système linéaire $|D|$. On déduit alors du §2,(v) que $\tilde{\varphi}$ contracte une droite sur un point, et est un plongement en dehors de cette droite. Il résulte alors de la proposition que α est un plongement.

Remarque : Par spécialisation, il est facile de décrire D(f) lorsque la droite f rencontre ℓ en un point x. En effet, les droites de F incidentes à ℓ et f sont d'une part les 4 droites de F passant par x distinctes de ℓ et f, et d'autre part la droite conjuguée de f par l'involution σ. On a donc $D(f) \equiv L - f + \sigma f$. Autrement dit, la restriction $\delta : \tilde{C} \longrightarrow S_0$ de D associe à un point x de $\tilde{C}$ l'unique diviseur effectif de $|L + \sigma x - x|$. Ceci prouve en particulier que $|L|$ est contenu dans S_1 (et $|\sigma L|$ dans S_0).

Un autre point clé est le calcul de l'application tangente à α. Soit T l'espace tangent à l'origine de JX ; l'espace tangent de JX en un point quelconque est canoniquement isomorphe à T. Pour tout f dans F, l'application $T_f(\alpha) : T_f(F) \longrightarrow T$ a pour image un sous-espace vectoriel de dimension 2 de T.

Proposition 4 ("théorème du fibré tangent") :

Il existe un isomorphisme de $\mathbb{P}(T)$ sur $\mathbb{P}^4$ tel que pour $f \in F$, l'image dans $\mathbb{P}(T)$ de $T_f(\alpha)$ s'identifie à la droite $f \subset \mathbb{P}^4$.

En d'autres termes, considérons sur F le fibré en droites projectives universel :

$$N = \{ (x,\ell) \in X \times F \mid x \in \ell \} ;$$

alors N s'identifie au fibré projectif tangent à F, de façon que l'application de ce fibré dans $\mathbb{P}(T)$ déduite de $T(\alpha)$ soit la première projection $N \longrightarrow \mathbb{P}^4$.

On peut déduire la prop.4 de la théorie des variétés de Prym, mais la démonstration est assez compliquée; je préfère renvoyer le lecteur à [T1], aux deux démonstrations de [C-G] et surtout à celle de [A-K], qui est la plus simple.

§ 5 LE CONE TANGENT DE Θ AU POINT SINGULIER

Proposition 5 :

Soit $\Phi : F \times F \longrightarrow JX$ l'application définie par $\Phi(x,y)=\alpha(x)-\alpha(y)$. L'image de Φ est un diviseur thêta de JX.

Démonstration :

Soit $(x,y) \in F \times F$. L'application tangente $T_{x,y}(\Phi) : T_x(F) \oplus T_y(F) \longrightarrow T$ est somme directe des applications $T_x(\alpha)$ et $-T_y(\alpha)$; il résulte aussitôt de la prop. 4 que $T_{x,y}(\Phi)$ est de rang 4 si (et seulement si) x et y ne se rencontrent pas. En particulier Φ est génériquement finie, et son image est un diviseur de JX.

D'après la prop.3, on a un diagramme commutatif

$$\begin{array}{ccccc} F_\ell \times F_\ell & \xrightarrow{D\times D} & S_o \times S_o & & \\ \downarrow{\scriptstyle \Phi o(\varepsilon,\varepsilon)} & & \downarrow{\scriptstyle \Psi} & \searrow^{\Psi^*} & \\ JX & \xleftarrow[u]{} & P & \xrightarrow[t]{} & P^* \end{array}$$

où $\Psi(A,B) = [A-B]$ $\quad \Psi^*(A,B) = [A + \sigma B]$

et où $t : P \longrightarrow P^*$ est la translation par $[\pi^* H]$.

Il est clair que l'image de Ψ^* est contenue dans Θ^* ; ainsi l'image de Φ est contenue dans un diviseur Θ de JX, donc lui est égale puisque Θ est irréductible (prop.2).

<u>Remarques</u> :

1) Contrairement à α qui n'est définie qu'à translation près, l'application Φ est canonique. On note dans la suite Θ son image ; c'est un diviseur thêta bien déterminé de JX (il correspond par l'isomorphisme $t \circ u^{-1}$ au diviseur Θ^* de P^*).

2) Soient s,t deux droites générales dans F, de sorte que l'hyperplan engendré par s et t coupe X suivant une surface lisse Σ. Soient $f_1,\dots,f_5$ les 5 droites de F incidentes à s et t ; notons t_i (resp. s_i) la droite résiduelle d'intersection du plan $< s, f_i >$ (resp. $< t, f_i >$) avec X. On a alors dans $\mathrm{Pic}(\Sigma)$:

$$s + t_i = t + s_i = h - f_i$$

(h désignant la classe d'une section hyperplane), d'où $s-t = s_i-t_i$. On a par conséquent $\Phi(s,t) = \Phi(s_i,t_i)$ pour $1 \leq i \leq 5$; ainsi <u>l'application</u>
<u>$\Phi : F \times F \longrightarrow \Theta$ est de degré</u> ≥ 6.

Nous allons voir que Φ est en fait de degré 6, de sorte que la fibre $\Phi^{-1}(\Phi(s,t))$ se compose de (s,t) et des cinq paires (s_i,t_i). La configuration formée par les droites $(s,s_1,\dots,s_5 ; t, t_1,\dots,t_5)$ de Σ est appelée classiquement un "<u>double six</u>".

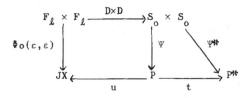

Proposition 6 :

Le diviseur Θ admet à l'origine un point triple ; lorsqu'on identifie $\mathbb{P}(T)$ à $\mathbb{P}^4$ par l'isomorphisme de la prop.4, la projectivisation du cône tangent à Θ en 0 est égale à X.

Démonstration :

Considérons le morphisme $\Phi : F \times F \longrightarrow JX$. Puisque α est un plongement (corollaire à la prop.3), la fibre $\Phi^{-1}(0)$ est la diagonale Δ de $F \times F$. Notons $\hat{F}^2$, $\hat{\Theta}$ et $\hat{J}$ les variétés obtenues en éclatant Δ dans $F \times F$, 0 dans Θ et 0 dans JX respectivement ; soient N, K et $\mathbb{P}(T)$ les diviseurs exceptionnels correspondants. Puisque $\Phi^{-1}(0) = \Delta$, on déduit de Φ un morphisme $\hat{\Phi} : \hat{F}^2 \longrightarrow \hat{J}$ qui se factorise à travers $\hat{\Theta}$, donnant lieu à un diagramme commutatif

$$\begin{array}{ccccc}
F^2 & \longrightarrow & \Theta & \hookrightarrow & JX \\
\uparrow & & \uparrow & & \uparrow \\
\hat{F}^2 & \longrightarrow & \hat{\Theta} & \hookrightarrow & \hat{J} \\
\cup & & \cup & & \cup \\
N & \longrightarrow & K & \hookrightarrow & \mathbb{P}(T)
\end{array}$$

On a par conséquent $\hat{\Phi}(N) = K$. Or N est le fibré projectif tangent à F et $K \subset \mathbb{P}(T)$ est la projectivisation du cône tangent à Θ en 0 ; l'application $N \longrightarrow \mathbb{P}(T)$ induite par $\hat{\Phi}$ est l'application projective associée à $T(\alpha)$. Identifions $\mathbb{P}(T)$ à $\mathbb{P}^4$ par l'isomorphisme de la prop.4 ; alors $\hat{\Phi}(N)$ est réunion des droites contenues dans X (prop.4), de sorte que K coïncide ensemblistement avec X. [1]

Il reste à prouver que K est réduit. Rappelons (prop.4) que N s'identifie au fibré en droites projectives universel sur F, de façon que l'application $N \longrightarrow \mathbb{P}(T)$ induite par $\hat{\Phi}$ soit la projection canonique ; en particulier, $\hat{\Phi}$ est non ramifié de degré 6 au point générique η de N. Soient $n : \tilde{\Theta} \longrightarrow \hat{\Theta}$ la normalisation de $\hat{\Theta}$ et $\tilde{K} = n^{-1}(K)$, de sorte que

[1] Ceci suffit pour prouver le théorème de Torelli.

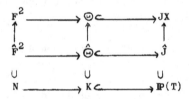

$\hat{\Phi}$ et sa restriction à N se factorisent comme suit :

$$\begin{array}{ccccccc}
\hat{F}^2 & \xrightarrow{g} & \tilde{\Theta} & \xrightarrow{n} & \hat{\Theta} & \hookrightarrow & \hat{J} \\
\cup & & \cup & & \cup & & \cup \\
N & \xrightarrow[\gamma]{} & \tilde{K} & \xrightarrow[\nu]{} & K & \hookrightarrow & \mathbb{P}(T)
\end{array}$$

Alors g est non ramifié en η, donc étale puisque $\tilde{\Theta}$ est normal ; on a $$\deg(\Phi) = \deg(g) = \deg(\gamma) \leq \deg(\nu\gamma) = 6.$$

Puisque $\deg(\Phi) \geq 6$ (remarque 2), on en déduit $\deg(\Phi) = 6$ et $\deg(\nu) = 1$. D'autre part on a $\hat{\Phi}^*(\mathbb{P}(T)) = N$, d'où $g^* \tilde{K} = N$; comme $\tilde{\Theta}$ est normal, ceci entraine que le diviseur $\tilde{K}$ est réduit. Notons A (resp. $\tilde{A}$) l'anneau local de $\hat{\Theta}$ (resp. $\tilde{\Theta}$) au point générique de K (resp. $\tilde{K}$), et $\mathfrak{m}$ (resp. $\tilde{\mathfrak{m}}$) son idéal maximal ; soit π une équation de K dans A. Alors $\tilde{A}$ est un anneau de valuation discrète, et π en est une uniformisante (puisque le diviseur $\nu^* K$ est réduit) ; de plus, comme $\deg(\nu) = 1$, l'extension $A/\mathfrak{m} \longrightarrow \tilde{A}/\tilde{\mathfrak{m}}$ est un isomorphisme. Ces propriétés entraînent $A = \tilde{A}$: en effet on a $\pi \in \mathfrak{m}$, d'où $\pi\tilde{A} = \mathfrak{m}\tilde{A} = \tilde{\mathfrak{m}}$; on en déduit que l'application $A/\mathfrak{m} \longrightarrow \tilde{A}/\mathfrak{m}\tilde{A}$ déduite de l'injection $A \longrightarrow \tilde{A}$ est bijective, et on conclut à l'aide du lemme de Nakayama. Par suite K est génériquement réduit, donc réduit puisque c'est une hypersurface, ce qui achève la démonstration.

Des propositions 2 et 6 résulte le théorème énoncé dans l'introduction :

Théorème :

Un diviseur thêta de JX admet un seul point singulier, qui est un point triple ; le cône tangent au diviseur en ce point est isomorphe au cône affine de base X.

Corollaire ("théorème de Torelli") :

La jacobienne intermédiaire JX, munie de sa polarisation principale, détermine la cubique X.

§ 6 COMPLEMENTS

D'autres résultats de [C-G] s'obtiennent facilement à l'aide des variétés de Prym.

Proposition 7 :

La classe de cohomologie de la surface $\alpha(F)$ dans $H^6(JX)$ est $\frac{\Theta^3}{3!}$.

(on note $H^*(JX)$ la cohomologie entière si $k = \mathbb{C}$, ℓ-adique sinon.)

On déduit cette proposition de la prop.3 et d'un calcul cohomologique, cf.[B2].

Proposition 8 :

L'application $\Phi : F \times F \longrightarrow \Theta$ est génériquement finie de degré 6.

Nous avons déjà obtenu ce résultat au cours de la démonstration de la prop.6. Nous indiquerons ici assez rapidement deux autres démonstrations possibles, qui n'utilisent pas le théorème du fibré tangent.

2ème Démonstration :

Il s'agit de montrer l'assertion analogue pour l'application $\Psi^* : S_o \times S_o \longrightarrow \Theta^*$. Notons que lorsqu'un diviseur E sur $\tilde{C}$ est de la forme $A + \sigma B$, avec $A,B \in S_o$, le diviseur $\pi_* E \in |2H|$ correspond à une conique dégénérée. Inversement, considérons un diviseur effectif E sur $\tilde{C}$ tel que $\pi_* E$ soit découpé par une conique dégénérée ; on peut écrire $E = A + B'$, avec A,B' dans S. Observons maintenant que $\Psi^*(S_i \times S_j)$ est contenu dans P^* si et seulement si $i \neq j$ (puisque $[\pi^*H] \in P^*$) ; on peut donc supposer $A \in S_o$, $B' \in S_1$, d'où $E = A + \sigma B$, avec $A,B \in S_o$. De plus cette écriture est unique si la conique dégénérée est de rang 2 et si son sommet n'est pas situé sur C.

Par conséquent la fibre $(\Psi^*)^{-1}(D)$, pour D assez général dans P^*, s'identifie à l'ensemble des diviseurs $E \in |D|$ pour lesquels $\pi_* E$ est découpé par une conique dégénérée. Or le système $\pi_*|D| \subset |2H|$ est découpé sur C par un pinceau quadratiques de coniques défini par une équation de la forme

$\lambda^2 p + \lambda\mu q + \mu^2 r = 0$, avec $(\lambda:\mu) \in \mathbb{P}^1$ et p, q, r $\in H^0(\mathbb{P}^2, \mathcal{O}_{\mathbb{P}}(2))$. Un tel pinceau, s'il contient une conique lisse, admet 6 coniques dégénérées (définies par l'équation det $(\lambda^2 p + \lambda\mu q + \mu^2 r) = 0$ sur $\mathbb{P}^1$), d'où le résultat.

3ème Démonstration :

En utilisant la prop. 7, on a dans $H^2(JX)$:

$$\Phi_* 1 = (\deg \Phi)\, \Theta = \frac{\Theta^3}{3!} * \frac{\Theta^3}{3!}$$

où $*$ désigne le produit de Pontrjagin dans $H^*(JX)$.

Par déformation, il suffit de calculer ce produit pour la jacobienne d'une courbe Γ de genre 5. Dans ce cas $\frac{\Theta^3}{3!}$ est la classe de $\varphi(\Gamma^{(2)})$, où $\varphi : \Gamma^{(2)} \longrightarrow J\Gamma$ est l'application canonique (définie à translation près). Considérons l'application $\Xi : \Gamma^{(2)} \times \Gamma^{(2)} \longrightarrow J\Gamma$ définie par $\Xi(A,B)=\varphi(A)+\varphi(B)$; son image est un diviseur thêta de $J\Gamma$, et on a donc

$$\Xi_* 1 = (\deg \Xi)\, \Theta = \frac{\Theta^3}{3!} * \frac{\Theta^3}{3!},$$

d'où
$$\deg \Phi = \deg \Xi = \binom{4}{2} = 6.$$

Proposition 9 :

L'application $A\ell b(F) \longrightarrow JX$ déduite de l'application d'Abel-Jacobi $\alpha : F \longrightarrow JX$ est un isomorphisme.

Démonstration (d'après [Mu3]) :

Nous admettons ici l'égalité $q(F) = 5$, déjà connue de Fano, pour laquelle on peut renvoyer par exemple à [A-K].

Compte tenu de la prop.3, il s'agit de montrer que l'application $a : A\ell b(S_o) \longrightarrow \mathbb{P}$ déduite de $\tilde{\varphi} : S_o \longrightarrow P$ est un isomorphisme. Considérons l'application $\delta : \tilde{C} \longrightarrow S_o$ qui associe à $x \in \tilde{C}$ l'unique diviseur de $|L+\sigma x-x|$ (cf. §4, remarque) ; soit $d : J\tilde{C} \longrightarrow A\ell b(S_o)$ l'homomorphisme correspondant. On a $\tilde{\varphi} \circ \delta(x) = [\sigma x - x + C^{te}]$, de sorte que l'homomorphisme composé

$$J\tilde{C} \xrightarrow{\ d\ } A\ell b(S_o) \xrightarrow{\ a\ } P$$

est égal à σ-1. Ceci entraîne que a est surjectif, donc une isogénie (puisque $q(S_0) = q(F) = 5$) ; de plus Ker $(a \circ d) = \pi^* JC$ est connexe, donc Ker(a) est connexe et a est un isomorphisme.

Bibliographie :

[A-K] A. Altman et S. Kleiman : Foundations of the theory of Fano Schemes. Compositio Mathematica 34 (1977) 3-47.

[B1] A. Beauville : Variétés de Prym et jacobiennes intermédiaires. Annales ENS 10 (1977) 309-391.

[B2] A. Beauville : Sous-variétés spéciales des variétés de Prym. Compositio Mathematica, à paraître.

[C-G] H. Clemens et P. Griffiths : The intermediate Jacobian of the cubic threefold. Annals of Math. 95 (1972) 281-356.

[M] D. Mumford : Prym varieties I. Contributions to Analysis, Academic Press, New-York (1974).

[Mu1] J.P. Murre : Algebraic equivalence modulo rational equivalence on a cubic threefold. Compositio Mathematica 25 (1972) 161-206.

[Mu2] J.P. Murre : Reduction of the proof of the non-rationality of a non-singular cubic threefold to a result of Mumford. Compositio Mathematica 27 (1973) 63-82.

[Mu3] J.P. Murre : Some results on cubic threefolds. Classification of algebraic varieties and compact complex manifolds, Springer-Verlag Lecture Notes 412 (1974) 140-164.

[T1] A.N. Tjurin : On the Fano surface of a nonsingular cubic in $\mathbb{P}^4$. Math. USSR Izvestija 4 (1970) 1207-1214.

[T2] A.N. Tjurin : The geometry of the Fano surface of a nonsingular cubic $F \subset \mathbb{P}^4$ and Torelli theorems for Fano surfaces and cubics. Math. USSR. Izvestija 5 (1971) 517-546.

[T3] A.N. Tjurin : Five Lectures on three-dimensional varieties. Russian Math. Surveys 27 (1972) 1-53.

THE FUNDAMENTAL GROUP OF THE FANO SURFACE, I.

by

Alberto Collino(*)

§ 0. Introduction. Let Γ be a rational normal curve of degree 4 in $\mathbb{P}_4$ The set covered by the lines which meet Γ in two points, secants, is a cubic threefold Y [7] . The "secant" threefold Y is singular exactly along Γ . Conversely, a cubic threefold which is singular along a rational normal quartic curve is projectively equivalent to Y. We study the degeneration of Hodge structure associated with a family of cubic threefolds with central fibre Y. It turns out that the associated family of intermediate Jacobians has good reduction and that the central fibre is the Jacobian of a hyperelliptic curve of genus 5. Further we study the related family of Fano surfaces. We find that the associated family of Albanese varieties has also good reduction, the central fibre being the Jacobian of the same curve. By this process one might recover some of the known results about the intermediate Jacobian of the cubic threefold [2] .

The new facts we prove are

(0.1) Theorem. Inside $\mathcal{A}_5$, the moduli space of principally polarized abelian varietes of dimension 5, the locus of Jacobians of hyperelliptic curves of genus 5 is contained in the closure of the set of intermediate Jacobians of cubic threefolds.

(0.2) Theorem. The commutator subgroup of the fundamental group of the Fano surface of a non singular cubic threefold is $\mathbb{Z}/2\mathbb{Z}$.

§ 1. The secant threefold. If Γ is the rational curve of parametric equations $x_0 = s^4$, $x_1 = s^3 t$, $x_2 = t^4$, $x_3 = st^3$, $x_4 = s^2t^2$, then the homogeneous equation of Y is

(1.1) $$E : (x_2 x_4 - x_3^2)\, x_0 - (x_4^3 - 2x_1 x_3 x_4 + x_1^2 x_2) = 0$$

and the singular locus of Y is exactly Γ . In [7] C. Segre shows that the surface F (Y) of the lines on Y splits in two components

(1.2) $$F(Y) = F_1(Y) \cup F_2(Y)$$

where $F_1(Y)$ = set of lines which are secant to Γ and $F_2(Y)$ = the set of lines l in $\mathbb{P}_4$ from which Γ is projected 2:1 onto a conic in a complementary plane. Both $F_1(Y)$ and $F_2(Y)$ are isomorphic to $\Gamma(2)$, the second symmetric product of Γ, and so to $\mathbb{P}_2$. The isomorphism $F_2(Y) \simeq \Gamma(2)$ is given by associating to l the pair of points on Γ which are the ramification points of the linear projection

(*) Lavoro svolto nell'ambito del gruppo G.N.S.A.G.A. del C.N.R.

from 1 . Further we have

(1.3) $\mathbb{P}_1 \simeq K = F_1(Y) \cap F_2(Y)$

where K is the set of lines tangent to Γ .

For later use, we note that under the identification $F_1(Y) = \mathbb{P}_2$, the curve K is a conic, because it corresponds to the diagonal in the symmetric product of Γ . A straight computation, which we shall outline later, shows that, as a scheme, F (Y) contains $F_2(Y)$ simply and $F_1(Y)$ with multiplicity four.

Let X be a non singular cubic threefold of equation G=0, X intersecting Γ transversally in $x_1, \dots, x_{12}$, and let tG + E = 0 be the pencil of cubic threefolds determined by X and Y. We plan to study the associated degeneration of Hodge structure, both for $H^3(X)$ and for $H^1(F(X))$, F(X) = Fano surface of the lines on X .

In order to have unipotent monodromy we base extend, setting $t = s^2$ we deal then with the pencil

(1.4) $s^2 G + E = 0$

We denote with X_s the threefold of equation 1.4, with F_s the associated Fano surface and with C the hyperelliptic curve of genus 5 which is the double covering of Γ ramified at the 12 points $x_1, \dots, x_{12}$.

We will show the following

(1.5) Proposition. The monodromy transformations on both $H^3(X_s)$ and $H^1(F_s)$ are the identity.

From this it follows that the Hodge structure stays pure also at the central fibre; or, equivalently, the family of intermediate Jacobians J (X_s) and of Albanese varieties A (F_s) both have good reduction at 0. We will compute $J_0 = J(C) = A_0$, where J(C) is the Jacobian of C. So, by Griffiths' extension theorem J(C) is a point in the closure of the set of intermediate Jacobians of non singular cubic threefolds. Varying X we obtain every set of 12 points on Γ, hence every hyperelliptic curve of genus 5 can appear as the curve C described above. From this (0.1) will follow.

In $\Delta \times \mathbb{P}_4$ let $\mathcal{X}$ be the hypersurface of equation (1.4), where s is the parameter in the disc Δ . We take the radius of Δ small, so that the fibre X_s, $s \neq 0$, is non-singular. Then $\mathcal{X}$ is singular exactly along the curve $\Gamma^* = \{0\} \times \Gamma$. A direct computation, which we outline in (1.8) below, shows the following. Blowing up $\Delta \times \mathbb{P}_4$ along Γ^* resolves the singularities of $\mathcal{X}$, i.e. $\mathcal{X}'$, the proper transform of $\mathcal{X}$, is non-singular. Let $p: \mathcal{X}' \longrightarrow \Delta$ be the natural projection, then

(1.6) $X'_0 = p^{-1}(0) = B \cup D$

where B is the desingularization of Y, obtained by blowing up Γ , and D is a non--singular threefold which is a "quadric bundle" over Γ . More precisely, there is a projection $q: D \to \Gamma$, the fibre $q^{-1}(g)$ is a quadric surface, if $g \notin X$ $q^{-1}(g)$ is non-singular, if $g \in X$ $q^{-1}(g)$ is an ordinary cone. Both B and D appear with multiplicity one in $p^{-1}(0)$ and they intersect transversally

(1.7) $B \cap D = S$

where S is a $\mathbb{P}_1$ bundle over Γ . In B the surface S is the exceptional divisor for $B \longrightarrow Y$.

(1.8) Outline of the computation. In affine coordinates the equation of Y is: $(x'z' - y'^2) - (z'^3 - 2y'z'w' + x'w'^2) = 0$ and Γ is the curve $(\beta^4, \beta^3, \beta^2, \beta)$. Setting new coordinates $w = w'$, $z = z' - w^2$, $y = y' - w^3$, $x = x' - w^4$, the equations of Γ are: $x = y = z = 0$, and the pencil $E + s^2 G = 0$ is given by

(+) $(xz - y^2 - z^3 - 3z^2w^2 + 2yzw) + s^2(a_0 + a_1 w + \dots) = 0.$

The condition that the origin is a point in the intersection $\Gamma \cap X$ is $a_0 = 0$ and, if $a_0 = 0$, the condition that X is not tangent to Γ at the origin is $a_1 \neq 0$.

Using $(s, x, y, z, w; \sigma, \xi, \eta, \zeta)$ as coordinates for $\Delta \times \mathbb{A}_4 \times \mathbb{P}_3$, the ideal of $\mathcal{X}'$ is generated by (+) and by the equations: $0 = s\xi - x\sigma = \dots = y\zeta - z\eta$. Then the polynomial $(\xi\zeta - \eta^2 - z\zeta^2 - 3w^2\zeta^2 + 2w\eta\zeta) + \sigma^2(a_0 + a_1 w + \dots)$ is also in the ideal of $\mathcal{X}'$. The stated properties of $p^{-1}(0)$ can now be checked by direct inspection.

(1.9) The cohomology of $X'_0 = B \cup D$ carries a mixed Hodge structure [4] , which is deduced from the Mayer - Vietoris sequence

(M.V.) $H^{q-1}(S) \to H^q(X'_0) \to H^q(B) \oplus H^q(D) \longrightarrow H^q(S)$

We need to compute $H^3(X'_0)$. By Tsen theorem D is a rational threefold, hence $0 = H_1(D, \mathbb{Z}) = H_5(D, \mathbb{Z}) = H^1(D, \mathbb{Z}) = H^5(D, \mathbb{Z})$ and there is no torsion in either homology or cohomology groups of D, [1] . Using the fibering $q: D \to \Gamma$ we see $\chi_{top}(D) = -4$, because $q^{-1}(g)$ is a quadric which is non-singular except when $g = x_1, \dots, g = x_{12}$.

(1.10) Lemma. (a) $R^0 q \mathbb{Z} = \mathbb{Z}$. (b) $R^1 q \mathbb{Z} = 0$.

Proof. (a) The fibres of q are connected, (b) $H^1(q^{-1}(g), \mathbb{Z}) = 0$ for every g.

(1.11) Lemma. $H^2(D, \mathbb{Z}) = \mathbb{Z} \oplus \mathbb{Z}$.

<u>Proof</u>. From the Leray spectral sequence one has the exact sequence

$$0 \to H^2(\Gamma, R^0q\mathbb{Z}) \to H^2(D, \mathbb{Z}) \to H^0(\Gamma, R^2q\mathbb{Z}) \to 0,$$

therefore it suffices to show $H^0(\Gamma, R^2q\mathbb{Z}) \simeq \mathbb{Z}$.

By Poincaré duality the surface S determines a class in $H^2(D, \mathbb{Z})$, hence a class σ in $H^0(\Gamma, R^2q\mathbb{Z})$. The intersection $S \cap q^{-1}(g)$ is the plane section in the quadric, hence $\sigma_g \in H^2(q^{-1}(g), \mathbb{Z})$ is the class of the plane section. Given $\alpha \in H^0(\Gamma, R^2q\mathbb{Z})$, the restriction α_g is invariant by monodromy on $\Gamma - \{x_1, \ldots, x_{12}\}$, therefore α_g is a multiple of the plane section, i.e. $\alpha_g = m_g\sigma_g$. Also m_g must be costant, by continuity.Hence $\alpha = m\sigma$ and $H^0(\Gamma, R^2q\mathbb{Z}) \simeq \mathbb{Z}\sigma$.

From the computations above it follows

(1.12) <u>Proposition</u>. $H^3(D, \mathbb{Z}) \simeq \mathbb{Z}^{10}$

(1.13) <u>Proposition</u>. $H^3(D, \mathbb{Z}) \simeq H^3(X_0', \mathbb{Z})$.

<u>Proof</u>. Since B is a $\mathbb{P}_1$ bundle over the symmetric product of Γ, i.e. over $\mathbb{P}_2$, $H^3(B, \mathbb{Z}) = 0$.

Also $H^3(S, \mathbb{Z}) = 0$, because S is a $\mathbb{P}_1$ bundle over Γ. Thus, (1.13) will follow from the M.V. sequence, once we show that $H^2(D) + H^2(B) \to H^2(S)$ is surjective. In $H^2(S, \mathbb{Z})$ consider the generators f and s, representing respectively the class of a fibre and of a section. Then f is in the image of $H^2(D)$ and f+s is the image of the cohomology class in $H^2(B)$ which represents the divisor covered by the lines passing through a fixed point of Γ.

Then the mixed Hodge structure on $H^3(X_0')$ is equal to the pure Hodge structure on $H^3(D)$. Consider now the sequence of mixed Hodge structures, exact over the rationals, see [4],

(1.14) $$H_5(X_0') \to H^3(X_0') \to H^3 \xrightarrow{N} H^3 \to H_3(X_0')$$

where $N = \log T$, T = the monodromy transformation, and $H^3 = H^3(X_s)$ carries the limit mixed Hodge structure. Proceeding as before we get $H_5(X_0') = 0$, hence $\ker N = H^3(X_0')$. Since $\dim H^3(X_0') = 10 = \dim H^3(X_s)$ (because X_s is a cubic threefold), $N = 0$ and T is the identity. This shows the following

(1.15) <u>Proposition</u>. The mixed Hodge structure on H^3 is pure and it is equal to the Hodge structure of $H^3(D)$.

Let $f^*: \Delta^* = \Delta - \{0\} \to \mathcal{a}_5$ be the period map, given by $f^*(s)$ = class $J(X_s)$. By Griffiths' extension theorem f^* can be extended to $f: \Delta \to \mathcal{a}_5$ and in our case $f(0)$ is the class of the intermediate Jacobian $J(D)$.

(1.16) <u>Proposition</u>. $J(D) \cong J(C)$, as principally polarized abelian varieties.

<u>Proof</u>. Let $L \hookrightarrow S$ be a section of $S \to \Gamma$, then L is also a section of $q: D \to \Gamma$, i.e. L meets each quadric $q^{-1}(g)$ once. If $q^{-1}(g)$ is non-singular there are two lines on it meeting L, if $q^{-1}(g)$ is a cone there is one line only meeting L, because the vertex of the cone does not lie on S. These lines are parametrized by C. Let $f: F \to C$ be the associated $\mathbb{P}_1$ bundle, there is a diagram

$$\begin{array}{ccc} F & \xrightarrow{k} & D \\ f \downarrow & & \downarrow q \\ C & \xrightarrow{h} & \Gamma \end{array}$$

where $kf^{-1}(c) = l_c$, the line represented by $c \in C$. Let $\{\gamma_i\}_{i=1\ldots 10}$ be representative cycles for a symplectic basis in $H_1(C, \mathbb{Z})$; $kf^{-1}(\gamma_i)$ are 3-cycles in D and $(\gamma_i \cdot \gamma_j) = (kf^{-1}(\gamma_i) \cdot kf^{-1}(\gamma_j))$, hence there is a symplectic embedding $k_*f^*: H_1(C, \mathbb{Z}) \longrightarrow H_3(D, \mathbb{Z})$. Since both lattices have the same rank ($=10$) k_*f^* is an isomorphism; this implies that the induced morphism $J(C) = (F^1H^1)^* / / H_1(C, \mathbb{Z}) \longrightarrow (F^2H^3)^* / H_3(D, \mathbb{Z}) = J(D)$ is also an isomorphism.

§ 2. <u>Degeneration of the Fano surface, with application to the computation of the fundamental group.</u>

We denote by $\varphi: \mathcal{F} \to \Delta$ the family of Fano surfaces associated with the pencil (1.4). Blowing-up $\Delta \times \mathbb{P}_4$ along $\{o\} \times \Gamma$ we desingularized $\mathcal{X}$. Similarly

(2.1) <u>Proposition</u>. Blowing-up the family of Grassmannians $\Delta \times G(1,4)$ along $\{o\} \times F_1(Y)$ resolves the singularities of $\mathcal{F}$. If $\mathcal{F}'$ is the proper transform of $\mathcal{F}$ and $\varphi': \mathcal{F}' \to \Delta$ the natural projection, then $\varphi'^{-1}(o) = F'_o$ is a reduced divisor with normal crossings. More precisely $F'_o = F_2(Y) \cup C(2)$, where $C(2)$ is the symmetric product of C, and $F_2(Y) \cap C(2) = K$, where K represents on $F_2(Y)$ and $C(2)$ respectively a conic and the unique g^1_2 on C.

<u>Proof</u>. We first explain informally why the central fibre should be $C(2) \cup F_2(Y)$ and then outline the computations needed to check the statements in (2.1).

(2.1.1) Under the blow-up of $\mathcal{X}$ along Γ^* (see 1.6), the "correct" transform as a cycle of a line l through a and b, points of Γ, is a reducible connected curve $l^+ = l + l'_a + l''_b$, where l'_a is a line in the quadric $q^{-1}(a)$, l''_b is a line in $q^{-1}(b)$. If $q^{-1}(a)$ is non singular then there are two choices for l'_a, otherwise there is only one choice, similarly for l''_b. Also the (Chow) variety of the l^+'s is just $C(2)$, and $F_2(Y)$ and $C(2)$ are glued together along a curve which is related to the curve of lines tangent to Γ. More precisely: if l is a line tangent to Γ at the point a and $q^{-1}(a)$ is smooth quadric then the cycle $l^{+} = l + l'_{a} + l''_{a}$, where l'_{a} and l''_{a} lie in different rulings, is represented by a point in $F_{2}(Y) \cap C(2)$, while $l + 2l'_{a}$ is represented by a point in C(2) which is not in $F_{2}(Y)$.

(2.1.2) We use $x_0, \dots, x_4$ as homogeneous coordinates on $\mathbb{P}_4$ and x, y, z, w as affine coordinates, where $x_1/x_0 = w$, $x_2/x_0 = x$, $x_3/x_0 = y$, $x_4/x_0 = z$. The parametric representation of Γ is then $(x,y,z,w) = (\beta^4, \beta^3, \beta^2, \beta)$ and the affine equation of Y is $(xz - y^2) - (z^3 - 2yzw + xw^2) = 0$. Following [2] 6.13 we set $u_j = (x_j/x_0)$, $z_j = (x_j/x_1)$ for $j = 2,3,4$, so that $(u_2, u_3, u_4, z_2, z_3, z_4)$ provide local affine coordinates in G(1,4) around the line $x_2 = x_3 = x_4 = 0$. The surface $F_1(Y)$ of the lines secant to Γ, (1.2), is given locally by the equations

(*) $$u_2 - u_4^2 - u_4 z_4^2 = u_3 - u_4 z_4 = z_2 - (z_4(z_3 + u_4)) = z_3 - (z_4^2 + u_4) = 0.$$

The family $\mathfrak{F}$ is locally the subscheme of $\Delta \times \mathbb{A}_6$ with ideal generated by the polynomials in (s, u., z.) which are the coefficients of the monomials in λ and μ inside

(+) $$E(\lambda(1,0,u_2,u_3,u_4) + \mu(0,1,z_2,z_3,z_4)) + s^2 G(\lambda(1,0,u_2,u_3,u_4) + \mu(0,1,z_2,z_3,z_4))$$

where E and G are the equations of Y and X respectively.

Setting, see (*), $w'_2 = u_2 - u_4^2 - u_4 z_4^2$, $w_3 = u_3 - u_4 z_4$, $y'_2 = z_2 - z_4(z_4^2 + 2u_4)$, $y_3 = z_3 - (z_4^2 + u_4)$ and $y_2 = y'_2 - 2y_3 z_4$, $w_2 = -w'_2 - y_3^2 - 2y_3 z_4 + 2w_3 z_4 + y'_2 z_4$, and using $w_2, w_3, u_4, y_2, y_3, z_4$ as new set of coordinates on G(1,4), we get that the coefficients of μ^3, $\lambda\mu^2$, $\lambda^2\mu$, λ^3 in (+) are in the order

$$e_1 : As^2 - y_2 \quad , \quad e_2 : Bs^2 + w_2$$

(++) $$e : Cs^2 + (-w_2 z_4 - y_3^2 z_4 + y_2 z_4^2 + y_2 u_4 - 2w_3 y_3)$$

$$e : Ds^2 - w_2 u_4 + y_2 u_4 z_4 - y_3^2 u_4 - w_3^2$$

where A,B,C,D are polynomials in the new variables depending on G.

The Fano scheme F(Y) is locally the complete intersection $s = e_1 = e_2 = e_3 = e_4 = 0$, hence $F_1(Y)$, which is the complete intersection $s = y_2 = y_3 = w_2 = w_3 = 0$, appears with multiplicity four in F(Y). $F_2(Y)$, which can be seen to be the surface of equations $s = y_2 = w_2 = 2w_3 + y_3 z_4 = 4u_4 + z_4^2 = 0$, appears simply in F(Y). F_1 and F_2 intersect along the curve $s = w_2 = w_3 = y_2 = y_3 = 4u_4 + z_4^2 = 0$, which is the curve of the lines tangent to $\{o\} \times \Gamma$.

It is convenient to write the equations of $\mathfrak{F}$, see (++), as follows

(+++) $e_1 = 0$, $e_2 = 0$, $(C + z_4 B + (z_4^2 + u_4^2)A)s^2 - y_3^2 z_4 - 2w_3 y_3 = 0$,
$(D + u_4 B + u_4 z_4 A)s^2 - y_3^2 u_4 - w_3^2 = 0$.

Then in $\Delta \times \mathbb{A}_6 \times \mathbb{P}_4$ with coordinates $(s; w_2, w_3, u_4, y_2, y_3, z_4; \sigma, \eta_2, \eta_3, \omega_2, \omega_3)$ the proper transform $\mathcal{F}'$ of $\mathcal{F}$ under the blow-up of $\Delta \times \mathbb{A}_6$ along $\{o\} \times F_1(Y)$ is given by (+++) together with the usual equations $\sigma y_2 - s\eta_2 = \ldots = w_2\omega_3 - \omega_2 w_3 = 0$.

Let $l = (0, \ldots) \in \{o\} \times F_1(Y)$, in order to compute the inverse image of l in $\mathcal{F}'$ we need to find $A(l), \ldots, D(l)$. Set $G = \sum_{i=0}^{3} G_i x_0^{3-i}$, where $G_i(x_1, x_2, x_3, x_4)$ is a homogeneous polynomial of degree i, and let l be the line secant to Γ through the origin and through the point of parameter ß, cf. (2.1.2). In local coordinates l is the point $0=s=y_2=y_3=w_2=w_3=u_4$, $z_4=\beta$; if l is the line tangent to Γ at the origin then $\beta = 0$. Using the notations above one has

$A(l) = G_3(1,\beta^3,\beta^2,\beta)$, $B(l) = G_2(1,\beta^3,\beta^2,\beta)$
$C(l) = G_1(1,\beta^3,\beta^2,\beta)$, $D(l) = G_0$.

Note $D(l)=0$ if the origin is a point in $\Gamma \cap X$, further if $D(l)=0$ then $G_1(1,0,0,0) \neq 0$ because X and Γ intersect transversally by hypothesis.

The inverse image of l in $\mathcal{F}'$ is then the subscheme of $\mathbb{P}_4$ determined by the equations: $\eta_2 = 0$, $\omega_2 = 0$, $G_0\sigma^2 - \omega_3^2 = 0$, $(G_1(1,\beta^3,\beta^2,\beta) + \beta G_2(1,\beta^3,\beta^2,\beta) + \beta^2 G_3(1,\beta^3,\beta^2,\beta))\sigma - \beta\eta_3^2 - 2\omega_3\eta_3 = 0$.

Counting multiplicities the inverse image of l is: (i) $4P$, if l is tangent to Γ in a point of X, (ii) $2P + P_1 + P_2$, if l is tangent to Γ in a point not lying on X, (iii) $P_1 + P_2 + P_3 + P_4$, if l is secant to Γ in two points which do not lie on X, (iv) $2P + 2Q$, if l is secant to Γ in two points of which one is on X, (v) $4P$, if l is secant to Γ in two points which lie on X.

In this way we see that the proper transform of $\{o\} \times F_1(Y)$ is C(2), as we have described it in (2.1.1). We leave to the reader the elementary and boring task of verifying the remaining statements in (2.1). For instance if l is the line tangent at the origin and the origin belongs to X, then $\mathcal{F}'$ is non-singular at P, because the tangent space to $\mathcal{F}'$ at P is the linear subspace of $\Delta \times \mathbb{A}_6 \times \mathbb{P}_4$ of equations $0=y_2=w_2=\eta_2=\omega_2=-2y_3=-u_4=s=w_3$.

(2.2) The same arguments used for (1.15) now yield

(2.2.1) Proposition. The monodromy action on $H^1(F_s)$ is trivial, the limiting mixed Hodge structure is pure and equal to the Hodge structure of $H^1(C(2))$. The family of associated Albanese varieties has good reduction and the special fibre is the Albanese variety of C(2), i.e. J(C).

Remark. From (1.15), (1.16), (2.2.1) it follows that the Albanese variety of F is isomorphic to J (X_s), which is one of the main results of [2].

(2.3) In this section we compute the commutator subgroup of the fundamental group of the Fano surface $F = F_s$, $s \neq 0$.

Recall the topological analysis of the structure of the degeneration $\mathcal{F}' \to \Delta$ as it is explained e.g. in [6] ch.II. The surface F is topologically obtained from F_0 by the following process. Remove from C(2) an open tubular neighborhood N_1 of the double curve K, $C(2)^+ = C(2) - N_1$ has for boundary the S^1 bundle D_1, boundary of N_1. Remove from $\mathbb{P}_2$ an open tubular neighborhood N_2 of K, $\mathbb{P}_2^+ = \mathbb{P}_2 - N_2$ has for boundary D_2, the S^1 bundle boundary of N_2. D_1 and D_2 are isomorphic bundles on K. The surface F is obtainable from $C(2)^+$ and $\mathbb{P}_2^+$ by glueing their boundaries together, via the isomorphism $D_1 \simeq D_2$. The fundamental group πF is therefore the product of $\pi C(2)^+$ and $\pi \mathbb{P}_2^+$ amalgamated along πD_1.

(2.3.1) Lemma. $\pi \mathbb{P}_2^+ \simeq (\mathbb{Z}/2)\alpha$, α = the normal loop on D_2.

Proof. $\mathbb{P}_2^+$ is homotopic to $\mathbb{P}_2^\circ = \mathbb{P}_2 - K$; the fundamental group of $\mathbb{P}_2^\circ$ is $\mathbb{Z}/2$, see 8.2 in [3].

(2.3.2) Lemma. $\pi D_1 = (\mathbb{Z}/4)\alpha$.

Proof. By the footnote at pg.11 of [5], it suffices to remark that D_1 is the S^1 bundle associated with the normal bundle of a conic in $\mathbb{P}_2$.

(2.3.3) Lemma. There is an exact sequence

$$\pi D_1 \longrightarrow \pi_1 C(2)^+ \longrightarrow \pi C(2) \longrightarrow 1 ,$$

Proof. $C(2) = C(2)^+ \cup \overline{N}$, hence $\pi C(2) = \pi C(2)^+ * \pi \overline{N}_1$ amalgamated along πD_1. Now $\overline{N}_1$ is simply connected, because it is homotopic to K, therefore the sequence is exact.

(2.3.4) Proposition. There is an exact sequence

$$(\mathbb{Z}/2)\alpha \longrightarrow \pi F \to \mathbb{Z}^{10} \to 0 ,$$

hence the commutator subgroup of πF is a quotient of $\mathbb{Z}/2$.

Proof. This follows from the preceding lemmas if we show $\pi C(2) \simeq \mathbb{Z}^{10}$. Indeed $\pi J(C) \simeq \mathbb{Z}^{10}$ and if $n \gg 0$ $\pi C(n) \simeq \mathbb{Z}^{10}$ because the symmetric product C(n) is a projective bundle over J. In addition if $n \geq 3$ $\pi C(n-1) \simeq \pi C(n)$ by the theorem of Lefschetz-Bott, because C(n-1) is an ample divisor in C(n), [3b] lemmma 2.7.

(2.3.5) Next we show that πF is not commutative, then (0.2) follows from (2.3.4)

(2.3.5.1) Let S be a connected surface such that the Albanese map $i: S \longrightarrow A$ is an inclusion (this is always true up to homotopy). Fixing a base point s_0 there is a commutative diagram of exact sequences

$$\begin{array}{ccccccccc} \pi_2(S, s_0) & \longrightarrow & \pi_2(A, s_0) & \longrightarrow & \pi_2(A, S, s_0) & \longrightarrow & \pi_1(S, s_0) & \longrightarrow & \pi_1(A, s_0) \\ \downarrow & & \downarrow & & \downarrow h & & \downarrow & & \downarrow \\ H_2(S, \mathbb{Z}) & \longrightarrow & H_2(A, \mathbb{Z}) & \xrightarrow{g} & H_2(A, S, \mathbb{Z}) & \longrightarrow & H_1(S, \mathbb{Z}) & \longrightarrow & H_1(A, \mathbb{Z}) \end{array}$$

Note: i) $\pi_2 A = 0$ because the universal covering space of A is contractible, ii) $\pi_1 A \simeq H_1 A$, hence $\pi_2(A, S, s_0)$ is a subgroup of $\pi_1 S$ containing the commutator subgroup. If $h(a) = g(b)$, $a \in \pi_2(A, S)$, $b \in H_2(A)$, then a is in the commutator. Since S and A are path connected and $\pi_1 S \longrightarrow \pi_1 A$ is surjective (because $H_1(S) \longrightarrow H_1(A)$ is surjective), we have $\pi_0 = \pi_1(A,S,s_0) = \{e\}$. By the generalized Hurewicz theorem, [8] ch.7 § 5, h: $\pi_2(A, S) \longrightarrow H_2(A, S)$ is an epimorphism, therefore $\pi_1(S, s_0)$ is not abelian if $i_*: H_2(S, \mathbb{Z}) \longrightarrow H_2(A, \mathbb{Z})$ is not surjective.

(2.3.5.2) We assume now that A is a principally polarized abelian variety of dimension g, that $\omega \in H^2(A, \mathbb{Z})$ is the polarization and that $w \in H_{2g-2}(A, \mathbb{Z})$ is the Poincaré dual of ω. Then $z = w^{g-1} / (g-1)!$ is an integral homology class, i.e. $z \in H_2(A, \mathbb{Z})$. In $H_2(S, Q)$ let $x = i^*w/m$, where m is an integer.

<u>Lemma</u>. If m does not divide g and if x is an integral class then $z \notin i_* H_2(S, \mathbb{Z})$.

<u>Proof.</u> If $z = i_* y$, $y \in H_2(S, \mathbb{Z})$, then $\langle y \cdot x \rangle_S = \langle z \cdot w/m \rangle_A = w^g / (g-1)!\, m = g/m$, since by Poincaré's formulas $w^g = g!$. This is a contradiction because $\langle y \cdot x \rangle$ should be integer.

(2.3.5.3) <u>Corollary</u>. πF is not abelian.

<u>Proof.</u> Apply the lemma in the case $S = F$, $g = 5$, $m = 2$, using the property, [2] (11.27), that $i^*w/2$ is an integral class.

(2.3.5.4) <u>Corollary.</u> The fundamental group of the non-singular Fano surface of the "lines" on a general quartic double solid is not abelian.

<u>Proof.</u> By [9] g=10. The lemma applies because of Welters' result that $i^*w/3$ is an integral class, see loc.cit. (6.1).

REFERENCES

1 M. Artin and D. Mumford: Some elementary examples of unirational varieties which are not rational, Proc. London Math. Soc., (3), 25 (1972), 75-95.

2 C.H. Clemens and P.A. Griffiths: The intermediate Jacobian of the cubic threefold, Ann. of Math. 95 (1972), 281-356.

3 W. Fulton and R. Lazarsfeld: Connectivity and its applications in Algebraic Geometry, Proceedings of the Midwest Alg.Geom.Conf., Springer Lec.Notes 862.

3b W. Fulton and R. Lazarsfeld: On the Connectedness of Degeneracy Loci and Special Divisors, Acta Math. 146 (1981), 271-283.

4 P. Griffiths and W. Schmid: Recent developments in Hodge theory: a discussion of technique and results, Proc. Of the Intern. Colloq. on Discrete Subgroups of Lie Groups and Applications to Moduli, Bombay, 1973, 31-127.

5 D. Mumford: The topology of normal singularities of an algebraic surface and a criterion for simplicity, I.H.E.S., Publ. Math. 9 (1961), 5-22.

6 V.Persson: On degenerations of algebraic surfaces, Memoirs of the American Mathematical Society, 189, vol.11, 1977.

7 C. Segre: Sulle varietà cubiche dello spazio a quattro dimensioni..., Memorie della R. Acc. Sc. di Torino, serie II, XXXIX, 1887, 3-48.

8 E.H. Spanier: Algebraic Topology, McGraw-Hill, 1966.

9 G.E. Welters: Abel-Jacobi isogenies for certain types of Fano threefolds, Thesis at Utrecht University, Mathematisch Centrum, Amsterdam, 1981, 1-141.

***** ***** ***** *****

Istituto di Geometria
Università
Via Principe Amedeo 8
10123 TORINO
ITALIA.

THE FUNDAMENTAL GROUP OF THE FANO SURFACE, II.

by

Alberto Collino(*)

In the preceding paper we have seen that the fundamental group of the Fano surface is not abelian and that there is an exact sequence

$$o \to \mathbb{Z}_2 \to \pi F \to \mathbb{Z}^{10} \to o ,$$

hence πF is a central extension of $\mathbb{Z}_2$ by $\mathbb{Z}^{10}$. Such central extensions are classified by $H^2(\mathbb{Z}^{10}, \mathbb{Z}_2)$. Recall that the cohomology of a group G is the cohomology of the corrisponding K(G,1) space, hence πF is determined by an element in $H^2(J, \mathbb{Z}_2)$, because the intermediate Jacobian J of the cubic threefold X is topologically $K(\mathbb{Z}^{10},1)$.

(1.1) <u>Theorem.</u> πF is determined by the class of the theta divisor in $H^2(J, \mathbb{Z}_2)$.

A geometric description of πF is given by the following theorem. Let $b,c \in \pi F$; let B,C be their images in $H_3(X, \mathbb{Z})$ via the composite map $\pi F \to H_1(F, \mathbb{Z}) \simeq H_3(X, \mathbb{Z})$; let $(B \cdot C)$ denote the intersection product on X, hence $(B \cdot C)$ is an integer.

(1.2) <u>Theorem.</u> $bcb^{-1}c^{-1} = (B \cdot C) \mod.(2)$.

In other words two loops on F commute in the fundamental group if and only if the intersection number of the corresponding 3-cycles on X is even. This result was conjectured to me by J.Steembrink. I thank him and J. Stienstra for asking me the right questions and for pointing out the relevant tools.

Proof of (1.1). It is known (see e.g. p. 195 in "Homotopy theory and differential forms" by Friedlander, Griffiths, Morgan, Seminario di Geometria, Firenze (1972)) that the 1-1 correspondence $\{\text{central extensions of A by G}\} \simeq H^2(K(G,1),A)$ can be given using obstruction theory: let $y \in H^2(K(G,1),A) = [K(G,1),K(A,2)]$ and let η be the corresponding morphism $\eta : K(G,1) \longrightarrow K(A,2)$. The pull back to K(G,1) via η of $\mathcal{P}K(A,2) \longrightarrow K(A,2)$ is a principal fibration B of fibre K(A,1). Then B is a K(Y,1) space, where Y is an extension $o \to A \to Y \to G \to 1$. The 1-1

(*) Partially supported by a grant Z.W.O.. The author is a member of G.N.S.A.G.A of CNR

correspondence associates Y to y.

Let now $G = \mathbb{Z}^{10}$, $A = \mathbb{Z}_2$, y = class of the theta divisor of J in $H^2(J, \mathbb{Z}_2)$, $i : F \to J$ the Abel-Jacobi map, which is also tha Albanese map. As before we denote by Y the extension associated with y. It is $o = i^*y$ in $H^2(F, \mathbb{Z}_2)$ because the pull back of the theta divisor to F is homologous to 2D, D = the curve of the lines on X which meet a given line on X. The corresponding map $F \to K(\mathbb{Z}_2, 2)$ is then homotopically trivial, because the correspondence $H^2(F, \mathbb{Z}_2) \simeq [F, K(\mathbb{Z}_2, 2)]$ is an isomorphism. Therefore the induced principal fibration $B_F \to F$ has a section $s: F \to B_F$, and $i: F \to J$ factors as pg, where $g : F \xrightarrow{1} B_F \hookrightarrow B$ and $p : B \to J$. The induced maps on the fundamental group give

$$\begin{array}{ccccccccc} o & \to & \mathbb{Z}_2 & \to & \pi F & \xrightarrow{i_*} & \mathbb{Z}^{10} & \to & o \\ & & & & \downarrow g_* & & \| & & \\ o & \to & \mathbb{Z}_2 & \to & Y & \xrightarrow{p_*} & \mathbb{Z}^{10} & \to & o \end{array}$$

Since Y is not the direct product, because the class of the theta divisor is not zero in $H^2(J, \mathbb{Z}_2)$, g_* is then an isomorphism.

Proof of (1.2). Let $e_1, \dots, e_{10}$ be a basis for $\mathbb{Z}^{10} = H_3(X, \mathbb{Z}) = H_1(F, \mathbb{Z}) = H_1(J, \mathbb{Z})$, the isomorphism $(\Lambda^2 \mathbb{Z}^{10})^* \otimes \mathbb{Z}_2 = H^2(K(\mathbb{Z}^{10}, 1), \mathbb{Z}_2) \simeq \{\text{Central extensions of } \mathbb{Z}_2 \text{ by } \mathbb{Z}^{10}\}$ can be given as follows. Let $y = \sum a_{ij} \cdot e_i^* \wedge e_j^*$, $i < j$, $a_{ij} \in \mathbb{Z}_2$, be a 2-form. The corresponding central extension Y is the group with generators $x_1, \dots, x_{10}$, ε, and with the relations $x_i x_j = a_{ij} x_j x_i$, $\varepsilon^2 = e$, $\varepsilon x_i = x_i \varepsilon$.

By the definition of principal polarization on the intermediate Jacobian, the form y corresponding to the theta divisor has coefficients $a_{ij} = (e_i \cdot e_j)$ mod.(2). Therefore the commutator $[x_i, x_j] = (e_i \cdot e_j)$ mod.(2). The statement follows from this by "linearity", recalling that for a central extension of abelian groups $[xy, z] = [x, z]\,[y, z]$.

ON THREEFOLDS WHOSE HYPERPLANE SECTIONS ARE ENRIQUES SURFACES

by

A. CONTE

As it is well known, Fano varieties, when embedded by the anticanonical system, are projective threefolds whose hyperplane sections are K3-surfaces. It is therefore natural to ask whether there exist projective threefolds having as hyperplane sections Enriques surfaces. This problem was solved by FANO himself in a paper of 1938 (with some gaps in the proofs; see [F]). J. P. MURRE and myself have started to fill in the gaps. This is a report of the first results obtained in this direction.

Let $W \subseteq P^N$ be a threefold such that:

(i) W is projectively normal;

(ii) if H is a sufficiently general hyperplane, then $F = W \cdot H$ is a smooth Enriques surface;

(iii) $g(W \cdot H \cdot H') = p$.

LEMMA 1.- W has isolated singularities.

Proof.- W has at most isolated singularities since $F = W \cdot H$ is smooth. Moreover, W has at least isolated singularities. For, assume by contradiction that W is smooth. Then, by the adjunction formula:

$$(K_W + F) \cdot F \sim K_F .$$

Put $T = K_W + F$. Then, since $2K_F \sim 0$, we will have $2T \cdot F \sim 0$ so that, by Weil'equivalence criterion, it will be $2T \sim 0$. It follows that $T \equiv 0$, i. e. $-K_W \equiv F$, so that, by Nakai's criterion of ampleness, $-K_W$ is ample and W is a Fano threefold. But then Pic W has no torsion, so that $T \sim 0$ and $K_F \sim 0$, which is a contradiction.

Let $P_1, \ldots, P_n$ be the singular points of W.

LEMMA 2.- $H^i(W, \mathcal{O}_W(n)) = 0$ for $i = 1, 2, 3$ and $n \geq 0$.

Proof.- From the exact sequence:

$$0 \longrightarrow \mathcal{O}_W(n) \longrightarrow \mathcal{O}_W(n+1) \longrightarrow \mathcal{O}_F(n+1) \longrightarrow 0$$

using Kodaira's vanishing theorem and decreasing induction on n.

COROLLARY.- The maps:

$$H^0(W, \mathcal{O}_W(n)) \rightarrow H^0(F, \mathcal{O}_F(n))$$

are surjective for all $n > 0$.

Let now $\Gamma = W \cdot H \cdot H' = F \cdot H'$, $g(\Gamma) = p$.

PROPOSITION 3.- $N = p$ and $\deg W = 2p - 2$, so that:

$$W = W_3^{2p-2} \subseteq P^p.$$

Proof.- From the general theory of Enriques surfaces it follows that $\Gamma^2 = 2p - 2$ and $h^0(\Gamma) = p - 1$. Moreover, form the assumption (i) and the Corollary to Lemma 2 it follows that the hyperplanes cut out a complete linear system on F, so that $\Gamma \subseteq P^{p-2}$, $F \subseteq P^{p-1}$ and $W \subseteq P^p$.

Note that Γ is not a canonically embedded curve since $\Gamma \subseteq P^{p-2}$. However, again from the theory of Enriques surfaces, one knows that there exists a smooth and irreducible curve $\Gamma' \in |\Gamma + K_F|$ such that $g(\Gamma') = p$, $\Gamma'^2 = 2p - 2$ and $h^0(\Gamma') = p + 1$. Moreover, Γ' is canonically embedded.

In the following we will always make the hypothesis that the points P_i's are "similar", i. e. that they all have the same properties. Under suitable hypotheses of generality for W one can therefore prove the following:

MAIN THEOREM.- Let W be not a cone and $p \geq 5$. Then, $n = 8$, i. e. W has exactly eight singular points $P_1, \ldots, P_8$. Each point P_i is a quadruple point and its tangent cone is a cone over the Veronese surface.

W carries a linear system $|\phi|$ of Weil divisors the general member of which is a K3-surface. This systema has dimension $p - 1$ and the base points of it are the P_i's, each of which is a rational double point on a general ϕ.

Let:

$$\lambda_{|\phi|} : W \dashrightarrow M \subseteq P^{p-1}$$

be the rational map defined by the system $|\phi|$. Then, $\lambda_{|\phi|}$ is a birational map and M spans a P^{p-1}, has degree $2p - 6$ and has K3-surfaces as (general) hyperplane sections (i. e., is a Fano variety in the classical sense). Furthermore, M contains 8 *planes* $\pi_1, \ldots, \pi_8$ *which are the "images" of the points* $P_1, \ldots, P_8$.

(Also M may have singularities; for instance, the points $\pi_i \cap \pi_j$ are, in general, double points of M).

Main ingredients of the proof:

(i) the construction of Φ (GODEAUX):

Let $F_\lambda = W\cdot H$ be a general pencil of Enriques surfaces on W. Let $\Gamma_o = F_{\lambda_1}\cdot F_{\lambda_2}$ be the axis of this pencil, so that $g(\Gamma_o) = p$. Let $A_1, \ldots, A_{p-1}$ be independent points on Γ_o, which determine uniquely $A_p, \ldots, A_{2p-2}$ such that $A_1 + A_2 + \ldots + A_{2p-2}$ is a canonical divisor. Let F be a generic member of the pencil. Then, there exists a unique $\Gamma'_\lambda \in |K_{F_\lambda} + \Gamma_o|$ going through $A_1, \ldots, A_{p-1}$ (and therefore also through $A_p, \ldots, A_{2p-2}$).

Letting F_λ vary in the pencil, the locus of Γ'_λ is a surface Φ on W. Note that $\Phi\cdot H = \Gamma'$ is a canonical curve. Both FANO and GODEAUX conclude from this fact that Φ is a K3-surface, but this is not correct, since (owing to the fact that $|\Phi|$ has fixed points in $P_1, \ldots, P_8$) Φ could be singular, and DU VAL has given examples of of *rational* singular surfaces with canonical curve sections. However, $h^1(\Phi) = 0$ and $h^2(\Phi) = 1$, i. e. Φ has the same invariants of a K3-surface.

(ii) Let $\widetilde{W} = B_{P_1,\ldots,P_n}(W)$ be the blow up of W in $P_1, \ldots, P_n$, $\pi : \widetilde{W} \longrightarrow W$ the coeesponding morphism and $E_i = \pi^{-1}(P_i)$ the corresponding exceptional divisor. We assume that both W and the E_i's are smooth and denote by $\widetilde{\Phi}$, $\widetilde{F}$ and $\widetilde{F}_i$ the proper transforms respectively of a general Φ, of a generale F and of a F_i, i. e. of a F going through P_i.

To study $\widetilde{\Phi}$ one needs the following lemma on surfaces:

LEMMA 4.- *Let S be a projective surface with one singular point P. Let:*

$$\begin{array}{ccc} C & \subseteq & \widetilde{S} \\ \downarrow & & \downarrow \pi \\ P & \in & S \end{array}$$

be a resolution of singularities. Assume $C = \pi^{-1}(P)$ *is smooth, irreducible and such that* $K_{\widetilde{S}} \equiv \rho C$, *with* $\rho \in \frac{1}{2}Z$. *Then, at most the following cases are possible:*

(i) $\rho = -2$, S *ruled and* $g(C) > 1$;

(ii) $\rho = -1$, S *ruled or rational*;

(iii) $\rho = -\frac{1}{2}$, S *rational*, $g(C) = 0$ *and* $C^2 = -4$;

(iv) $\rho = 0$, $g(C) = 0$, $C^2 = -2$;

(v) $\rho = 1$, $g(C) = 0$, $C^2 = -1$.

We have the following situation:

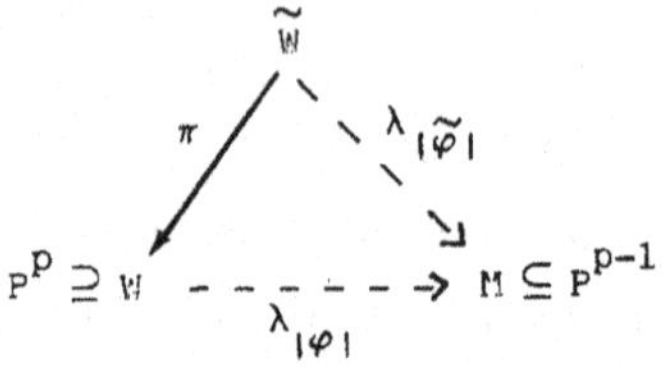

As a consequence of COSSEC [C] and LETIZIA [L] results on projective embeddings of Enriques surfaces one can show that, if W is not a cone and $p \geq 5$, then $M = \lambda_{|\varphi|}(W)$ is always threedimensional. Moreover, using SAINT-DONAT [SD] and REID [R] results on projective embeddings of K3-surfaces, one can prove that, under the same hypotheses, $\lambda_{|\varphi|}$ is always birational. It follows that the generic $\tilde{\varphi}$ (hence φ) is a K3 surface. Therefore, the hyperplane sections of M are K3-surfaces spanning a P^{p-2}, so that $\deg M = 2(p-2) - 2 = 2p - 6$.

If we put $\Delta_i = F_i \cdot E_i$, one can show that Δ_i is smooth and irreducible and that $|\Delta_i|$ is very ample. Moreover, from the lemma on surfaces it follows that in our case $\rho = -\frac{1}{2}$, so that on F_i one has $\Delta_i^2 = 4$ and $g(\Delta_i) = 0$, which implies that the P_i's are rational quadruple points.

From the Picard-Del Pezzo theorem ($|\Delta_i|$ very ample and $g(\Delta_i) = 0$) it follows also that the E_i's are rational. Other coputations give that, on W, one has $\Delta_i^2 = 4$ and $K_{E_i}^2 = 9$, so that the E_i's are planes, the Δ_i's are conics and the tangent cone to W in P_i is the cone over the Veronese surface.

Moreover, on W, from the relations $F \cdot F^* = \Gamma$, $\varphi \cdot F^* = \Gamma' = K_{F^*} + \Gamma$, it follows that $(F - \varphi) \cdot F^* \sim K_{F^*}$, so that $2(F - \varphi) \cdot F^* \sim 0$ and therefore, by Weil's equivalence criterion, $2F \sim 2\varphi$. Since the kernel of the map π_*: Pic $\tilde{W} \longrightarrow$ Pic W is generated by the E_i's, it follows that:

$$2\tilde{F} \sim 2\tilde{\varphi} + \sum_{i=1}^{n} t_i E_i .$$

On the other hand, one can show that all the t_i's are equal to 1, so that, by taking degrees and remembering that the E_i's are planes, one gets that $2(2p - 2) = 2(2p - 6) + n$, so that $n = 8$.

In his paper FANO gives also a classification of these varieties W which is based on the classification of the Fano varieties M (note however that M may have double points, so that strictly speaking it is not a Fano variety in the modern sense used by ISKOHVSHIH, which implies smoothness). We have not yet come across Fano's arguments, so that for us, at the moment, the following are only examples of the varieties we are studyng and not necessarily the only existing ones.

FANO's classification goes as follows: first of all, as to the two exceptional cases:

$p = 3$ W_3^4 is a 4-tuple P^3;

$p = 4$ $W_3^6 \subseteq P^4$ is the famous "Enriques threefold", the double cover of P^3 branched over an Enriques sextic surface plus four planes, having one quadruple point, four triple lines and six double planes, whose equation is:

$$x_1x_2x_3x_4\left[x_o^2 + 2x_of_1(x_1, \ldots, x_4) + f_2(x_1, \ldots, x_4)\right] +$$
$$+ \varphi_2(x_2x_3x_4, x_3x_4x_1, x_4x_1x_2, x_1x_2x_3) = 0,$$

where f_1 is linear and f_2 and φ_2 are quadratic.

FANO thought that it was not even unirational. ROTH proved that it is unirational and thought he had shown that it is not rational by proving that it has Severi torsion (in the $H_1(W, Z)$). But SERRE pointed out that it is simply connected. TYRRELL explained this apparent contradiction by showing that W has not ordinary singularities (there are some more hidden non-ordinary quadruple points!). L. PICCO BOTTA and A.VERRA have recently proved [PB-V] that it is not rational (W is a conic bundle over an open set of a quadric).

$p = 5$ There exists no such threefold (projection from the line $l_{ij} = \overline{P_iP_j}$ would give a P^3 with six double points!).

$p = 6$ $M_3^6 \subseteq P^5$ is the intersection of a quadric and a cubic hypersurface (i. e., a cubic complex of lines, if one identifies the quadric with the Grassmannian $G(1,3)$ of the lines of P^3). Since it contains 8 planes, one can show that it is the complex of the lines lying on some quadric of a net in P^3 (the planes correpond to the lines going through the 8 base points of the net).

To se what W_3^{10} is, let $\check{P}^9$ be the projective space parametrising the dual quadrics of P^3 and let:

$$\check{P}^9 \supseteq \Delta_8^4 \supseteq V_6^{10} \supseteq V_3^8$$

be the filtration by the rank, where Δ_8^4 corresponds to the dual quadrics of rank ≤ 3, V_6^{10} to the quadrics of rank ≤ 2 (couples of points) and V_3^8 to the quadrics of rank 1 (points counted twice). Note that V_3^8 has multiplicity 4 inside V_6^{10}. Then, if $\check{P}^6$ is a 6-dimensional linear subspace of $\check{P}^9$, $W_3^{10} = V_6^{10} \cdot \check{P}^6$ is a threefold having 8 quadruple points in the intersection $V_3^8 \cdot \check{P}^6$. The hyperplane sections of W_3^{10} are the couples of points contained into a 5-dimensional linear system of dual quadrics so, by apolarity, they are the couples of points conjugate with respect to all quadrics of a 3-dimensional web of quadrics of P^3. The lines joining these points give rise to what is called a "Reye congruence". The "Reye congruences" are Enriques surfaces of special type making up a 9-dimensional family; they have been characterized recently by COSSEC in his thesis in the following way:

THEOREM 5.- An Enriques surface S is a "Reye congruence" if and only if there exist on it an elliptic pencil $|P|$ and two rational curves C_1, C_2 such that $PC_1 = 6$ and $PC_2 = 2$.

The K3-surface which is the universal covering of a Reye congruence is a quartic surface in P^3 with 10 nodes whose enveloping cones split in two cubic cones.

p = 7 M_3^8 is the intersection of three quadrics with 8 planes in common; W_3^{12} is the image of P^3 by the linear system of all sextic surfaces going doubly through the edges of a tethraedron and containing one plane cubic.

p = 8 There exists no such threefold.

p = 9 Here we have the following situation: let Q_1, Q_2 be two quadrics in P^5 having equations:

$$A_1(y_o, y_1, y_2) + B_1(y_o, y_1, y_2) = 0,$$

$$A_2(y_o, y_1, y_2) + B_2(y_o, y_1, y_2) = 0,$$

where the A_i's and B_i's are quadratic polynomials. Let $Y = Q_1 \cdot Q_2$ and $X = Y/i$, where i is the involution defined by:

$$i[(y_o, \ldots, y_5)] = (-y_o, -y_1, -y_2, y_3, y_4, y_5).$$

Then, there are two suitable embeddings:

$$\begin{array}{ccc} & Y & \\ & \swarrow \quad \searrow & \\ P^9 \supseteq W_3^{16} & \dashrightarrow & M_3^{12} = s(P^2 \times P^2)\cdot Q \subseteq P^8, \end{array}$$

where s is the Segre embedding, Q is a quadric of P^8 and M_3^{12} contains eight planes. Here the hyperplane sections of W_3^{16} are Enriques surfaces obtained modulo i from the intersection $S = Q_1 \cdot Q_2 \cdot Q_3$, where Q_3 is a third quadric of P^5 of equation:

$$A_3(y_0, y_1, y_2) + B_3(y_3, y_4, y_5) = 0.$$

It is well known that the generic Enriques surface can be obtained in this way. This type of surfaces has been charcterized recently by A. VERRA [V] in the following way:

THEOREM 6.- An Enriques surface S is of the type described above if and only if the following two equivalent conditions hold:

(i) there exists an irreducible and non-singular curve C on S such that $p_a(C) = 3$ and the generic member of $|C|$ is not hyperelliptic;

(ii) there exists a cartesian square:

$$\begin{array}{ccc} S & \longrightarrow & P^2 \\ \varphi \downarrow & & \downarrow \psi \\ P^2 & \longrightarrow & P^2, \end{array}$$

where φ and ψ are morphisms of degree 4.

Every Reye congruence satisfies this condition.

p = 10, 11, 12 There exist no such threefold.

p = 13 $W_3^{24} \subseteq P^{13}$ is the image of P^3 by the linear system of all sextic surfaces going doubly through the edges of a tethraedron; $M_3^{20} \subseteq P^{12}$ is the image of P^3 by the linear system of all quartic surfaces going simply through the edges of the same tethraedron.

$p \geq 14$ There exist no such threefold.

All these varieties are rational, with the only exception of the Enriques threefold. Most of them are conic bundles. For instance, the $M_3^6 = Q\cdot C \subseteq P^5$ is a conic bundle over the Del Pezzo rational double plane branched over a quartic plane curve.

For detailed proofs of the above statements, see [C-M].

REFERENCES

[C] F. COSSEC, Projective models of Enriques surfaces and Reye congruences, thesis Yale University, 1981.

[C-M] A. CONTE and J. P. MURRE, Three-dimensional algebraic varieties whose hyperplane sections are Enriques surfaces, Mittag-Leffler Institut, Report n. 10, 1981.

[F] G. FANO, Sulle varietà algebriche a tre dimensioni le cui sezioni iperpiane sono superficie di genere zero e bigenere uno, Mem. Soc. it. Sci. (detta dei XL), 24 (1938), 41-66.

[L] M. LETIZIA, Sistemi lineari completi su superficie di Enriques, Ann. Mat. Pura App., CXXVI (1980), 267-82.

[R] M. REID, Hyperelliptic linear systems on K3-surfaces, J. London Math. Soc., 13 (1976), 425-31.

[SD] B. SAINT-DONAT, Projective embeddings of K3-surfaces, Am. J. Math., 96 (1975), 602-39.

[PB-V] L. PICCO BOTTA and A. VERRA, The non-rationality of the generic Enriques threefold, to appear in Comp. Math.

[V] A. VERRA, Superficie di Enriques e reti di quadriche, to appear in Ann. Mat. Pura App.

Ample Divisors on 3-folds

by Andrew John Sommese

In this article I will discuss without proofs my recent work $[S_1, S_2]$ about ample divisors on 3-folds. I work over $\mathbb{C}$. A *divisor* A on a connected projective manifold X is said to be *ample* if [A], the associated line bundle is ample. A *line bundle* L on such an X is said to be *ample* if there exists an $n > 0$ such that L^n is very ample. A *line bundle* L on such an X is said to be *very ample* if there exists an embedding $\phi : X \to \mathbb{P}_{\mathbb{C}}$ with $\phi^* \mathcal{O}_{\mathbb{P}_{\mathbb{C}}}(1) \approx L$; in this case an element $A \in |L|$, the linear system of zero sets of sections of not identically zero sections of L is often called a *hyperplane section* of X.

All the results here came from a study of the following question which I will make precise and then completely answer.

Question A. *Let X be a smooth connected projective 3-fold. Let S be a smooth ample divisor on X. When can (X,S) be 'replaced' by a new pair (X',S') with S' a minimal model in the sense of Zariski* [Z]?

The following definition is useful.

Definition. *Let L be a holomorphic line bundle on a smooth connected 3-fold, X. A reduction (X',L') of the pair (X,L) consists of a holomorphic line bundle, L', on a smooth connected 3-fold, X', and a holomorphic surjection* $\pi : X \to X'$ *such that:*

a) π *expresses X as X' with a finite set F blown up,*

b) $\pi^* L' \otimes [\pi^{-1}(F)]^{-1} \approx L$.

Note that:

1) if $S \in |L|$ is smooth then $\pi(S)$ is smooth and $\pi(S) \in |L' - F|$, where $|L' - F|$ denotes the linear system of elements of $|L'|$ that contain F,

2) sending a smooth $S' \in |L' - F|$ to its proper transform on X sets up a (1-1) correspondence of smooth $S' \in |L' - F|$ and smooth $S \in |L|$,

3) if L is ample then so also is L'.

Question A can now be made precise.

<u>Question A'</u>. <u>Let</u> L <u>be an ample line bundle on a smooth connected projective threefold</u> X. <u>Does there exist a reduction</u> (X',L') <u>of</u> (X,L) <u>such that all smooth</u> $S' \in |L'|$ <u>are minimal models in the sense of Zariski</u>?

The first result gives a large class of pairs (X,L) for which Question A' has an affirmative answer.

<u>Theorem I</u>. <u>Let</u> L <u>be an ample line bundle on a smooth connected projective threefold</u>, X. <u>Assume that</u> $h^0(K_X^n \otimes L^n) \neq 0$ <u>for some</u> $n > 0$. <u>Then all smooth</u> $S \in |L|$ <u>are not birational to</u> $\mathbb{P}^1_{\mathbb{C}} \times C$ <u>for any curve</u>. <u>Further a reduction</u> (X',L') <u>of</u> (X,L) <u>exists such that smooth</u> $S' \in |L'|$ <u>are minimal models</u>.

If $|L|$ contains at least one smooth element, (X',L') is unique. Further if there is an $S \in |L|$, then $h^0(K_X^n \otimes L^n) \neq 0$ for some $n > 0$, says by definition that X - S has non-negative logarithmic Kodaira dimension in the sense of Iitaka [I].

The above result is proven in $[S_1]$ under the stronger condition that X has non-negative Kodaira dimension, i.e. that $\Gamma(K_X^n) \neq 0$ for some $n > 0$. That proof is easy to generalize to a proof of the above. At the end of this paper I sketch this proof. Most of the remaining results below that do not assume $h^0(K_X^n) \neq 0$ for some $n > 0$, require a much more elaborate method of proof. This other method is a modification of the Fano-Morin adjunction process [R, pg. 66]. It is discussed in $[S_2$, §1].

Theorem II. Let X, L, X', and L' be as in theorem I. Assume that there is a smooth $S' \in |L'|$. Then:

a) $K_{X'} \otimes L'$ is arithmetically effective, i.e. given an effective divisor D,

$$(K_X' + L') \cdot C > 0 < (K_{X'} + L') \cdot (K_{X'} + L') \cdot D$$

b) $H^i(K_{X'}^n \otimes L'^n) = 0$ for all $i > 0$ and $n > 0$

c) if X has non-negative Kodaira dimension, i.e. $\Gamma(K_X^n) \neq 0$ for some $n > 0$, then smooth $S' \in |L'|$ are of general type and:

$$K_{S'} \cdot K_{S'} \geq K_{S'} \cdot L_{S'} \geq L_{S'} \cdot L_{S'}$$

and

$$K_S \cdot K_S \geq L_S \cdot L_S$$

for smooth $S \in |L|$. Further $K_S \cdot K_S = L_S \cdot L_S$ if and only if $K_{S'} \cdot K_{S'} = L_{S'} \cdot L_S$ if and only if $K_{X'}^t \approx \mathcal{O}_{X'}$ for some $t \neq 0$.

Note that by Nakai's criterion, a) implies that $K_{X'}^n \otimes L'^{n+1}$ is ample for all $n \geq 0$. This combined with Kodaira's vanishing theorem implies b). Also b) implies the important fact that

$$\chi(X', K_{X'}^n \otimes L'^n) = h^0(K_{X'}^n \otimes L'^n) \qquad \text{for } n > 0.$$

Note that since $\pi^*(K_{X'} \otimes L') \approx K_X \otimes L \otimes [\pi^{-1}(F)]^{-1}$, it follows from Hartog's theorem that:

$$h^0(K_X^n \otimes L^n) = h^0(K_{X'}^n \otimes L'^n).$$

This yields a very surprising corollary of $[S_2]$.

Corollary II-A. *Let S be a smooth ample divisor on a smooth connected projective threefold, X. Then there exists a polynomial q(n) such that*

$$h^0(K_X^n \otimes [S]^n) = q(n) \qquad \textit{for } n > 0.$$

The following two results are proved in $[S_1]$.

Corollary II-B. *Let S be a smooth ample divisor on a smooth connected projective threefold X. If the Kodaira dimension of X is non-negative then* $h^{2,0}(S) \geq 2 \leq \chi(\mathcal{O}_S)$.

Corollary II-C. *Fix* $M > 0$. *Let* S_M *be the set of smooth projective surfaces S that satisfy* $\chi(\mathcal{O}_S) \leq M$. *Then the set of smooth projective threefold of non-negative Kodaira dimension containing an* $S \in S_M$ *as an ample divisor is finite up to algebraic deformation (and hence diffeomorphism).*

The next result is an easy consequence of part a) of theorem II and the generalization of the Ramanujam-Kodaira vanishing theorem, proved independently by Kawamata [K_2] and Viehweg [V].

Corollary II-D. *Let X, L, X', and L' be as in theorem II. Assume that the degree of $\chi(X', K^n_{X'} \otimes L'^n)$ is three. Then there is an $N > 0$, such that $\Gamma(K^N_{X'} \otimes L'^N)$ spans $K^N_{X'} \otimes L'^N$ and smooth $S' \in |L'|$ are of general type. If L is very ample then N = 5 will suffice; if further the Kodaira dimension of X is non-negative, then N = 3 will suffice.*

The above can be used to describe thc −2 rational curves on smooth $S' \in |L'|$; this will be done elsewhere. The following question suggested by the above corollary would if answered affirmatively have some very nice consequences.

Question B. *Let X, L, X', and L' be as in theorem II. Is there an $N > 0$ such that $(K^N_{X'} \otimes L'^N)$ spans $K^N_{X'} \otimes L'^N$?*

If one smooth $S' \in |L'|$ is an elliptic surface, and $H^1(S',\mathbb{C}) \neq 0$ then it is shown in [S_2] that the above is true. This result shows also that for such S' with $H^1(S',\mathbb{C}) \neq 0$, the map from S' to its image under the Albanese map is an elliptic fibration without multiple fibres.

What happens when $h^0(K^n_X \otimes L^n) = 0$ for all $n > 0$?

Theorem III. *Let S be a smooth ample divisor on a connected smooth projective 3-fold, X. Then $h^0(K^n_X \otimes [S]^n) = 0$ for $n > 0$ if and only if $h^0(K^n_X \otimes [S]^{n+1}) = 0$ for all $n \gg 0$, if and only if either:*

a) X *is a* $\mathbb{P}^1_{\mathbb{C}}$ *bundle over* $\tilde{S}$ *a smooth projective surface, and* S *is a meromorphic section of* X *over* $\tilde{S}$,

or,

b) S *is birational to* $\mathbb{P}^1_{\mathbb{C}} \times C$ *for some curve* C.

Note that there is a one-to-one correspondence of triples $(X,S,\tilde{S})$ as in a) and triples $(E,s,\tilde{S})$ where:

1) E is an ample rank two vector bundle on $\tilde{S}$,

and

2) s is a holomorphic section of E whose graph in E is transverse to the zero section.

The situation of case b) is generalized by the following, due to Morin-Fano [R, pg. 66] if S is a very ample divisor.

Theorem IV. Let S *be a smooth ample divisor on a smooth connected projective threefold,* X. *Assume that* S *is birational to* $\mathbb{P}^1_{\mathbb{C}} \times C$ *for some curve* C. *Then there is a reduction* (X',L') *of* $(X,[S])$ *such that either:*

a) $(X',L') \approx (\mathbb{P}^3_{\mathbb{C}}, \mathcal{O}_{\mathbb{P}^3_{\mathbb{C}}}(e))$ *for* $e = 1, 2,$ *or* 3,

or,

b) X' *is holomorphic to a smooth quadric in* $\mathbb{P}^4_{\mathbb{C}}$ *and* L' *is the restriction of* $\mathcal{O}_{\mathbb{P}^4_{\mathbb{C}}}(e)$ *for* $e = 1$ *or* 2,

or,

c) $K_{X'} \approx L'^{-2}$, i.e. X' is a Fano threefold of the principal series (these are described in [F, I+S]).

or,

d) X' fibres holomorphically, $\phi : X' \to C$ over a smooth curve C and a generic fibre F of ϕ is biholomorphic to $\mathbb{P}^2_{\mathbb{C}}$ with $L'_F \approx \mathcal{O}_{\mathbb{P}^2_{\mathbb{C}}}(e)$ for e = 1 or 2,

or,

e) X fibres holomorphically $\phi : X' \to C$ over a smooth curve C, a generic fibre F of ϕ is biholomorphic to a smooth quadric $\mathbb{P}^2_{\mathbb{C}} \times \mathbb{P}^1_{\mathbb{C}}$ of $\mathbb{P}^3_{\mathbb{C}}$, and L_F is isomorphic to the restriction of $\mathcal{O}_{\mathbb{P}^3_{\mathbb{C}}}(1)$.

or,

f) (X,L) is as in a) of theorem III. If e = 1 in d), c), a) or b) then $X \approx X'$ and $L \approx L'$.

The following surprising corollaries follow from the above results.

Corollary IV-A. Let S be a smooth ample divisor on a connected smooth projective 3-fold X. Assume that S is not birational to $\mathbb{P}^1_{\mathbb{C}} \times C$ for any curve C. Assume that the map from S to its minimal model S' has a fibre f such that $S \cdot f \geq 2$, e.g. assume that f is not irreducible. Then X is a $\mathbb{P}^1_{\mathbb{C}}$ bundle over a smooth surface $\tilde{S}$ and S is a meromorphic section of X over $\tilde{S}$ (S cannot be a holomorphic section).

Corollary IV-B. Let S be a smooth ample divisor on a smooth connected projective threefold, X. Assume that S is birational to a surface S' which is a $K(\pi,1)$, i.e. the universal cover of S' is contractible. Then X is a $\mathbb{P}^1_{\mathbb{C}}$ bundle over a smooth surface $\tilde{S}$ and S is a meromorphic section of X over $\tilde{S}$ (S can never be a holomorphic section).

The proof of theorem I is based on the following two lemmas.

Lemma A. Let S be an ample divisor on a smooth connected projective 3-fold X. Let E be a smooth rational curve where $E \subseteq S_{reg}$, the smooth points of S. Assume that $E \cdot E = -1$ on S and $h^0(K_X^n \otimes [S]^n) \neq 0$ for some $n > 0$. Then the closure $\mathcal{D}$ of the union of all deformations of E is a normal divisor on X such that S meets $\mathcal{D}$ transversely in $E \subseteq \mathcal{D}_{reg}$.

Sketch of proof. Since $(K_X + S) \cdot E = K_S \cdot E = -1$, and since $n(K_X + S)$ is effective $\mathcal{D} \neq X$. Since $E \cdot E = -1$ on S and $L \cdot E \geq 1$, $H^1(E, N_E) = 0$ where E is the normal bundle of E in X. Also $h^0(N_E) \geq 2$. By Kodaira-Spencer theory $\dim \mathcal{D} \geq 2$. Using the projectivity of a family over a component of the Hilbert scheme of X, it is clear $\mathcal{D}$ is an irreducible divisor.

If N_E was spanned, then by Kodaira-Spencer theory the deformations of E in X would fill out an open set of X contradicting the fact that $\mathcal{D}$ is a divisor. It is straightforward to show that $N_E \approx \mathcal{O}_{\mathbb{P}^1_{\mathbb{C}}}(-1) \oplus L_E$. This shows that a union U of small deformations of E in X gives a complex manifold that meets S transversely in E.

By ampleness of S the pullback of S to the normalization of $\mathcal{D}$ is connected. This shows that $\mathcal{D}$ meets S transversely in E. Since the intersection is transverse and E is smooth the singularities of $\mathcal{D}$ are in $\mathcal{D} - S$. Since $\mathcal{D} - S$ is affine and therefore contains no compact subvarieties except points, the singular set of $\mathcal{D}$ is finite. Since $\dim X = 3$ and X is smooth this implies that $\mathcal{D}$ is normal. □

It is an easy argument [cf. S_1, §0] to see that $\mathcal{D}$ is either $\mathbb{P}^2_{\mathbb{C}}$, F_r or $\tilde{F}_2$. Here F_r for $r \geq 0$ is the unique $\mathbb{P}^1_{\mathbb{C}}$ bundle over $\mathbb{P}^1_{\mathbb{C}}$ with a section E having self intersection $-r$ on E; E is unique if $r \neq 0$. Also $\tilde{F}_2$ is F_2 with E blown down.

__Lemma B__. $\mathcal{D}$ __is biholomorphic to__ $\mathbb{P}^2_{\mathbb{C}}$.

__Proof__. Since $-1 = (K_X + S) \cdot E = (K_X + S)_{\mathcal{D}} \cdot E$ and since intersection of Cartier divisors on $\tilde{F}_2$ is always even, it follows that only $\mathbb{P}^2_{\mathbb{C}}$ or F_r are possible.

I will now show that $\mathcal{D}$ is not biholomorphic to F_r. Assume to the contrary that it is. Let f denote a fibre of the projection $F_r \to \mathbb{P}^1_{\mathbb{C}}$ transverse to E.

Since $L_{\mathcal{D}}$ is ample and E is rational it follows that $L_{\mathcal{D}} \approx [E] \otimes [f]^{r+k}$ with $k > 0$, e.g. [H, pg. 369 ff]. It follows that there is a smooth rational curve $C \in |L_{\mathcal{D}} - f|$ satisfying $C \cdot C \geq 0$.

Note that:

*) $$\mathcal{D} \cdot f < 0.$$

To see this assume that *) is false. Note that $f \cdot f = 0$ on F_r and that $[\mathcal{D}]_{\mathcal{D}} \approx N_{\mathcal{D}}$, the normal bundle of $\mathcal{D}$ on X. Since f is a smooth rational curve it follows from $\mathcal{D} \cdot f \geq 0 \leq f \cdot f$ that $\Gamma(N_f)$ spans N_f, the normal bundle of f in X. It also follows that $H^1(f, N_f) = 0$. This implies by Kodaira-Spencer theory that the deformations of f in X fill out an open set of X. Since $n(K_X + S)$ is effective it follows that $(K_X + S) \cdot f \geq 0$. Since $S \cdot f = \text{degree } L_{\mathcal{D},f} = 1$ this implies that $K_X \cdot f \geq -1$. Since it is assumed that $\mathcal{D} \cdot f \geq 0$ this yields $(K_X + \mathcal{D}) \cdot f \geq -1$. This contradicts $(K_X + \mathcal{D}) \cdot f = -2$ which follows from the adjunction formula.

By *) for $C \in |L_{\mathcal{D}} - f|$ it follows that:

**) $$\mathcal{D} \cdot C = \mathcal{D} \cdot E - \mathcal{D} \cdot f > \mathcal{D} \cdot E.$$

By the adjunction formula $(K_X + \mathcal{D} + S) \cdot E = -2$. Using this, the fact that $(K_X + S) \cdot E = -1$ and **) it follows that:

***) $$\mathcal{D} \cdot E = -1 \quad \text{and} \quad \mathcal{D} \cdot C \geq 0$$

Therefore by the same argument used with f in the last paragraph, $(K_X + S) \cdot C \geq 0$. This implies that:

$$\begin{aligned} -1 = (K_X + S) \cdot E &= (K_X + S) \cdot C + (K_X + S) \cdot f \\ &\geq (K_X + S) \cdot f \geq K_X \cdot f + 1 \end{aligned}$$

or

$$K_X \cdot f \leq -2.$$

This together with *) yields:

$$(K_X + \mathcal{D}) \cdot f \leq -3$$

contradicting the adjunction formula.

Therefore $\mathcal{D} \approx \mathbb{P}^2_{\mathbb{C}}$. □

A simple argument shows that $L_{\mathcal{D}} \approx \mathcal{O}_{\mathbb{P}^2_{\mathbb{C}}}(1)$ and $N_{\mathcal{D}} \approx \mathcal{O}_{\mathbb{P}^2_{\mathbb{C}}}(-1)$. Therefore $\mathbb{P}^2_{\mathbb{C}}$ can be smoothly blown down to yield a projective three-fold X' with $\pi : X \to X'$ the blowing down map. $(X',[\pi(S)])$ is easily checked to be a reduction of $(X,[S])$.

The above is the key argument in proving theorem I. For full details in proceeding along these lines see [S_1] and [S_2].

References

[F] T. Fujita, On the structure of polarized manifolds of total deficiency, 1, J. Math. Soc. Japan 32 (1980), 709-725.

[H] R. Hartshorne, Algebraic Geometry, Springer Verlag, New York (1977).

[I] S. Iitaka, On logarithmic Kodaira dimension of algebraic varieties, Complex Analysis and Algebraic Geometry, ed. by W. L. Baily, Jr. and T. Shioda (1977), 175-189, Iwanami Shoten.

[I+S] V. A. Iskovskih and V. V. Šokurov, Biregular theory of Fano 3-folds, Proceedings Algebraic Geometry Conference, Copenhagen 1978, ed. by K. Lønsted, Lect. Notes in Math. 732 (1979), 171-182.

[K] Y. Kawamata, A generalization of Kodaira-Ramanujam's vanishing theorem, preprint.

[R] L. Roth, Algebraic Threefolds, Springer Verlag, Heidelberg (1953).

[S_1] A. J. Sommese, On the minimality of hyperplane sections of projective threefolds, to appear in Journal für die reine und angewandte Mathematik.

[S_2] A. J. Sommese, The birational theory of hyperplane sections of projective threefolds, preprint.

[V] E. Viehweg, Vanishing Theorems, preprint.

[Z] O. Zariski, Introduction to the Problem of Minimal Models in the Theory of Algebraic Surfaces, Pub. Math. Soc. of Japan 4 (1958).

REVETEMENTS CYCLIQUES

par Hélène Esnault et Eckart Viehweg

Soient Y une variété algébrique, projective, lisse sur le corps des nombres complexe et Z une désingularisation d'un revêtement galoisien de Y. Nous décrivons ici les faisceaux des formes différentielles à pôles logarithmiques le long du diviseur de ramification dans Z dans le cas où le discriminant de Z sur Y n'est "pas trop mauvais" (§1). Dans le cas où Z est cyclique sur Y, on peut déterminer la filtration de Hodge de la structure de Hodge mixte affectée à la partie ouverte de Z complémentaire du diviseur de ramification, sans pour autant pouvoir déterminer la filtration par le poids. Dans (§3), on applique cette construction aux revêtements cycliques de $\mathbb{P}^2$ et l'on obtient ainsi par des méthodes algébriques certains invariants topologiques de la fibre de Milnor d'un cône sur une courbe plane singulière. Les détails sont dans [4]. La symétrie des nombres de Hodge sur Z permet d'identifier la cohomologie de certains faisceaux inversibles sur Y avec celle de faisceaux de formes différentielles méromorphes, de sorte que des théorèmes d'annulation pour ces derniers s'interprètent en termes de théorèmes d'annulation pour ces faisceaux inversibles. Comme application, nous donnons dans (§2) une forme arithmétique du théorème d'annulation de Kodaira. Ceci fut le propos de l'exposé à Varenna du deuxième auteur. Les détails sont dans [7]. Le même théorème fut prouvé indépendemment et parallèlement par Y. Kawamata [5].

§1. REVETEMENTS

(1.1) Notations

(1.1.1) Un diviseur effectif D sur une variété complexe projective lisse Y est dit à croisements normaux si toutes ses composantes sont lisses et se coupent transversalement.

(1.1.2) Soient (Y,D) comme dans (1.1.1), $\tau : Y' \to Y$ un revêtement fini et galoisien tel que Y' soit normale, le discriminant $\Delta(\tau)$ soit contenu dans D, et $d : Z \to Y'$ une désingularisation de Y' telle que si $f : Z \to Y'$ désigne le morphisme composé, $f^{-1}(D) = D'$ soit un diviseur à croisements normaux. Un tel triplet $((Z,D'),Y',(Y(D))$ est dit bon revêtement.

(1.1.3) On dénote par $\Omega_Y^{\cdot}\langle D\rangle$ le complexe des formes différentielles logarithmiques le long de D, c'est-à-dire le complexe des formes différentielles holomorphes sur $Y-D$ qui ont au plus des pôles logarithmiques le long de D et par $W_n\Omega_Y^p\langle D\rangle$ la filtration par le poids de $\Omega_Y^p\langle D\rangle$ définie par les p-formes ayant au plus n pôles le long de D.

Lemme 1.2 [7] - Soit $((Z,D'),Y',(Y,D))$ un bon revêtement. On a les inclusion $f^* \Omega^p_Y <D> \hookrightarrow \Omega^p_Z <D'>$ et égalité $f_* \Omega^p_Z <D'> = \Omega^p_Y <D> \otimes f_* \mathcal{O}_Z$

Lemme 1.3 - Soit $((Z,D'),Y',(Y,D))$ un bon revêtement. Alors $R^q d_* \Omega^p_Z <D'> = 0$ et $R^q f_* \Omega^p_Z <D'> = 0$ pour $q > 0$. En particulier $R^q d_* \mathcal{O}_Z = 0$ pour $q > 0$ et Y' n'a que des singularités rationnelles.

Ce dernier fait est bien connu et la démonstration de (1.3) se fait selon la ligne expliquée dans [8] .

Corollaire 1.4 - Le morphisme $\tau : Y' \to Y$ est plat.

Afin de prouver (1.3), on utilise le

Lemme 1.5 - Soit $((Z,D'),Y',(Y,D))$ un bon revêtement tel que $Y' = Y$. Alors $R^q f_* \Omega^p_Z <D'> = 0$ pour $q > 0$.

Ne connaissant pas d'autres références, nous donnons une démonstration de (1.5) qui se ramène à un résultat de P. Deligne [3] .

Remarque - Le paragraphe 2 n'utilise pas (1.3), au contraire du paragraphe 3. Dans ce cas particulier d'une surface, la démonstration peut se faire plus aisément à l'aide d'un "bon choix" de Z [4] .

Démonstration de (1.5) - Soit H un diviseur très ample tel que, en notant $H' = f^{-1}H$, $((H',H' \cap D'), \tau^{-1}H , (H,H \cap D))$ soit un bon revêtement. Notons F_p le conoyau de l'inclusion $f^* \Omega^p_Y < D > \hookrightarrow \Omega^p_Z < D' >$. On a $R^q f_* \Omega^p_Z < D' > = R^q f_* F_p$ pour $q > 0$. On se ramène au cas où les $R^q f_* F_p$ sont des faisceaux gratte-ciel en "coupant par H" de la façon suivante.
De la suite exacte

$$0 \to \Omega^p_Z < D' > \otimes f^* \mathcal{O}(-H) \longrightarrow \Omega^p_Z <D' > \to \Omega^p_Z < D' > |_{H'} \to 0$$

on tire

$$R^q f_* (\Omega^p_Z <D' > |_{H'}) = R^q f_* \Omega^p_Z < D' > |_H \ .$$

La suite exacte

$$0 \to \mathcal{O}_{H'}(-H') \to \Omega^1_Z < D' > |_{H'} \to \Omega^1_{H'} < D' \cap H' > \to 0$$

induit la suite exacte

$$0 \to \Omega^{p-1}_{H'} < D' \cap H' > \otimes \mathcal{O}_{H'}(-H') \to \Omega^p_Z < D' > |_{H'} \to \Omega^p_{H'} < D' \cap H'> \to 0 .$$

Fixons q et soit p_0 le plus petit p tel que $R^q f_* \Omega^p_Z \langle D'\rangle \neq 0$. Bien sûr $p_0 > 0$, ce qui en fait importe peu. On a $R^q f_* \Omega^p_{H'} \langle D' \cap H'\rangle = 0$ pour $p < p_0$ et $R^q f_* \Omega^{p_0}_{H'} \langle D' \cap H' \rangle = R^q f_* \Omega^{p_0}_Z \langle D' \rangle |_{H'}$.

On peut donc supposer que les $R^q f_* F_p$ sont concentrés en des points. Mais alors, si $H^q(F_p) = H^0(R^q f_* F_p) \neq 0$, on a $h^q(\Omega^p_Z \langle D' \rangle) > h^q(\Omega^p_Y \langle D \rangle)$, ce qui contredit [3, théorème 3.2.5] .

Démonstration de (1.3) - Le morphisme τ étant affine, il suffit de montrer la première égalité. L'assertion étant locale, on peut supposer que Y = Spec A et $D_{réd} = \sum_1^r D_i$ est donné par les r premiers éléments $\langle f_1 \ldots f_r \rangle$ d'un système régulier de paramètres de A . Notons m_i l'ordre de ramification en D_i de τ . La variété Spec $A[f_i^{1/m_i}]_{1 \leq i \leq r}$ est régulière, de même, d'après le lemme d'Abhyankar, que la normalisée W de Y' dans $\mathbb{C}(Y')[f_i^{1/m_i}]_{1 \leq i \leq r}$. De plus, W est étale sur $Y' - \tau^{-1}D$. Soient W' la normalisée de Z dans $\mathbb{C}(W)$ et X une désingularisation de W' . On obtient le diagramme suivant

$$\begin{array}{ccccc} X & \xrightarrow{\delta} & W' & \xrightarrow{h} & W \\ & & \downarrow{\scriptstyle g'} & & \downarrow{\scriptstyle g} \\ & & Z & \xrightarrow{d} & Y' \end{array}$$

On peut supposer que $\Delta = \delta^{-1} g'^{-1} D'$ est un diviseur à croisements normaux. Alors $((X,\Delta),W',(Z,D'))$ est un bon revêtement, de même que $((X,\Delta),W',(W,g^{-1}\tau^{-1}D))$. Supposons maintenant par récurrence que pour tout bon revêtement $((Z,D'),Y',(Y,D))$ on ait $R^i d_* \Omega^p_Z \langle D' \rangle = 0$ pour $0 < i < q$ et un p fixé, ou bien que $R^1 d_* \Omega^p_Z \langle D' \rangle \neq 0$ (si q = 1). Dans tous les cas, on a une inclusion provenant de la suite spectrale de Leray

$R^q h_* (\delta_* \Omega^p_X \langle \Delta \rangle) \hookrightarrow R^q (h \circ \delta)_* \; \Omega^p_X \langle \Delta \rangle$, dont le deuxième membre est nul d'après (1.5). Comme g et g' sont affines, on a $g_* R^q h_* (\delta_* \; \Omega^p_X \langle \Delta \rangle) = R^q d_* (g'_* (\delta_* \; \Omega^p_X \langle \Delta \rangle)) = 0$. D'après (1.2), $\Omega^p_Z \langle D' \rangle$ est un facteur direct de $g'_* \delta_* \Omega^p_X \langle \Delta \rangle$. Donc $R^q d_* \; \Omega^p_Z \langle D' \rangle = 0$.

Corollaire 1.6 - Soit $((Z,D'),Y',(Y,D))$ un bon revêtement.
Alors $H^q(Z, \Omega^p_Z \langle D' \rangle) = H^q(Y, \Omega^p_Y \langle D \rangle \otimes f_* \mathcal{O}_Z)$

(1.7)
Notons $D = \sum B_k + \sum \nu_j E_j$ la décomposition de D en composantes irréductibles de multiplicités 1 et ν_j . On pose $B = \sum B_k$, $E = \sum \nu_j E_j$, $M = \mathcal{O}(B)$ et on suppose qu'il existe un faisceau inversible L tel qu'une puissance positive N-ième vérifie $M = L^N \otimes \mathcal{O}(-E)$. L'inclusion $L^{-N} \hookrightarrow \mathcal{O}_Y$ correspondante à D définit sur le

faisceau de modules $\overset{N-1}{\underset{0}{\oplus}} L^{-i}$ une structure de $\mathcal{O}_Y$-algèbre. On considère par la suite des bons revêtements $((Z,D'),Y',(Y,D))$ pour lesquels Y' est la normalisée de $\mathrm{Spec}_Y(\overset{N-1}{\underset{0}{\oplus}} L^{-i})$. Un tel bon revêtement $f : Z \to Y$ est dit <u>extraction</u> N-<u>ième de</u> D. Le groupe de Galois de Y' sur Y est alors le groupe cyclique d'ordre N. Une racine primitive N-ième de l'unité, e, définit un automorphisme semi-simple sur $f_*\mathcal{O}_Z = \tau_*\mathcal{O}_{Y'}$, compatible à l'inclusion $\overset{N-1}{\underset{0}{\oplus}} L^{-i} \hookrightarrow \tau_*\mathcal{O}_{Y'}$. Ainsi $\tau_*\mathcal{O}_{Y'}$ admet une décomposition en somme directe de sous-faisceaux propres associés aux valeurs propres e^i, dont on peut supposer qu'ils contiennent L^{-i}, et qui sont inversibles. Appelons-les $L^{(i)^{-1}}$. Pour chaque nombre réel x, dénotons par $[x]$ sa partie entière.

<u>Lemme 1.8</u> [4] - <u>Soit</u> $f : Z \to Y$ <u>une extraction</u> N-<u>ième de</u> D.
<u>Alors</u> $f_*\mathcal{O}_Z = \overset{N-1}{\underset{0}{\oplus}} L^{(i)^{-1}}$ <u>où</u> $L^{(i)^{-1}} = L^i \otimes \mathcal{O}(-[\nu_j i/N]E_j)$.

En général, l'identification de la filtration par le poids des $\Omega_Z^p\langle D'\rangle$ en fonction de certaines filtrations sur la base Y est difficile. (Voir le cas des surfaces au paragraphe 3). Pour le terme W_0, on a des renseignements plus précis.

<u>Lemme 1.9</u> [7] - <u>Soit</u> $f : Z \to Y$ <u>une extraction</u> N-<u>ième de</u> D.
<u>Alors</u>,

i) <u>on a une inclusion</u>

$$f_*\Omega_Z^p \hookrightarrow \Omega_Y^p + \overset{N-1}{\underset{1}{\oplus}} \Omega_Y^p\langle D\rangle \otimes L^{(i)^{-1}} .$$

ii) <u>Si</u> D <u>est un diviseur lisse, donc en particulier</u> $E = 0$ <u>et</u> $L^{(i)} = L^i$, <u>cette inclusion est un isomorphisme</u>.

La démonstration de i) est donnée dans [7]. Pour ii), il suffit de remarquer qu'alors f est affine, D est isomorphe à D' et donc que

$$f_*\Omega_Z^p = \mathrm{Ker}(\overset{N-1}{\underset{0}{\oplus}} \Omega_Y^p\langle D\rangle \otimes L^{(i)^{-1}} \to \Omega^{p-1}{}_D) .$$

<u>Théorème 1.10</u>. <u>Soit</u> $f : Z \to Y$ <u>une extraction</u> N-<u>ième de</u> D.

<u>Alors</u>, <u>pour</u> $1 \le i \le N-1$

i) $h^p(L^{(N-i)^{-1}}) \le h^0(\Omega_Y^p\langle D\rangle \otimes L^{(i)^{-1}})$

ii) <u>Si</u> D <u>est lisse, alors</u>

$h^q(\Omega_Y^p \langle D\rangle \otimes L^{-i}) = h^p(\Omega_Y^q \langle D\rangle \otimes L^{-N+i})$

Démonstration - ii) On peut choisir Z de telle sorte que $Z = \mathrm{Spec}(\overset{N-1}{\underset{0}{\oplus}} L^{-i})$. Une racine primitive N-ième e de l'unité définit un automorphisme de Z, qui fait de $f_* \Omega_Z^p$ (resp. $H^{p+q}(Z,\mathbb{C})$) un $\langle e\rangle$ -faisceau (resp. un $\langle e\rangle$ -module), dont on désigne par $(\ )_i$ le sous-faisceau (resp. sous-module) propre associé à e^i. On a $(f_* \Omega_Z^p)_i = \Omega_Y^p \langle D\rangle \otimes L^{-i}$. Par conjugaison, $\overline{e^i \text{ et } e^{N-i}}$ sont échangés, donc de même $(H^{p+q}(Z,\mathbb{C}))_i$ et $(H^{p+q}(Z,\mathbb{C}))_{N-i}$. Donc $\overline{(H^q(Z,\Omega_Z^p))_i} = (H^p(Z,\Omega_Z^q))_{N-i}$ et $\overline{H^q(Y,\Omega_Y^p \langle D\rangle \otimes L^{-i})} = H^p(Y,\Omega_Y^q \langle D\rangle \otimes L^{-N+i})$

i) Bien que n'opérant que birationnellement sur Z, e opère sur les groupes de cohomologie $H^o(\Omega_Z^p)$ et $H^p(\mathcal{O}_Z)$, indépendants du modèle Z choisi. Donc $\overline{H^o(\Omega_Z^p))_i} = (H^p(\mathcal{O}_Z))_{N-i}$ et (1.9 i)) $h^p(L^{(N-i)^{-1}}) \leq h^o(\Omega_Y^p \langle D\rangle \otimes L^{(i)^{-1}})$.

§2. THEOREMES D'ANNULATION

Théorème 2.1 [7] - Soit $f : Z \to Y$ une extraction N-ième de D. Si la dimension de Kodaira de $L^{(N-i)}$ vérifie $\kappa(L^{(N-i)}) = \dim Y = n$, alors $H^p(Y,L^{(i)^{-1}}) = 0$ pour $p < n$.

Démonstration - C'est une conséquence directe de (1.10 i)) et de l'annulation dans ce cas de $H^o(\Omega_Y^p \langle D\rangle \otimes L^{(N-i)^{-1}})$, que F. Bogomolov [2] prouve en utilisant que les formes différentielles globales à pôles logarithmiques sont fermées.

Remarque - De (2.1), on tire la forme classique du théorème d'annulation de Kodaira lorsque le diviseur D est ample.

En raisonnant par récurrence sur la dimension de Y et en utilisant (2.1), on obtient le

Lemme 2.2 [7] - Soient K et M deux faisceaux inversibles sur Y tels que pour toute courbe C sur Y, la première classe de Chern de M vérifie $c_1(M).C \geq 0$. Il existe alors des réels a_q strictement positifs vérifiant

$h^q(K \otimes M^m) \leq a_q . m^{n-q}$ pour $m > 0$.

A l'aide de (2.1) et (2.2) on peut démortrer le résultat principal de ce paragraphe.

Théorème 2.3 [7] - Soit M un faisceau inversible sur Y de dimension n, dont la première classe de Chern vérifie $c_1(M)^n > 0$ et $c_1(M).C \geq 0$ pour toute courbe C sur Y. Alors $H^q(Y,M^{-1}) = 0$ pour $q < n$.

Démonstration - D'après (2.2) et le théorème de Riemann-Roch, M vérifie $\kappa(M) = \dim Y = n$. Un faisceau inversible ample H étant donné, il existe une puissance $N_1 > 0$ telle que $M^{N_1} = H \otimes \mathcal{O}(\sum \nu_j E_j)$, pour un diviseur effectif $\sum \nu_j E_j$ dont on peut supposer, après éclatement éventuel de Y, que c'est un diviseur à croisements normaux. En fait, pour une suite d'éclatements $\rho : X \to Y$, $\rho^* H^m \otimes \mathcal{O}(-F)$ est ample pour m grand et F un diviseur contenu dans le lieu exceptionnel de ρ. Cela dit, pour chaque $N_2 > 0$, $M^{N_2} \otimes H$ est ample et pour N_3 grand, on peut poser $(M^{N_2} \otimes H)^{N_3} = \mathcal{O}(B)$ pour un diviseur B régulier tel que $B + \sum N_3 \nu_j E_j$ soit un diviseur à croisements normaux. Il suffit alors d'appliquer (2.1) à $M = L$, $N = (N_1+N_2)N_3$ et de prendre N_2 suffisamment grand afin que $[N_3\nu_j/N] = 0$ et donc $L^{(1)^{-1}} = M^{-1}$.

§3. FIBRE DE MILNOR D'UN CONE SUR UNE COURBE PLANE

Soit C une courbe plane réduite d'équation f et X la fibre de Milnor correspondante : $X = \{(x,y,z) \in \mathbb{C}^3 / f(x,y,z) = 1\}$. Nous montrons ici brièvement comment utiliser le paragraphe 1 dans l'étude des invariants topologiques suivants de X : nombres de Betti $b_k(X)$, rang, signature de la matrice intersection sur $H^2(X,\mathbb{C})$. Ceux-ci sont exprimables en termes du degré N, du nombre de composantes r et d'invariants locaux de C que l'on définit en liaison avec la structure de Hodge mixte portée par les $H^k(X,\mathbb{C})$. Ces derniers sont lisibles sur la désingularisation plongée de C au voisinage d'un point singulier. Nous mentionnons aussi comment à notre sens on pourrait utiliser le paragraphe 1 dans la recherche de courbes planes C pour lesquelles, degré, nombre de composantes et singularités locales étant fixées, $b_1(X)$ et donc aussi le groupe fondamental du complémentaire de C dans $\mathbb{P}^2$, "sautent" en fonction de la position des singularités, c'est-à-dire, en d'autres termes, d'exemple "à la Zariski".

(3.1) - Soit $\sigma : Y \to \mathbb{P}^2$ la désingularisation plongée de C. On a $\sigma^* C = D = \sum_1^r B_k + \sum \nu_j E_j$, où le diviseur $B = \sum_1^r B_k$ est la normalisation de C et D est à croisements normaux. On pose $L = \sigma^* \mathcal{O}(1)$ et Y' la normalisation de $\mathrm{Spec}_Y(\bigoplus_0^{N-1} L^{-i})$ où la structure de $\mathcal{O}_Y$-algèbre du faisceau de modules $\bigoplus_0^{N-1} L^{-i}$ est donnée par la section de L^N correspondante à D. On construit alors comme dans (1.7) une extraction N-ième de D, $f : Z \to Y$. En fait, Z est une compactification lisse de la fibre de Milnor X dont le bord $D' = Z - X$ est un diviseur à croisements normaux.

roposition 3.2 [4] - On a $b_2(X) = b_1(X) + (N-1)^3 - N \sum \mu_p$, où μ_p est le nombre e Milnor de C au point p .

émonstration - On applique le théorème suivant de P. Deligne [3 , page 35] . Il xiste une suite spectrale de terme $E_1^{pq} = H^q(Z,\Omega_Z^p < D' >)$ qui converge vers $^{p+q}(X,\mathbb{C})$ et dégénère en E_1 . Au regard de (1.6) et (1.8) on obtient donc $_2(X) = \sum_{p+q=2} \sum_0^{N-1} h^q(\Omega_Y^p < D > \otimes L^{(i)^{-1}})$. De plus $b_k(X) = 0$ si $k \geq 3$. On applique e théorème de Riemann-Roch. Il faut identifier $h^0(\Omega_Y^1 < D >)$, ce que l'on fait à 'aide de la suite exacte courte

$$0 \longrightarrow \Omega_Y^1 \longrightarrow \Omega_Y^1 < D > \longrightarrow n_* \mathcal{O}_{\tilde{D}} \longrightarrow 0 \text{ , où } \quad n : \tilde{D} \longrightarrow D_{réd}$$

st la normalisation du diviseur réduit $D_{réd}$; ce qui donne $h^0(\Omega_Y^1 < D >) = r-1$. n obtient la formule

$$b_2(X) = b_1(X) + (N-1)^3 + N/2. \sum (\nu_j - a_j - 1)E_j \sum (\nu_j - 1)E_j - \sum \delta_p$$

ù $\sigma^* \mathcal{O}(-3) \otimes \mathcal{O}(\sum a_j E_j)$ est le diviseur canonique de Y et δ_p le conducteur e C en p défini par la suite exacte

$$0 \longrightarrow \mathcal{O}_C \longrightarrow \sigma_* \mathcal{O}_B \longrightarrow \sum \mathbb{C}^{\delta_p} \longrightarrow 0 .$$

eci permet de conclure en utilisant la formule de N.A'Campo [1] exprimant le nomre de Milnor de C en p en fonction des ν_j .

Pour décrire la structure de Hodge mixte sur $H^2(X,\mathbb{C})$, on applique le théorème e P. Deligne [3 , page 38] . Il existe une suite spectrale de terme $_1^{-n,k+n} = H^k(Z,Gr_n^W \Omega_Z^p < D' >)$ qui dégénère en E_2 et converge vers $H^k(Z,\Omega_Z^p < D' >)$. e qui donne ici, en utilisant la suite exacte

$$0 \longrightarrow \Omega_Z^1 \longrightarrow \Omega_Z^1 < D' > \longrightarrow n_* \mathcal{O}_{\tilde{D}'} \longrightarrow 0 \text{ , où}$$

$: \tilde{D}' \longrightarrow D'_{réd}$ est la normalisation du diviseur réduit $D'_{réd}$:

3.3)

$$_0 H^2(X,\mathbb{C}) = H^0(\Omega_Z^2) + H^1(\Omega_Z^1)/\mathrm{Im}(H^0(n_* \mathcal{O}_{\tilde{D}'}) \rightarrow H^1(\Omega_Z^1)) + H^2(\mathcal{O}_Z)$$

$$ir_1{}^W H^2(X,\mathbb{C}) = \mathrm{Ker}(H^0(Gr_1{}^W \Omega_Z^2 < D' >) \rightarrow H^1(\Omega_Z^2)) + \mathrm{Ker}(H^1(n_* \mathcal{O}_{\tilde{D}'}) \rightarrow H^2(\Omega_Z^1))$$

$$ir_2{}^W H^2(X,\mathbb{C}) = \mathrm{Ker}(H^0(Gr_2{}^W \Omega_Z^2 < D' >) \rightarrow H^1(Gr_1{}^W \Omega_Z^2 < D' >))$$

3.4) Le terme $H^{11} = H^1(\Omega_Z^1)/\mathrm{Im}(H^0(n_* \mathcal{O}_{\tilde{D}'}) \rightarrow H^1(\Omega_Z^1))$ de $W_0 H^2(X,\mathbb{C})$ est inscrit ans la suite exacte

$$0 \longrightarrow H^{11} \longrightarrow H^1(\Omega_Z^1 < D' >) \longrightarrow H^1(n_* \mathcal{O}_{\tilde{D}'}) \longrightarrow H^2(\Omega_Z^1) \longrightarrow 0$$

De même le terme $H^{10} = \mathrm{Ker}(H^1(n_* \mathcal{O}_{\tilde{D}'}) \to H^2(\Omega^1_Z))$ de $Gr_1^W H^2(X,\mathbb{C})$ est inscrit dans la suite exacte

$$0 \longrightarrow H^{10} \longrightarrow H^1(n_* \mathcal{O}_{\tilde{D}'}) \longrightarrow H^2(\Omega^1_Z) \longrightarrow 0 .$$

De sorte qu'il suffit d'identifier $H^1(n_* \mathcal{O}_{\tilde{D}'})$ pour obtenir la W-filtration de $H^2(X,\mathbb{C})$.

(3.5) Posons $d_j = \langle N,\nu_j \rangle$ le dénominateur commun à N et ν_j et $d_{jk} = \langle N,\nu_j,\nu_k \rangle$ celui à N, ν_j et ν_k . Introduisons les invariants suivants.

$\beta_1 = \sum (d_j-1)E_j \sum(\nu_j-1)E_j$

$\beta_2 = \sum (d_j-1)(E_j^2+2)$

$\beta_3 = \sum_{j<k} E_j.E_k(d_{jk}-1) - \sum \varepsilon_{B.E_j} B_j$

avec

$\varepsilon_{B.E_j} = 1$ si $B.E_j = 0$ et 0 sinon

B_j = cardinal $\{i/1 \leq i \leq d_j-1$ et $i\nu_k/d_j$ est entier pour tout k tel que $E_j.E_k \neq 0\}$.

<u>Lemme 3.6</u> 4 - <u>Avec les notations de (3.5), on a</u>

$h^1(n_* \mathcal{O}_{\tilde{D}'}) = r-1 + (N-1)(N-2)/2 - \sum \delta_p - (\beta_1+\beta_2)/2 - \beta_3$.

<u>Remarque</u> - En fait, les composantes exceptionnelles de $d : Z \to Y'$ ne jouent aucun rôle pour ce qui concerne la structure de Hodge mixte de $H^2(X,\mathbb{C})$. Elles n'apparaissent ni dans le H^{11} de $W_0H^2(X,\mathbb{C})$ puisqu'elles sont à la fois dans $H^1(\Omega^1_Z)$ et dans $\mathrm{Im}(H^0(n_* \mathcal{O}_{\tilde{D}'}) \longrightarrow H^1(\Omega^1_Z))$, ni dans $H^1(n_* \mathcal{O}_{\tilde{D}'})$.

<u>Démonstration</u> - Le terme $H^1(\mathcal{O}_B)$ étant connu, il suffit en fait d'évaluer quel type de revêtement de la composante E_k donne l'extraction N-ième de D . Pour cela, on remarque qu'à une ramification totale près, on extrait la racine d_k-ième du diviseur $B + \sum_{j\neq k} \nu_j E_j$. On applique alors (1.8), puis le théorème de Riemann-Roch aux faisceaux obtenus. Les invariants β_i introduits permettent d'exprimer la trivialité de ces faisceaux qui sont négatifs.

(3.7) On définit sur la cohomologie à support compact $H^2_c(X,\mathbb{C})$ la matrice intersection q de la façon usuelle, dont rang et signature s'expriment en fonction des nombres de Hodge sur les gradués $Gr^W_n H^2(X,\mathbb{C})$ [6] .

Théorème 3.8 [4] - On a

$$\dim W_0H^2(X,\mathbb{C}) = \text{rang } q = (N-1)(N^2-3N+3) + 2b_1(X) - 2(r-1) - (N-1)\sum \mu_p+(\beta_1+\beta_2+\beta_3)$$

$$\text{signature } q = -(N-1)(N^2+N-3)/3 + (N-1)\sum \mu_p + 2\sum_1^{N-1}(\sum [\nu_j i/N]E_j)^2 +$$

$$+ (N-1)\sum(\nu_j-1)(E_j^2+2) - \beta_1 - \beta_3 .$$

Remarque - On voit donc que tous les invariants topologiques calculés dépendent, outre de N, r et d'invariants locaux de C, du premier nombre de Betti $b_1(X)$. Ce dernier est l'objet de ce qui suit.

Proposition 3.9 [4] - Si la dimension de Kodaira de $L^{(i)}$ vérifie $\kappa(L^{(i)} = 2$ pour $1 \leq i \leq N-1$, alors $b_1(X) = r-1$.

Démonstration - On applique (1.6), (1.8), (2.1), puis il faut calculer $h^0(\Omega_Z^1 < D' >)$ en fonction de $h^0(\Omega_Z^1)$, ce que l'on fait en calculant le nombre de composantes algébriquement indépendantes à support dans D', qui donnent une contribution non nulle à $\text{Im}(H^0(n_* \mathcal{O}_{\tilde{D}'}) \longrightarrow H^1(\Omega_Z^1))$. On a $h^0(\Omega_Z^1 < D' >) = h^0(\Omega_Z^1) + r-1$.

Lemme 3.10 [4] - En particulier, si tous les ν_j sont premiers à N, alors $b_1(X) = r-1$.

Remarque - C'est aussi une conséquence de la dualité de Serre appliquée à $H^2(\Omega_Y^1 < D > \otimes L^{(i)^{-1}}) = 0$.

Sous les hypothèses de (3.10), on obtient une forme particulièrement simple de rang q.

(3.11) En général on a $b_1(X) = r-1 + 2\sum_1^{N-1} h^1(L^{(i)^{-1}})$. En appliquant la suite spectrale de Leray, on trouve

$$h^1(L^{(i)^{-1}}) = \text{Coker}(\sum \mathbb{C}^{m_p} \longrightarrow H^0(\mathcal{O}(i-3))) \text{, avec}$$

$$-m_p = [\nu_j i/N] + \sum [\nu_j i/N]([\nu_j/N]+1)/2.E_j^2 + \sum_{j<k} [\nu_j i/N][\nu_k i/N]E_j.E_k ,$$

cette somme étant évaluée sur les E_j au-dessus du point singulier p.
En particulier, $b_1(X)$ dépend du nombre de courbes de degré i-3, pour $1 \leq i \leq N-1$, passant par les points p avec la multiplicité m_p.

(3.12) Prenons l'exemple de Zariski. C'est une courbe de degré 6 avec 6 cusps comme singularités. On a

$$R^1\sigma_* L^{(i)^{-1}} = 0 \text{ pour } 1 \leq i \leq 4 \text{ et } (R^1\sigma_* L^{(5)^{-1}})_p = \mathbb{C} .$$

Donc $h^1(L^{(5)^{-1}}) = \mathrm{Coker}(\sum_1^6 \mathbb{C} \longrightarrow H^0(\mathcal{O}(2))$ et $b_1(X) = 0$ si les 6 cusps ne sont pas une conique

$= 2$ s'ils le sont

Références

[1] A'Campo, N — La fonction zêta d'une monodromie, Comment. Math. Helvetici, 50, (1975), 233-248.

[2] Bogomolov, F — Unstable vector bundles and curves on surfaces, Proc. Int. Congress of Maths, Helsinki, (1978), 517-524.

[3] Deligne, P — Théorie de Hodge II, Pub. Math. I.H.E.S., 40, 5-57.

[4] Esnault, H — Fibre de Milnor d'un cône sur une courbe plane singulière, manuscrit.

[5] Kawamata, Y — A generalization of Kodaira-Ramanujam's Vanishing theorem, Manuscrit.

[6] Steenbrink, J — Intersection form for quasi-homogeneous singularities, Comp. Math., 34, fasc. 2 (1977)

[7] Viehweg, E — Vanishing theorems. Manuscrit .

[8] Viehweg, E — Rational singularities of higher dimensional schemes, Proc. of the A.M.S., 63, (1977).

Hèlène Esnault
Université de Paris VII
U.E.R. de Mathématiques
Tour 45-55, 5ème étage
2, Place Jussieu
75251 Paris Cedex 05
France

Eckart Viehweg
Institut für Mathematik
und Informatik
A 5, Seminargebäude
D-68 Mannheim
République Fédérale Allemande

Some questions on the canonical ring of threefolds of general type.

P.M.H.Wilson, Department of Pure Mathematics,
University of Cambridge,
16 Mill Lane, Cambridge, England.

Introduction.

This paper is a sequel to the paper [13], in which we considered the problem of whether the canonical ring $R(V) = R(V,K_V) = \bigoplus_{n\geq 0} H^0(V,nK_V)$ of a compact complex algebraic manifold V is finitely generated or not as an algebra over $k=\mathbb{C}$. We shall summarize some of the results of [13], omitting proofs, in Section 1 of this paper. In particular, for threefolds of general type we see that one cannot hope for a simple counterexample with R(V) not finitely generated, i.e. an example where for n sufficiently large, the fixed components of $|nK_V|$ are isolated.

For threefolds of general type, one is led to consider the case when K_V is arithmetically effective (denoted a.e.); i.e. where $K_V \cdot C \geq 0$ for all curves C on V. It would be a major advance to know whether the canonical ring is finitely generated in this case. By the results of Section 1, this is so if and only if $|nK_V|$ has no fixed points for some positive integer n.

The remainder of this paper is devoted to studying the case when V is a smooth complex algebraic threefold of general type for which the m-canonical system $|mK_V|$ has no fixed components for some positive integer m. This is in fact a special case of K_V being a.e., and so we wish to know whether there exists a positive integer n for which $|nK_V|$ is base point free.

Following a suggestion of Bombieri, Reid in [10] asked whether we could construct a threefold V for which the canonical ring was not finitely generated by arranging for there to be a single base curve C of $|mK_V|$ for some m, such that $\mathcal{O}_C(K_V)$ is a non-torsion element of $\mathrm{Pic}^0(C)$. In view of the Zariski construction of a divisor D on a surface S for which the ring R(S,D) is not finitely generated (see [15]), this at first sight might seem plausible.

We note in passing the result of Zariski for D any divisor on a smooth surface, that if $|D|$ has no fixed components, then $|nD|$ is base point free for some positive n.

We shall see in Section 2 of this paper that if $|mK_V|$ has no fixed components, we need only consider those base curves C for which $K_V \cdot C = 0$. Moreover, any base curve C with $K_V \cdot C = 0$ has arithmetic genus 0 or 1. For such a curve, we would like to be able to find an integer n for which C was not a base curve of $|nK_V|$. If this is possible for all such curves, then it follows easily that $|NK_V|$ is without fixed point for some N.

In Section 3, we show that if C is a smooth base curve of $|mK_V|$ with $K_V \cdot C = 0$, then C is isomorphic to $\mathbb{P}^1$. In particular we have the following result.

Theorem. Suppose V is a non-singular complex projective threefold of general type such that for some positive m, the linear system $|mK_V|$ has no fixed components. If C is a base curve of $|mK_V|$ with $K_V \cdot C = 0$, then C is rational.

This result has been proved independently by Henry Pinkham. Note that if C were singular, then it would be isomorphic to a nodal or cuspidal cubic in $\mathbb{P}^2$. Using similar methods as in the non-singular case (i.e. Section 3), it is believed that the singular case can be shown not to occur, and thus the proposed counterexample of [10] will not then exist.

In Section 4, we return to the question of whether a base curve of $|mK_V|$ remains a base curve of $|nK_V|$ for all n. In particular, we study the case when C does not meet any other base curves of the linear system. Here we shall wish to use the recently proved Kawamata-Viehweg form of Kodaira vanishing.

The author would like to thank Eckart Viehweg, Shigefumi Mori and Henry Pinkham for the benefit of various written and verbal communications.

1. General results.

Let D be a divisor on a compact complex algebraic manifold V. Following Zariski [15], we define the ring $R(V,D) = \bigoplus_{n\geq 0} H^0(V,nD)$, which may or may not be finitely generated as an algebra over $k=\mathbb{C}$. In any case, we can define the D-dimension of V by $\kappa(V,D) =$ tr. $\deg._k R(V,D)$ if the right hand side is non-negative, and $\kappa(V,D) = -\infty$ otherwise. The ring R(V,D) is of particular interest when $\kappa(V,D) = d = \dim(V)$.

As is well known, R(V,D) is not always finitely generated, and in the case of surfaces the question was investigated by Zariski in [15]. We can moreover translate the condition that R(V,D) is finitely generated into a geometric criterion concerning the fixed locus of $|nD|$ for n large.

Theorem 1.1. Let D be a divisor on a compact algebraic manifold V, where $\kappa(V,D) = d = \dim(V)$. The ring R(V,D) is finitely generated if and only if there exists a positive integer n and a smooth modification $f : \tilde{V} \to V$ such that the map ϕ_{nf^*D} on $\tilde{V}$ (defined by the linear system $|nf^*D|$) is a birational morphism which contracts to a lower dimension every fixed component of $|nf^*D|$.

Proof. See (1.2) of [13].

In the case when $D=K_V$, we see that the canonical ring of an algebraic manifold V of general type is finitely generated if and only if there is a positive integer n and a birationally equivalent smooth model $\tilde{V}$ on which $\phi_{nK_{\tilde{V}}}$ is a birational morphism contracting every fixed component of $|nK_{\tilde{V}}|$. Using this criterion, we can prove :

Theorem 1.2. Let V be a smooth threefold of general type, and suppose that for n sufficiently large, the mobile part of $|nK_V|$ is base point free and the fixed components are normal and disjoint. Then the canonical ring R(V) is finitely generated.

Proof. See (5.2) of [13].

This result may be contrasted with the Zariski example of a divisor D on a surface S with R(S,D) not finitely generated [15]. Here $\kappa(S,D) = \dim(S) = 2$, and for all $n \geq 1$ the mobile part of $|nD|$ is very ample and there is just one (smooth) fixed component. The example depends on the observation that if we have a fixed component E of $|nD|$ with positive but bounded multiplicity as $n \to \infty$, then R(V,D) is not finitely generated (see [15], page 562). We may also contrast (1.2) with the examples given in Section 4 of [13]. The first example is of a normal Gorenstein threefold V for which $R(V,K_V)$ is not finitely generated. One should remark here that this example has $\kappa(V,K_V)=2$, not $\kappa(V,K_V)=3$ as stated in [13]. The second example is a non-algebraic compact complex manifold for which the canonical ring $R(M) = \bigoplus_{n \geq 0} H^0(M,\omega_M^{\otimes n})$ is not finitely generated. This example has $a(M) = 2 = \kappa(M)$, where a(M) denotes the algebraic dimension of M. Both these examples utilize the idea in the Zariski example mentioned above.

In the case of algebraic surfaces, it is well known that the canonical ring is finitely generated. The proofs of this fact all require the existence of a smooth model on which the canonical divisor is a.e. Unfortunately in dimensions 3 and higher, such a smooth model does not always exist. However, it is reasonable to ask what may be said if such a model does exist. This is all the more reasonable for threefolds on general type in view of the recent results of Mori concerning the case when the canonical divisor is not a.e. (see [8]).

Let us consider the general case when V is any compact complex algebraic variety of dimension d, and D is an arithmetically effective Cartier divisor on V. We say that the fixed locus of $|nD|$ is numerically bounded if for every birational morphism $f : \tilde{V} \to V$, the fixed components of $|nf^*D|$ have bounded multiplicities as $n \to \infty$. The following result gives a suitable generalization to higher dimensions of Theorem 10.1 of [15].

Theorem 1.3. If the fixed locus of $|nD|$ is numerically bounded, then D is a.e. Furthermore, if $\kappa(V,D) = \dim(V)$, then the converse holds.

Proof. See (2.2) of [13].

Corollary 1.4. If D is a.e. with $\kappa(V,D) = \dim(V)$, then the ring $R(V,D)$ is finitely generated if and only if $|nD|$ has no fixed points for some positive integer n.

Proof. The only if part follows from (1.3) and the observation on page 562 of [15]. The if part is a special case of (1.1). □

Using the above and the results of Mori [8], it is easy to deduce the following fact in dimension three.

Theorem 1.4. Let V be a smooth complex projective threefold of general type. The following conditions are equivalent :

(a) K_V is a.e.

(b) The fixed locus of $|nK_V|$ is numerically bounded.

(c) V is relatively minimal and $h^2(nK_V)=0$ for all $n>1$.

If the canonical ring is finitely generated, then we have a further equivalent condition :

(d) $|nK_V|$ has no fixed points for some positive n.

Proof. See (6.2) of [13].

This last result may be compared with the result for a surface S of general type that the following conditions are equivalent : (i) K_S is a.e. (ii) $|nK_S|$ has no fixed points for n sufficiently large. (iii) S is minimal. (iv) $h^1(nK_S)=0$ for $n>1$.

From (1.4) we have seen that for a smooth threefold of general type with K_V a.e., any part of the fixed locus of $|nK_V|$ will appear with bounded multiplicity. This in turn yields useful numerical conditions. For the case of fixed components, compare with (6.3) of [13]. We shall deal in the remainder of this paper with the case when for some positive integer m, the linear system $|mK_V|$ has no fixed components.

2. Preliminaries on case with no fixed components.

Throughout the rest of this paper V will denote a smooth complex projective threefold of general type, and m a positive integer such that $|mK_V|$ has no fixed components.

Lemma 2.1. K_V is arithmetically effective.

Proof. If there exists a curve C on V with $K_V \cdot C < 0$, then by the theory of Hilbert schemes (see for instance [7]), we deduce that C moves in an algebraic family. This then would yield a fixed component of $|mK_V|$ (one can of course say more ; see [8]). □

Proposition 2.2. If $K_V \cdot C > 0$ for all base curves C of $|mK_V|$, then for some n the linear system $|nK_V|$ is base point free. If C is a base curve with $K_V \cdot C = 0$, then C has arithmetic genus $p_a(C) \leq 1$.

Proof. Let D, D' $\in |mK_V|$ be general elements, and $Z = D'|D$ denote the scheme theoretic intersection. For $n \geq 1$, we have the exact sequence of sheaves

$$0 \to \mathcal{O}_D(nD) \to \mathcal{O}_D((n+1)D) \to \mathcal{O}_Z((n+1)D) \to 0.$$

From the Kawamata-Viehweg form of Kodaira vanishing ([6] or [11]), we see that $h^i(nD) = 0$ for $i>0$ and $n>1$. Thus from the exact sequence

$$0 \to \mathcal{O}_V(nD) \to \mathcal{O}_V((n+1)D) \to \mathcal{O}_D((n+1)D) \to 0$$

we deduce that $h^i(\mathcal{O}_D(nD)) = 0$ for $i>0$ and $n>2$. Thus from the former exact sequence, we see that $h^1(\mathcal{O}_Z(nD)) = 0$ for $n>3$.

Suppose first that C is a base curve with $K_V \cdot C = 0$. We have a surjection $\mathcal{O}_Z(nD) \to \mathcal{O}_C(nD)$ of sheaves supported on Z, and hence we deduce that $h^1(\mathcal{O}_C(nD)) = 0$ for $n>3$. Since $D \cdot C = 0$, we see that $\chi(\mathcal{O}_C) = \chi(\mathcal{O}_C(nD)) \geq 0$. Thus $p_a(C) \leq 1$ as required.

Suppose therefore that $K_V \cdot C > 0$ for all base curves C of $|mK_V|$. We see then that for all curves C underlying Z, we have $D \cdot C > 0$. Hence D is ample on Z (Propositions 4.2 and 4.3 of [4]). Thus for n sufficiently large, $\mathcal{O}_Z(nD)$ is generated by its global sections.

From the first exact sequence and the fact that $h^1(\mathcal{O}_D(nD)) = 0$ for large n, we deduce that $\mathcal{O}_D(nD)$ is also generated by its global sections for n sufficiently large. Hence, using the second exact sequence, we see that the same is also true of the sheaf $\mathcal{O}_V(nD)$. Thus the linear system $|nD|$ is base point free for large n. □

Remark. We may compare the above Proposition with the result that if X is any smooth projective threefold with $\kappa(X) > 0$ and $K_X \cdot C > 0$ for all curves C on X, then K_X is ample (see [12], Proposition 2.3).

We see from the above that it is therefore vital to investigate those base curves C of $|mK_V|$ for which $K_V \cdot C = 0$. First however, let us fix some notation.

Let C be a smooth base curve of $|mK_V|$ with $K_V \cdot C = 0$ and genus g. From (2.2), g = 0 or 1. From the adjunction formula, we have that $\deg(\Lambda^2 N_{V/C}) = 2g-2$, where $N_{V/C}$ denotes the normal bundle of C in V. We now put $N^{\vee}_{V/C} = L \otimes E$ where L is a line bundle of degree a on C, E a rank 2 vector bundle with $\deg(\Lambda^2 E) = -e$ such that $h^0(C, E) > 0$, but that $h^0(C, E \otimes L') = 0$ for any line bundle L' with $\deg(L') < 0$ (cf. notation of [5], Chapter V, Section 2). Here $N^{\vee}_{V/C}$ denotes the dual of $N_{V/C}$. We therefore have the indentity $e = 2g-2+2a$.

3. Rationality of base curves.

Proposition 3.1. If C is a smooth base curve of $|mK_V|$ with $K_V \cdot C = 0$, then C is isomorphic to $\mathbb{P}^1$.

Proof. We blow up the curve C, say $f_1 : V_1 \to V$, with exceptional surface E_1. Thus, with the notation of Section 2, we obtain $E_1 = \mathbb{P}(N^{\vee}_{V/C}) = \mathbb{P}(\mathcal{E})$, and a morphism $\pi : E_1 \to C$. We have a section C_1 of π with $\mathcal{O}_{E_1}(C_1) \cong \mathcal{O}_{\mathbb{P}(\mathcal{E})}(1)$.

Hence $\mathcal{O}_{E_1}(-E_1) \simeq \mathcal{O}_{\mathbb{P}(N^{\vee})}(1) \simeq \mathcal{O}_{\mathbb{P}(\mathcal{E})}(1) \otimes \pi^* \mathcal{L}$. Therefore $-E_1|E_1 \equiv C_1 + aF_1$, where F_1 denotes a fibre of π and $\equiv$ denotes algebraic equivalence.

The general surface $D \in |mK_V|$ is irreducible, and non-singular outside the base locus (Bertini's theorem). Let the strict transform of D under f_1 be denoted by D_1, and suppose that $f_1^* D = D_1 + r_o E_1$ (i.e. $|D|$ has multiplicity r_o along C).

Since $K_V \cdot C = 0$, we know that $0 \equiv D_1|E_1 + r_o E_1|E_1$. Hence

$$D_1|E_1 \equiv r_o C_1 + a r_o F_1 .$$

I claim first that $e \geq 0$; suppose not, then from the results of [5], Chapter V or [9], we have that $g = 1$, $e = -1$ and $\mathcal{E}$ is indecomposable (since by (2.2) $g \leq 1$). This however is impossible, since $e = 2g-2+2a$.

Thus $e \geq 0$, and so by Proposition 2.20 of [5], Chapter V, and the fact that $D_1|E_1$ is effective, we deduce that $a \geq 0$. Also note that $(D_1|E_1)\cdot C_1 = r_o(C_1+aF_1)\cdot C_1 = r_o(a-e)$. We show in the lemma below that if $e-a > 0$, then we must have $g = 0$. Hence we need only consider the cases when $e-a = 2g-2+a \leq 0$. These yield the following three possibilities :

(a) $g=0$, $a=1$, $e=0$, $N_{V/C} = \mathcal{O}_C(-1)\oplus\mathcal{O}_C(-1)$

(b) $g=0$, $a=2$, $e=2$, $N_{V/C} = \mathcal{O}_C(-2)\oplus\mathcal{O}_C$

(c) $g=1$, $a = e = 0$.

Let us therefore show that the third possibility cannot occur.

Since $a=0$, we have $D_1|E_1 \equiv r_oC_1$. Since $e=0$, we deduce that any fixed curve of $|D_1|$ on E_1 is arithmetically equivalent to sC_1 for some $s>0$, and that the mobile part cuts out on E_1 a curve arithmetically equivalent to tC_1 for some t. Since $C_1^2 = 0$, the mobile part has no fixed points on E_1.

Let B be any fixed curve of $|D_1|$ on E_1 , say $B \equiv sC_1$. Since $C_1^2 = 0$, we see easily that B is a non-singular elliptic curve. Moreover, on B we have ([3],16.2.7) an exact sequence of sheaves

$$0 \to \mathcal{O}_{E_1}(-E_1)\otimes\mathcal{O}_B \to N^{\vee}_{V_1/B} \to \mathcal{O}_B(-B) \to 0 \quad .$$

Thus $N^{\vee}_{V_1/B}$ is an extension of line bundles of degree 0, and so $N^{\vee}_{V_1/B}$ has basic invariants $a'=e'=0$.

Therefore, if we blow up B, we obtain $f' : V' \to V_1$ with exceptional divisor E', minimal section C'. If r' denotes the multiplicity of $|D_1|$ along B, and D' the strict transform of D_1 , we see that $D'|E' \equiv r'C'$, and so we may repeat the argument on E' if necessary.

Eventually we reach a stage where the strict transform of D moves in a linear system which has no fixed points on the exceptional surfaces ; say $g : \tilde{V} \to V$, with $\tilde{D}$ the strict transform of D, and $|\tilde{D}|$ having no fixed points on $g^{-1}(C)$.

In particular we deduce that C does not meet any of the other fixed curves of $|D|$, and hence that $g^{-1}(C)$ does not meet the fixed locus $|\tilde{D}|$. Moreover, the rational map $\phi_{\tilde{D}}$ corresponding to the linear system $|\tilde{D}|$ is birational and contracts every component of $g^{-1}(C)$ to lower dimensions.

We can now apply Proposition 1.1 of [13] to deduce that if $|mg^*K_V| = |\tilde{D}| + Z$, then for all positive n we have $|nmg^*K_V| = |n\tilde{D}| + nZ$. Since $Z > 0$, this gives an immediate contradiction to (1.3) and the fact that K_V is a.e. □

The proof of Proposition 3.1 has now been reduced to proving the following lemma.

Lemma 3.2. With the notation as in (3.1), if $e-a > 0$, then $g = 0$.

Proof. Suppose therefore that $g>0$ and $e>a$. We saw in the proof of (3.1) that $(D_1|E_1)\cdot C_1 = r_o(e-a) < 0$. Thus C_1 is a fixed curve of $|D_1|$, with multiplicity $r_1>0$ say. We also note that $g(C_1) = g$.

As with C, we write $\check{N}_{V_1/C_1} = \mathcal{E}_1 \otimes \mathcal{L}_1$ with $\deg(\mathcal{L}_1) = a_1$ and $\deg(\Lambda^2 \mathcal{E}_1) = -e_1$. We note from [3], 16.2.7 that we have an exact sequence

$$0 \to \mathcal{O}_{E_1}(-E_1)\otimes\mathcal{O}_{C_1} \to \check{N}_{V_1/C_1} \to \mathcal{O}_{C_1}(-C_1) \to 0$$

and hence that $\deg(\check{N}_{V_1/C_1}) = a$. Thus $e_1 = 2a_1-a$. We now blow up C_1 , say $f_2 : V_2 \to V_1$, with exceptional surface E_2, minimal section C_2.

We may then blow up C_2 on V_2 and continue the procedure by induction ; at the i'th step we blow up the curve C_{i-1} on $E_{i-1} \subset V_{i-1}$, say $f_i : V_i \to V_{i-1}$. We show by induction that for all $i \geq 0$, we have $e_i - a_i \geq 0$, $e_i \geq 0$, $a_i \geq 0$ and $r_i > 0$ (where $e_o = e$, $a_o = a$ and r_i denotes the multiplicity of C_i in $|D_i|$).

This being so gives us our required contradiction since the curve C_i on E_i will always be a fixed curve of $|D_i|$ (i.e. we cannot resolve the base locus), which is clearly in contradiction to the results say of [14].

Suppose therefore that the inequalities are true for $i \le k-1$, where $k \ge 1$. I claim that they are the also true for $i=k$.

We blow up the curve C_k, and let D_{k+1} denote the strict transform of D_k under f_{k+1}. If F_{k+1} denotes a fibre of E_{k+1}, it is straightforward to check that

$$D_{k+1}|E_{k+1} \equiv r_k C_{k+1} + \{a_k r_k - r_{k-1}(e_{k-1}-a_{k-1}) - \ldots - r_o(e-a)\}F_{k+1}$$

(noting that $D_i \cdot C_i = r_{i-1}(a_{i-1}-e_{i-1})$ for all $i \ge 1$). Since $D_{k+1}|E_{k+1}$ is effective, $e_i - a_i \ge 0$ for all $i<k$ and $e-a > 0$, we see that $r_k=0$ is impossible.

We now show that $e_k \ge 0$. Suppose this were not so ; then using Proposition 2.21 of [5], Chapter V again, we deduce that

$$2a_k r_k - 2\sum_{i=o}^{k-1} r_i(e_i-a_i) \ge r_k e_k \ .$$

Since $e_k = 2a_k - a_{k-1}$, we deduce that

$$2\sum_{i=o}^{k-1} r_i(e_i-a_i) - r_k a_{k-1} \le 0.$$

Noting now that $r_o \ge r_1 \ge \ldots \ge r_k$, we see that for all $i \ge 1$,

$$r_{i-1}(e_{i-1}-a_{i-1}) - r_i a_{i-1} \ge r_{i-1}(e_{i-1}-2a_{i-1}) = -r_{i-1}a_{i-2} \ .$$

Thus we deduce by induction that

$$\sum_{i=o}^{k-1} r_i(e_i-a_i) - r_k a_{k-1} \ge (e-2a)r_o \ge 0 \quad \text{since } g > 0.$$

This would then imply that

$$\sum_{i=o}^{k-1} r_i(e_i-a_i) \le 0,$$

which contradicts our induction hypotheses taken with the assumption that e-a > 0. We therefore deduce that $e_k \geq 0$.

Since $D_{k+1}|E_{k+1}$ is effective, we can now use Proposition 2.20 of [5], Chapter V to deduce that

$$r_k a_k \geq \sum_{i=0}^{k-1} r_i(e_i - a_i) > 0.$$

Therefore $a_k > 0$ as required.

Finally note that

$$r_k(e_k - a_k) = r_k(a_k - a_{k-1}) \geq \sum_{i=0}^{k-1} r_i(e_i - a_i) - a_{k-1}r_k$$

$$\geq (e-2a)r_o \text{ as above.}$$

Hence we deduce that $e_k - a_k \geq 0$, and the induction is now complete.

Combining the results (2.2) and (3.1), we see now that any base curve of $|mK_V|$ with $K_V \cdot C = 0$ must be rational.

4. Isolated base curves.

Let us now consider the case when C is a smooth base curve of $|mK_V|$ not meeting any other base curve. If $K_V \cdot C = 0$, then by the previous section we know that C is rational. In this case we shall also assume that $a-e \geq 0$, i.e. that $-E_1|E_1$ is a.e. on E_1. Hence the divisor $mf_1^*K_V-E_1$ is a.e. on V_1. If on the other hand $K_V \cdot C > 0$, then we may clearly choose m sufficiently large so that $mf_1^*K_V-E_1$ is a.e. on V_1, and also that $(2m+1)K_V \cdot C > 2g(C)-2$.

Proposition 4.1. With the notation as above, we have that C is not a base curve of $|(2m+1)K_V|$.

Proof. The divisor $mf_1^*K_V-E_1$ is a.e. and it is easily checked that $(mf_1^*K_V-E_1)^3 > 0$. Therefore, by the Kawamata-Viehweg form of Kodaira vanishing ([6] or [11]), we deduce that

$$H^1(V_1, \mathcal{O}_{V_1}(K_{V_1}+2mf_1^*K_V-2E_1)) = 0.$$

But $K_{V_1} = f_1^*K_V+E_1$, and thus we have an exact sequence

$$0 \to H^0((2m+1)f_1^*K_V-E_1) \to H^0((2m+1)f_1^*K_V) \to H^0(\mathcal{O}_{E_1}((2m+1)f_1^*K_V)) \to 0.$$

In the case when $K_V \cdot C = 0$, we know that C is rational, and so $h^0(\mathcal{O}_{E_1}((2m+1)f_1^*K_V)) = h^0(\mathcal{O}_{E_1}) = 1$. In the case when $K_V \cdot C > 0$, we have chosen m sufficiently large so that $h^0(\mathcal{O}_{E_1}((2m+1)f_1^*K_V)) > 0$.

From the above exact sequence, we see therefore that E_1 is not a fixed component of $|(2m+1)f_1^*K_V|$, and hence that C is not a base curve of $|(2m+1)K_V|$ on V. □

In the cases considered above therefore, Theorem 6.2 of [15] now yields that for some positive n, the linear system $|nK_V|$ has no fixed points on C. In particular, if C is the only base curve, then we have that $|nK_V|$ is base point free for some n, and so the canonical ring is finitely generated.

We are left therefore with the question of what happens in the case when $K_V \cdot C = 0$ and $e-a > 0$. With the notation of the previous section, if we know that for all $i \geq 1$ the normal bundle of C_i splits along E_i and the normal direction,

i.e. $N_{V_i/C_i} = (N_{V_i/E_i}|C_i) \oplus N_{E_i/C_i} = \mathcal{O}_{C_i}(e_{i-1}-a_{i-1}) \oplus \mathcal{O}_{C_i}(-e_{i-1})$,

then we would obtain an immediate contradiction. For if such a splitting always occurred, then we could deduce by induction that for all $i \geq 0$, we have $e_i = (i+1)e - ia$, $a_i = ie - (i-1)a$, and $D_{i+1} \cdot C_{i+1} = -(r_o + \ldots + r_i)(e-a) < 0$. Thus $r_i > 0$ for all $i \geq 0$, which as we noted in (3.2) is impossible.

Unfortunately, despite (1.8) of [1], it appears that an example of Laufer provides a case when such a splitting does not occur. It is believed however by the author that it is still true that C is not a base curve of $|nK_V|$ for some integer n.

The general case with $|mK_V|$ having no fixed components will involve further complications, mainly associated with combinatorial type problems involving questions as to which configurations of rational curves can appear in the base locus.

References.

1. A. Fujiki : On the minimal models of complex manifolds. Math. Ann. 253 (1980), 111-128.

2. H. Grauert and O. Riemenschneider : Verschwindungssätz für analytische Kohomologiegruppen auf komplexen Räumen. Inventiones math. 11 (1970), 263-292.

3. A. Grothendieck and J. Dieudonné : Eléments de Géométrie Algébrique IV. Publ. Math. IHES. 1967.

4. R. Hartshorne : Ample subvarieties of algebraic varieties. Lecture Notes in Mathematics 156. Berlin, Heidelberg, New York; Springer 1977.

5. R. Hartshorne : Algebraic Geometry. Graduate Texts in Mathematics 52. Berlin, Heidelberg, New York; Springer 1977.

6. Y. Kawamata : A Generalization of Kodaira-Ramanujam's Vanishing Theorem : preprint.

7. S. Mori : Projective manifolds with ample tangent bundles. Annals of Math. 110 (1979), 593-606.

8. S. Mori : Threefolds whose canonical bundles are not numerically effective. Proc. Nat. Acad. Sci. USA. 77 (1980), 3125-6.

9. M. Nagata : On the self-intersection number of a section on a ruled surface. Nagoya Math. J. 37 (1970), 191-196.

10. M.A. Reid : Canonical 3-folds ; in Journées de Géométrie Algébrique, Juillet 1979, edited by A. Beauville. Sijthoff & Noordhoff 1980.

11. E. Viehweg : Vanishing theorems; appearing ibid.

12. P.M.H. Wilson : On complex algebraic varieties of general type. Symposia Math. XXIV (1981), 65-73.

13. P.M.H. Wilson : On the canonical ring of algebraic varieties ; to appear in Compositio Math. (1981).

14. O. Zariski : Resolution of the singularities of algebraic three dimensional varieties. Ann. of Math. 45 (1944), 472-547.

15. O. Zariski : The theorem of Riemann-Roch for high multiples of an effective divisor on an algebraic surface. Ann. of Math. 76 (1962), 560-615.

Deformations of Irregular Threefolds

by Marc Levine*
University of Pennsylvania

For purely topological reasons, the classification of curves by Kodaira dimension is stable under smooth deformation. In addition, it is quite easy to show that the pluri-genera remain constant under smooth deformation. The analogous facts for compact complex surfaces were proved by Iitaka [4], using much of the information provided by the classification of surfaces.

Here, we consider the generalizations of these facts to the case of threefolds. As the classification of the regular threefolds is still mainly a mystery, we will restrict our attention to the irregular threefolds. For technical reasons, we will also restrict ourselves to projective threefolds and projective deformations.

Let V be a smooth threefold with positive irregularity $q(V) = h^1(V,\mathcal{O}_V)$, and Kodaira dimension zero, embedded in a projective space $\mathbb{P}^N_{\mathbb{C}}$. Let V' be a smooth deformation of V in $\mathbb{P}^N_{\mathbb{C}}$. We will show that V' also has Kodaira dimension zero and in fact the pluri-genera of V and V' are the same (Corollary 10). We then use results of Fujiki [2], Ashikaga and Ueno [1], and ourselves [7], to show that the Kodaira dimension of an irregular threefold W in $\mathbb{P}^N_{\mathbb{C}}$ is invariant under smooth deformation in that projective space (Corollary 11).

We use the following notations and conventions:

We fix an algebraically closed field k of characteristic zero. Unless mentioned otherwise, all scheme, morphisms and rational maps are over k. We also fix a discrete valuation ring $\mathcal{O}$ with quotient field K and residue field k.

---

*Partially supported by NSF Grant

A polarized variety is a pair (X,D) consisting of a smooth complete variety X and an ample divisor D on X, where divisors D and D' give equivalent polarizations on X if there are positive integers n and n' such that nD and $n'D'$ are numerically equivalent on X. If P is the polarization given by an ample divisor D on X, we let $\mathrm{Aut}_p(X)$ denote the group of polarized automorphisms of X. These notions can be relativised in the obvious fashion.

If $f: V \to \mathrm{Spec}(\mathcal{O})$ is a smooth projective fiber space, $\pi: \mathrm{Alb}(V/\mathcal{O}) \to \mathrm{Spec}(\mathcal{O})$ will denote the relative Albanese scheme of $f: V \to \mathrm{Spec}(\mathcal{O})$. If f admits a section $s: \mathrm{Spec}(\mathcal{O}) \to V$, we let $a_V: V \to \mathrm{Alb}(V/\mathcal{O})$ denote the associated Albanese morphism.

<u>Proposition 1</u>. Let V and S be smooth varieties and let $f: V \to S$ be a fiber bundle (in the complex topology) with smooth complete connected fiber F. Let L be a relatively ample invertible sheaf on V, let 0 be a point of S. Fix an isomorphism of F with $f^{-1}(0)$, and let P denote the polarization on F induced by the restriction of L to $g^{-1}(0)$. Suppose that F is not a ruled variety. Then there is a finite irreducible Galois etale cover $r: S' \to S$, with group G, and a homomorphism $\sigma: G \to \mathrm{Aut}_p(F)$ such that V is isomorphic over S to the quotient $S \times F/G$. Here, G acts on $S \times F$ by

$$g(s,x) = (g(s),\sigma(g)(x))$$

<u>Proof</u>. Let $h: \mathrm{Iso}(V,S\times F) \to S$ be the subscheme of $\mathrm{Hom}_S(V,S\times F)$ representing the functor of S-schemes $T \to S$

$$T \to \{T\text{-isomorphisms } q: T\times_S V \to T\times F\} \quad .$$

$h: \mathrm{Iso}(V,S\ F) \to S$ is just the principal $\mathrm{Aut}(F)$ bundle associated to the fiber bundle $g: V \to S$.

Let $h_p: Iso_p(V,S\times F) \to S$ be the S-subscheme of $Iso(V,S\times F)$ representing the functor of S-schemes $T \to S$

$$T \to \{\text{polarized } T \text{ isomorphisms } q: T\times_S V \to T\times F\} .$$

These two S-schemes exist by the results of section 4 of [3].

The existence of $Iso_p(V,S\times F)$ implies that the group of the bundle $f: V \to S$ can be reduced to $Aut_p(F)$. Also, $h_p: Iso_p(V,S\times F) \to S$ is just the principal $Aut_p(F)$ bundle associated to $g: V \to S$, so it is enough to find a cover $r: S' \to S$ of the desired form such that $S'\times_S Iso_p(V,S\times F)$ is the trivial $Aut_p(F)$ bundle over S'. Denote $Iso_p(V,S\times F)$ by X.

Since F is not ruled, the results ([8] and [9] cor.2) imply that $Aut_p(F)$ is a compact algebraic group, whose connected component of the identity is an abelian variety A. Let $\bar{h}_p: \bar{X} \to S$ be the principal bundle with group $Aut_p(F)/A$ canonically associated to $h: X \to S$. As $Aut_p(F)$ is algebraic, $\bar{h}_p$ is a finite etale Galois map; letting S_1 be an irreducible component of $\bar{X}$, and replacing $f: V \to S$ with $f_1: V\times_S S_1 \to S_1$, reduces us to the case in which X contains a principal A subbundle. The result then follows from a theorem of Safarevic ([13],section 6).

q.e.d.

The next proposition is the main structure result for irregular threefolds of Kodaira dimension zero. The hard parts (i) and (ii) are proved in [15].

<u>Proposition 2.</u> Let V be a smooth threefold with very ample invertible sheaf L. Suppose

a) $q(V) = q > 0$

b) $\kappa(V) = 0$.

Then V is birational over $Alb(V) = Alb(V')$ to a projective threefold V' such that

(i) the Albanese map $a_{V'}: V' \to \mathrm{Alb}(V')$ is smooth and surjective with connected fibers of dimension $3-q$.

(ii) if $q = 1$ or 2, then $a_{V'}: V' \to \mathrm{Alb}(V')$ is a fiber bundle (in the etale topology). The fiber F' has $\kappa(F') = 0$, and is an absolute minimal model.

Let $f: V \to V'$ be the birational map given by the above. Then

(iii) there is an open subset U of $\mathrm{Alb}(V)$ such that $a_V^{-1}(U)$ is smooth over U , and f restricted to $a_V^{-1}(U)$ is a morphism.

Let s be in U and let F denote the fiber $a_V^{-1}(s)$. Let D be a hyperplane section of F induced by L . Fix an isomorphism of F' with $a_{V'}^{-1}(s)$. Then

(iv) $D' = f_*(D)$ is ample on F' , inducing a polarization p

(v) there is a finite irreducible etale Galois cover $r: S' \to \mathrm{Alb}(V')$, with group G , and a homomorphism $\sigma: G \to \mathrm{Aut}_p(F')$ such that V' is isomorphic over S to the quotient $S' \times F'/G$.

Proof: As mentioned in Ueno's talks (and proved in [15]) , V is birational over $\mathrm{Alb}(V)$ to a smooth projective threefold V'' such that $a_{V''}: V'' \to \mathrm{Alb}(V'')$ is a fiber bundle in the etale topology with fiber, say, F'' . Furthermore, if $q = 3$, then $a_{V''}$ is an isomorphis if $q = 1$ or 2 then $\kappa(F'') = 0$. This completes the proof in case $q = 3$, we now consider the case $q < 3$.

As $\kappa(F'') = 0$, F'' is not ruled. By proposition 1, there is a finite irreducible etale Galois cover

$$r: S'' \to \mathrm{Alb}(V'')$$

with group G, and a homomorphism $\sigma_1: G \to \mathrm{Aut}(F'')$ such that V'' is isomorphic to $S'' \times F''/G$.

Next, consider the case $q = 1$, i.e., F'' is a surface. Let F' be the absolute minimal model of F''. Let g be in G. Then $\sigma_1(g)$ induces a birational automorphism of F', which is in fact a biregular automorphism as F' is minimal. Thus there is a homomorphism

$$\sigma_2: G \to \mathrm{Aut}(F')$$

such that $V' = S'' \times F'/G$ is birational to V via a map $f: V \to V'$ over $\mathrm{Alb}(V) = \mathrm{Alb}(V')$. This proves (i) and (ii) in case $q = 1$. Assertion (iii) follows from the minimality of F', and the fact that f is a map over $\mathrm{Alb}(V) = \mathrm{Alb}(V')$.

To prove (iv), we consider a curve C on F'. Then

$$C \cdot D' = C \cdot f_*(D) = f^*(C) \cdot D > 0$$

and $$D'^{(2)} = f_*(D)^{(2)} \geq D^{(2)} > 0 ,$$

so by the Nakai-Moishezon criteria ([10],[11]), D' is ample on F'.

In case $q = 2$ (F'' a curve), we take $F' = F''$ and $V' = V''$. The assertions (i), (ii), and (iii) are then obvious; as $f: F \to F'$ is an isomorphism, (iv) is also clear.

We now prove (v). Let H be an irreducible hyperplane section of V induced by L, H' the proper transform of H under f, and H'' the pullback of H' to $S'' \times_{\mathrm{Alb}(V')} V' \cong S'' \times F'$. By the above argument, H' restricts to an ample divisor on the fiber $a_{V'}^{-1}(s)$ for all s in an open subset of $\mathrm{Alb}(V')$, hence H'' restricted to $s'' \times F'$ is ample for all s'' in an open subset of S''. Clearly $p_2(H'' \cdot (s_1'' \times F'))$ and $p_2(H'' \cdot (s_2'' \times F'))$ are algebraically equivalent divisors on F' for each s_1'', s_2'' in S'', thus H'' restricts to an ample divisor on $s'' \times F'$ for all s'' in S''. Therefore H'' is relatively ample on $S'' \times_{\mathrm{Alb}(V')} V'$, and hence H' is relatively ample on V' .

By construction, if s is in U , we have

$$\begin{aligned} H'\cdot a_{V'}^{-1}(s) &= f_*(H)\cdot a_{V'}^{-1}(s) \\ &= f_*(H\cdot a_V^{-1}(s)) \\ &= f_*(D) \\ &= D' \end{aligned}$$

Thus H' induces the polarization P on F' . Applying proposition 1 proves (v) . q.e.d.

Our main aim (Theorem 6) is to show that the structure result of proposition 2 carries over in the context of a family of Kodaira dimension zero threefolds. In other words, we would like to replace an arbitrary family of such threefolds with a family of fiber bundles, which are much easier to study. In fact, we shall be able to do this assuming only that the generic member of the original family is of Kodaira dimension zero (and irregular). Our main tools will be two results of Matsusaka and Mumford [9] which we recall here and rephase slightly for the convenience of the reader.

Theorem I: Let $f: X \to \mathrm{Spec}(\mathcal{O})$, $g: Y \to \mathrm{Spec}(\mathcal{O})$ be smooth and proper fiber varieties. Suppose $X \otimes k$ is not a ruled variety. Suppose further that there is a birational map

$$h_K: X\otimes K \to Y\otimes K$$

defined over K . Then there is a birational $\mathcal{O}$-map $h: X \to Y$ such that $h \otimes K$ is h_K , $h \otimes k$ is defined, and $h\otimes k: X\otimes k \to Y\otimes k$ is birational.

Theorem II: Let $f: X \to \mathrm{Spec}(\mathcal{O})$, $g: Y \to \mathrm{Spec}(\mathcal{O})$ be smooth and projective fiber spaces such that $X \otimes k$ is not ruled. Let L_X (resp. L_Y) be an f-ample (resp. g-ample) invertible sheaf on X (resp Y) . Suppose there is an isomorphism of the polarized varieties

$$h_K: (X\otimes K, L_X\otimes K) \to (Y\otimes K, L_Y\otimes K)$$

defined over K. Then there is an $\mathcal{O}$-isomorphism of the relatively polarized $\mathcal{O}$-schemes

$$h: (X, L_X) \to (Y, L_Y)$$

such that $h\otimes K = h_K$.

The next lemma shows how a fiber bundle over K can be extended to a family of fiber bundles over $\mathrm{Spec}(\mathcal{O})$. Theorem II is used in an essential manner.

<u>Lemma 3</u>. Let $q: F \to \mathrm{Spec}(\mathcal{O})$ be a smooth and projective fiber variety, with relatively ample invertible sheaf L on F. Let $p: S \to \mathrm{Spec}(\mathcal{O})$ be an abelian scheme over $\mathcal{O}$. Suppose

a) $F \otimes k$ is not ruled

b) there is a finite irreducible etale Galois cover

$$r_K: S_K^* \to S\otimes K$$

with group G, and a homomorphism

$$\sigma: G \to \mathrm{Aut}_p(F\otimes K) .$$

Then there is a local extension $\mathcal{O}'$ of $\mathcal{O}$, with quotient field K' and residue field k, a finite irreducible etale Galois cover

$$r: S' \to S\otimes_{\mathcal{O}}\mathcal{O}'$$

with group G, and a homomorphism

$$\sigma': G \to \mathrm{Aut}_p(F\otimes_{\mathcal{O}}\mathcal{O}'/\mathcal{O}')$$

satisfying

(i) the quotient $V' = S'\times_{\mathcal{O}'}(F\otimes_{\mathcal{O}}\mathcal{O}')/G$ is a smooth fiber space over $S \otimes_{\mathcal{O}}\mathcal{O}'$

(ii) $V' \otimes K'$ is isomorphic to $S_K^*\times(F\otimes K)/G$ over K'

(iii) $V' \otimes k$ is a fiber bundle with base $S \otimes k$ and fiber $F \otimes k$.

Proof. Suppose that $r_K: S_K^* \to S\otimes K$ is of degree d. Then there is a finite, etale map

$$\tilde{r}_K: S\otimes K \to S_K^*$$

such that $r_K \circ \tilde{r}_K: S\otimes K \to S\otimes K$ is multiplication by d. Let H_K denote the kernel of $\tilde{r}_K$. Replacing $\mathcal{O}$ by a local extension, and changing notation, we may assume that H_K is defined over K. Then the k-closure H of H_K is a subgroup scheme of S, etale over $\mathcal{O}$, and $H\otimes K = H_K$. Let S^* be the quotient S/H. Then S^* is the desired cover of $dS = S$.

Let K' be a common field of definition for the automorphisms $\sigma(g)$, g in G. Let $\mathcal{O}'$ be a local extension of $\mathcal{O}$ with quotient field K' and residue field k. Denote $F \otimes_{\mathcal{O}} \mathcal{O}'$ by F' and $S^* \otimes_{\mathcal{O}} \mathcal{O}$ by S'. By Theorem II, there is a polarized $\mathcal{O}'$-automorphism

$$\sigma'(g): F' \to F'$$

such that $\sigma'(g)\otimes K': F'\otimes K' \to F'\otimes K'$ is $\sigma(g)$.

This also implies that

$$\sigma': G \to \mathrm{Aut}_p(F'/\mathcal{O}')$$
$$g \to \sigma'(g)$$

is a homomorphism.

Property (i) is immediate, as G acts without fixed points on S', hence also on $S' \times_{\mathcal{O}'} F'$. By construction, we have $S'\otimes K'$ is isomorphic over K' to S_K^*, and

$$\sigma'(g): F'\otimes K' \to F'\otimes K'$$

is isomorphic over K' to

$$\sigma(g): F\otimes K \to F\otimes K \quad ,$$

which proves (ii). Finally, $V'\otimes k$ is the quotient $(S'\otimes k) \times (F\otimes k)/G$, proving (iii).

q.e.

For technical reasons, it is somewhat easier to work with minimal models of surfaces in our construction of the family of fiber bundles. The next lemma enables us to replace certain families of surfaces with the family of minimal models.

Lemma 4. Let $p : F \to \mathrm{Spec}(\mathcal{O})$ be a smooth projective family of surfaces such that $\kappa(F \otimes K) = 0$. Then $\kappa(F \otimes k) = 0$. Furthermore, there is a local extension $\mathcal{O}'$ of $\mathcal{O}$, with quotient field K' and residue field k , a smooth projective family of surfaces $p' : F' \to \mathrm{Spec}(\mathcal{O}')$, and a birational $\mathcal{O}$-morphism $f: F \otimes_{\mathcal{O}} \mathcal{O}' \to F'$ such that $F' \otimes K'$ and $F' \otimes k$ are the absolute minimal models of $F \otimes K$ and $F \otimes k$ respectively. Finally, let D be a relatively ample divisor on F and let D^* denote the divisor $p_1^*(D)$ on $F \otimes_{\mathcal{O}} \mathcal{O}'$. Then $f_*(D)$ is relatively ample on F' .

Proof. The proof of the first two assertions appears in [4]; we sketch the proof here for the convenience of the reader.

By ([6], thm. 3.5) $F \otimes k$ is not ruled. We first show the existence of a smooth family $p': F' \to \mathrm{Spec}(\mathcal{O}')$, and a birational $\mathcal{O}'$-morphism $f: F \otimes_{\mathcal{O}} \mathcal{O}' \to F'$ such that $F' \otimes k$ is absolutely minimal.

Let $\bar{F}$ be the minimal model of $F \otimes k$. Then the birational morphism $F \otimes k \to \bar{F}$ can be written as a sequence of monoidal transformations,

$$F \mathcal{O} k = F_r \to F_{r-1} \to \dots \to F_0 = \bar{F} \quad .$$

By induction, we may assume $r = 1$. Let E be the exceptional curve in $F \otimes k$. By ([5], Thm. 5), E is a stable subvariety of $F \otimes k$; thus there is a local extension $\mathcal{O}'$ of $\mathcal{O}$, with quotient field K' and residue field k , and a subscheme $\tilde{E}$ of $F \otimes_{\mathcal{O}} \mathcal{O}'$ such that $\tilde{E}$ is smooth over $\mathcal{O}'$, $\tilde{E} \otimes k$ is E , and $\tilde{E} \otimes K'$ is an exceptional curve on $F \otimes K'$. We then blow down $\tilde{E}$ by $f: F \otimes_{\mathcal{O}} \mathcal{O}' \to F'$ as desired.

Suppose now that $F' \otimes K'$ contained an exceptional curve. Let $\overline{F}'$ be the minimal model of $F' \otimes K'$. Then ([23],page 6) $K_{\overline{F}'}$ is numerically equivalent to zero, hence $(K_{F'\otimes K'})^{(2)}$ is strictly negative. As intersection products are preserved under specialization, we have that $(K_{F'\otimes k})^{(2)}$ is also negative. As $F' \otimes k$ is minimal and not ruled, this is impossible (see [12], page 6). Thus $F' \otimes K'$ is minimal. In particular, $12K_{F'\otimes K'}$ is linearly equivalent to zero, hence $12K_{F'\otimes k}$ is also linearly equivalent to zero, and $\kappa(F'\otimes k) = 0$. This proves the first two assertions. Arguing as in the proof of Proposition 2(iv) proves the final assertion.

q.e.d.

We isolate a few useful facts in the next lemma to avoid encumbering the later argument.

Lemma 5. Let $f: V \to \mathrm{Spec}(\mathcal{O})$ be a smooth projective fiber space with a section, let $a_V: V \to \mathrm{Alb}(V/\mathcal{O})$ be the associated Albanese map, and let $\pi: \mathrm{Alb}(V/\mathcal{O}) \to \mathrm{Spec}(\mathcal{O})$ denote the structure morphism. Suppose $a_{V\otimes K}: V\otimes K \to \mathrm{Alb}(V\otimes K)$ is surjective with connected fibers. Then

(i) $a_{V\otimes k}: V\otimes k \to \mathrm{Alb}(V\otimes k)$ is also surjective with connected fibers

(ii) there is an open subset U of $\mathrm{Alb}(V/\mathcal{O})$ such that $U\otimes k \neq \emptyset$ and $a_V^{-1}(U)$ is smooth over U.

Proof: We note that $\mathrm{Alb}(V/\mathcal{O})$ is an abelian scheme over $\mathcal{O}$, hence smooth (characteristic zero). Let

$$\begin{array}{ccc} V & \xrightarrow{g} & W \\ & \searrow^{a_V} & \downarrow h \\ & & \mathrm{Alb}(V/\mathcal{O}) \end{array}$$

be the Stein factorization of a_V. By assumption $a_V\otimes K = a_{V\otimes K}$ has connected fibers, hence h is birational and finite. As $\mathrm{Alb}(V/\mathcal{O})$ is normal, h is an isomorphism. Noting that $a_{V\otimes k} = a_V\otimes k$ completes the proof of (i).

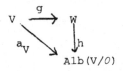

To prove (ii), we have that $a_{V\otimes k}$ is smooth over an open sub-
et of $V \otimes k$ by Bertini's theorem. As $a_{V\otimes k} = a_V \otimes k$, this implies
he existence of an open subset U of $\mathrm{Alb}(V/\mathcal{O})$ such that $U\otimes k \neq \emptyset$
nd $a_V^{-1}(x)$ is smooth and of the proper dimension for each x in U.
ince both V and $\mathrm{Alb}(V/\mathcal{O})$ are smooth, this gives the smoothness
f $a_V^{-1}(U)$ over U by the usual Jacobian criterion.

q.e.d.

We are now ready to prove the main structure result for our
amilies of threefolds.

<u>Theorem 6</u>. Let $g\colon V \to \mathrm{Spec}(\mathcal{O})$ be a projective fiber variety of
iber dimension three. Suppose

(a) $q = q(V\otimes K) > 0$

(b) $\kappa(V\otimes K) = 0$

hen there is a local extension $\mathcal{O}'$ of $\mathcal{O}$, with quotient field K'
nd residue field k, a smooth, projective fiber variety
$'\colon V' \to \mathrm{Spec}(\mathcal{O}')$ and a birational $\mathcal{O}'$-map $h\colon V\otimes_{\mathcal{O}}\mathcal{O}' \to V'$ such that

(i) $h\otimes K'\colon V\otimes_{\mathcal{O}}K' \to V'\otimes K'$ and $h\otimes k\colon V\otimes k \to V'\otimes k$ are birational

(ii) $a_{V'}\colon V' \to \mathrm{Alb}(V'/\mathcal{O}')$ is smooth and surjective, with connected fibers of dimension $3-q$

(iii) if $q < 3$, then there is a smooth projective fiber variety $p'\colon F' \to \mathrm{Spec}(\mathcal{O}')$, of fiber dimension $3-q$,

uch that $a_{V'\otimes K'}\colon V'\otimes K' \to \mathrm{Alb}(V'\otimes K')$ (resp. $a_{V'\otimes k}\colon V'\otimes k \to \mathrm{Alb}(V'\otimes k)$)
s a fiber bundle in the etale topology with fiber $F' \otimes K'$ (resp.
$'\otimes k$). Furthermore, $F' \otimes K'$ and $F' \otimes k$ are both absolute minimal
odels and $\kappa(F'\otimes K') = \kappa(F'\otimes k) = 0$.

<u>Proof</u>. We may assume that g admits a section; let
$_V\colon V \to \mathrm{Alb}(V/\mathcal{O})$ be the associated relative Albanese map, and let
$\colon \mathrm{Alb}(V/\mathcal{O}) \to \mathrm{Spec}(\mathcal{O})$ be the structure morphism. By proposition 2,

$a_V \otimes K$: $V \otimes K \to \mathrm{Alb}(V \otimes K)$ is surjective with connected fibers generically of dimension $3-q$. By lemma 5, $a_V \otimes k$ is also surjective with connected fibers. Thus, if $q=3$, we may take $V' = \mathrm{Alb}(V/\mathcal{O})$, and $h = a_V$, which completes the proof in this case. We consider the case $q < 3$.

Let L be a relatively invertible ample sheaf on V. We apply propostion 2 to $V \otimes K$, constructing a fiber bundle $a_{V'_K}$: $V'_K \to \mathrm{Alb}(V'_K)$ with fiber F'_K, and a birational map h_K: $V \otimes K \to V'_K$. We recall that h_K is a map over $\mathrm{Alb}(V \otimes K) = \mathrm{Alb}(V'_K)$, and there is an open subset U_K of $\mathrm{Alb}(V \otimes K)$ such that $a^{-1}_{V \otimes K}(U_K)$ is smooth over U_K and h_K restricted to $a^{-1}_{V \otimes K}(U_K)$ is a morphism.

By lemma 5, there is an open subset U of $\mathrm{Alb}(V/\mathcal{O})$ such that

(1) $a_V^{-1}(U)$ is smooth over U

(2) $U \otimes k \neq \emptyset$

(3) $U \otimes K \subseteq U_K$

After making a local extension of $\mathcal{O}$, and changing notation, we may assume there is a section s: $\mathrm{Spec}(\mathcal{O}) \to U$ to the map $\pi \circ i_u$: $U \to \mathrm{Spec}(\mathcal{O})$. Let p: $F \to \mathrm{Spec}(\mathcal{O})$ be the $\mathcal{O}$-scheme

$$F = a_V^{-1}(U) \times_U \mathrm{Spec}(\mathcal{O})$$
$$= V \times_{\mathrm{Alb}(V)} \mathrm{Spec}(\mathcal{O}) ,$$

where $\mathrm{Spec}(\mathcal{O})$ is a U-scheme via s. F is smooth and projective over $\mathcal{O}$, and is in a natural way an $\mathcal{O}$-subvariety of V.

We now consider the case $q=1$, so p: $F \to \mathrm{Spec}(\mathcal{O})$ is a family of surfaces. By proposition 2, we have $\kappa(F \otimes K) = 0$. After making a local extension, and changing notation again, we may assume there is a family p': $F' \to \mathrm{Spec}(\mathcal{O})$, and a birational $\mathcal{O}$-morphism f: $F \to F'$ satisfying the conclusion of lemma 4.

Let H be a divisor on V such that $L \cong \mathcal{O}_V(H)$. We may suppose that $H \cap (F \otimes K)$ and $H \cap (F \otimes k)$ are proper intersections. Let D be the divisor $H\cdot F$ on F . Then D is p-ample on F ; $f_*(D)$ is p'-ample on F' by lemma 4.

Let P_K be the polarization on F'_K given by proposition 2. We claim that the polarized varieties $(F'\otimes K, f_*(D)\otimes K)$ and (F'_K, P_K) are isomorphic. Indeed, both $F'\otimes K$ and F'_K are absolute minimal model of $F\otimes K$. Thus there is an isomorphism $\tau\colon F'\otimes K \to F'_K$ making the diagram

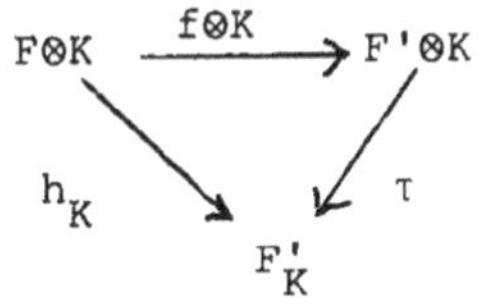

commute. By proposition 2(iv), the polarization P_K on F'_K is given by the ample divisor $h_{K*}(D\otimes K)$. As $f_*(D)\otimes K = (f\otimes K)_*(D\otimes K)$, we see that τ gives the desired isomorphism of $(F'\otimes K, f_*(D)\otimes K)$ with (F'_K, P_K) . We may assume that τ is defined over K .

Next, by proposition 2(v), there is a finite irreducible etale Galois cover,

$$r_K\colon S_K \to \mathrm{Alb}(V_K)$$

with group G , and a homomorphism

$$\sigma_K\colon G \to \mathrm{Aut}_{P_K}(F'_K) = \mathrm{Aut}_{f_*(D)\otimes K}(F'\otimes K)$$

such that

$$V'_K \cong S'_K \times F'_K/G \cong S'_K \times (F'\otimes K)/G$$

Making another local extension, and changing notation, we apply lemma 3, and construct the fiber variety $a'\colon V' \to \mathrm{Alb}(V/\mathcal{O})$, having the properties

(4) $V'\otimes K \cong V'_K$

(5) $V'\otimes k$ is a fiber bundle over $\mathrm{Alb}(V\otimes k)$ with fiber $F'\otimes k$.

In particular

$$\kappa(V'\otimes k) \geq \kappa(F'\otimes k) + \kappa(\mathrm{Alb}(V\otimes k)) = 0$$

by the easy addition formula ([14], Thm.6.12) . Thus $V'\otimes k$ is not ruled.

By Theorem I, the birational map h_K extends to a birational $\mathcal{O}$-map $h: V \to V'$ with the properties desired in (i). This completes the proof in case $q = 1$.

If $q = 2$, so that $p: F \to \mathrm{Spec}(\mathcal{O})$ is a family of elliptic curves, we take $F' = F$ and proceed as above. This completes the proof of the theorem.

q.e.d.

Corollary 7. Let $g: V \to \mathrm{Spec}(\mathcal{O})$ be a smooth projective fiber space, of fiber dimension three. Suppose

(a) $q = q(V\otimes K) > 0$

(b) $\kappa(V\otimes K) = 0$

Then $\kappa(V\otimes k) = 0$.

Proof. We retain the notation of Theorem 6. If $q = 3$, then $V \otimes k$ is birational to the abelian variety $\mathrm{Alb}(V\otimes k)$, completing the proof in this case. If $q < 3$, then $V \otimes k$ is birational to the fiber bundle $a_{V'\otimes k}: V'\otimes k \to \mathrm{Alb}(V'\otimes k)$, which has fiber $F'\otimes k$. By the addition formula for locally trivial fiber bundles ([14],Thm.15.1),

$$\begin{aligned}\kappa(V'\otimes k) &= \kappa(\mathrm{Alb}(V'\otimes k)) + \kappa(F'\otimes k)\\ &= 0\end{aligned}$$

As κ is a birational invariant, this gives $\kappa(V\otimes k) = 0$, as desired.

q.e.d.

Theorem 6 can also be used to prove a finer statement than that of corollary 7: the pluri-genera also remain constant. I am indebted to Pehham Wilson for suggesting the use of the following lemma to aid in the proof of corollary 9.

Lemma 8. Let $g: W \to \mathrm{Spec}(\mathcal{O})$ be a smooth projective fiber space. Suppose the canonical sheaf $K_{W\otimes K}$ of $W \otimes K$ is an element of finite order n in $\mathrm{Pic}(W\otimes k)$. Then $K_{W\otimes k}$ is also of order n in $\mathrm{Pic}(W\otimes k)$.

Proof. The canonical sheaf K_W of W is an invertible sheaf on W, and as such defines a section $s: \mathrm{Spec}(\mathcal{O}) \to \mathrm{Pic}(W/\mathcal{O})$ of the relative Picard scheme of W over $\mathcal{O}$. By the adjunction formula, we have

$$s(\mathrm{Spec}(K)) = K_{W\otimes K}$$

and

$$s(\mathrm{Spec}(k)) = K_{W\otimes k} \quad .$$

By assumption, $s(\mathrm{Spec}(K))$ lands in the n-torsion subgroup $\mathrm{Pic}(W\otimes K)_n$. Thus $s(\mathrm{Spec}(\mathcal{O}))$ generates a finite subgroup scheme G of $\mathrm{Pic}(W/\mathcal{O})$. As the characteristic is zero, $G \otimes k$ is reduced, of order n, and is generated by $s(\mathrm{Spec}(k))$. This completes the proof.

q.e.d.

Corollary 9. Let V and $V \otimes K$ be as in Corollary 7. Then

$$P_m(V\otimes K) = P_m(V\otimes k)$$

for all $m > 0$.

Proof. We retain the notation of theorem 6. As the plurigenera are birational invariants, we have

$$P_m(V\otimes k) = P_m(V'\otimes k)$$

and

$$P_m(V\otimes K) = P_m(V'\otimes K) \quad .$$

We claim that the canonical sheaf $K_{V'\otimes K}$ is a torsion element of $\mathrm{Pic}(V'\otimes K)$. Indeed, if $q = 3$, then $V'\otimes K$ is an abelian variety and $K_{V'\otimes K} = \mathcal{O}_{V'\otimes K}$. If $q < 3$, then there is a finite etale cover

$$r: S \to \mathrm{Alb}(V\otimes K)$$

such that

$$(V'\otimes K)_S \overset{\text{def}}{=} S\times_{\mathrm{Alb}(V'\otimes K)}(V'\otimes K) \cong S\times(F'\otimes K) .$$

As S is an abelian variety, and $F'\otimes k$ is either an elliptic curve, or a minimal surface of Kodaira dimension zero, it follows that $K_{(V'\otimes K)_S}$ is a torsion element of $\mathrm{Pic}((V'\otimes K)_S)$. Since $(V'\otimes K)_S \to V'\otimes K$ is etale, our claim is proved.

We now apply lemma 8. If n is the common order of $K_{V'\otimes K}$ and $K_{V'\otimes k}$, we have

$$P_m(V'\otimes k) = P_m(V'\otimes K) = \begin{cases} 0 & m\nmid n \\ 1 & m|n \end{cases} , \quad m > 0 ,$$

proving the corollary.

q.e.d.

We now prove the main results on the deformation invariance of κ and P_m .

<u>Corollary 10</u>. Let V and M be varieties, $g: V \to M$ a smooth projective fiber space of fiber dimension three, and 0 a point of M .

Suppose

$$q(g^{-1}(0)) > 0$$

and

$$\kappa(g^{-1}(0)) = 0$$

Then

$$P_m(g^{-1}(x)) = P_m(g^{-1}(0))$$

and

$$\kappa(g^{-1}(x)) = 0$$

for each x in M .

<u>Proof</u>. It is enough to prove the statement on the pluri-genera. We reduce immediately to the case in which M is a smooth curve. Let $K = k(M)$. By the upper-semi continuity of the pluri-genera, we have

$$P_m(V\otimes K) \le P_m(g^{-1}(0)) \le 1$$

for all m . Thus either $\kappa(V\otimes K) = -\infty$, or $\kappa(V\otimes K) = 0$. As $q(V\otimes K) = q(g^{-1}(0)) > 0$, we have by ([15]) that

$$\kappa(V\otimes K) = -\infty \iff V\otimes K \text{ is uni-ruled.}$$

On the other hand, as uni-ruledness is preserved under specialization ([7]) we have

$$\kappa(V\otimes K) = -\infty \implies \kappa(g^{-1}(0)) = -\infty .$$

Thus $\kappa(V\otimes K) = 0$.

We apply corollary 9 twice to find

$$P_m(V\otimes K) = P_m(g^{-1}(0))$$

and

$$P_m(V\otimes K) = P_m(g^{-1}(x)) .$$

as desired. q.e.d.

<u>Corollary 11</u>. Let V and M be varieties. Let $g: V \to M$ be a smooth projective fiber space of fiber dimension three, and let 0 be a point of M . Suppose

$$q(g^{-1}(0)) > 0 .$$

Then

$$\kappa(g^{-1}(x)) = \kappa(g^{-1}(0))$$

for all x in M .

<u>Proof</u>. The case $\kappa(g^{-1}(0)) = -\infty$ is done in [2] and [7] and the $\kappa(g^{-1}(0)) = 0$ is corollary 10.

Ashikaga and Ueno [1], have shown the following:

<u>Theorem:</u> Let $\bar{f}: \bar{V} \to \mathrm{Spec}(\mathcal{O})$ be a smooth projective fiber variety of fiber dimension three. If $\kappa(\bar{V}\otimes K) > 0$, then $\kappa(\bar{V}\otimes K) = \kappa(\bar{V}\otimes k)$.

Using this result, and arguing as in corollary 10, completes the proof.

q.e.d.

References

[1] Ashikaga, with appendix by K. Ueno, "The relative Kodaira dimension of fiber spaces", preprint.

[2] A. Fujiki, "Deformation of uni-ruled manifolds", preprint.

[3] A. Grothendieck, "Les schemas de Hilbert", Sem. Bourb. No. 221, 1960/1961.

[4] S. Iitaka, "Deformations of compact complex surfaces II", J. Math. Soc. Japan 22 (1971), 247-261.

[5] K. Kodaira, "On stability of compact submanifolds of complex manifolds", Am. J. Math. 85 (1963), 79-84.

[6] __________ and D. C. Spencer, "On deformations of complex analytic structures I", Am. J. Math. 67 (1958), 328-401.

[7] M. Levine, "Deformation of uni-ruled varieties", Duke Math J. 48 (1981), 467-473.

[8] T. Matsusaka, "Polarized varieties, fields of moduli and generalized Kummer varieties of polarized Abelian varieties", Am. J. Math. 80 (1958), 45-88.

[9] ____________ and D. Mumford, "Two fundamental theorems on deformations of polarized varieties", Am. J. Math. 86 (1964), 668-684.

[10] B. Moishezon, "The criterion of projectivity of complete algebraic abstract varieties", Doklady Annsr. Math. Ser. 28, 179-224.

[11] Y. Nakai, "A criterion of an ample sheaf on a projective scheme", Am. J. Math. 85 (1963), 14-26.

[12] I. R. Shafarevie, Algebraic Surfaces, Proc. Steklov Inst. Math. 75 (1965).

[13] ________________, "Principal homogeneous spaces defined over function fields", Am. Math. Sec. Transl. (2) 63 (1967).

[14] K. Ueno, Classification Theory of Algebraic Varieties and Compact Complex Spaces, LNM No. 439, Springer-Verlag.

[15] E. Viehweg, "Klassifikationtheorie algebraischer varietaten der dimension drei", Comp. Math. 41 (1980), 361-400.

Plane Forms and Multiple-Point Formulas

by Steven L. Kleiman[1]

1. Introduction.

By definition, a plane form is a 3-fold Z in $Y = \mathbb{P}^4$ swept out (or ruled) by an irreducible 1-parameter family of planes such that a general point of Z lies in exactly one plane. The ground field k can be of any characteristic. In fact, the characteristic will not enter the present discussion except in one minor sidelight, (4.1). For convenience of exposition, k will be taken algebraically closed, but this hypothesis is of no essential importance.

Since Z is of codimension 1 in Y, it is reasonable to expect that the locus of i-fold points is of pure codimension i in Y (for locally it ought to be the intersection of i branches). Let d_i denote the degree of this locus, viewed as an i-cycle. Thus, d_1 is the degree of Z and d_2 is the degree of the double surface, d_3 is the degree of the triple curve and d_4 is the number of quadruple points.

Let p denote the geometric genus of a general plane-section of Z. Then the degrees d_2, d_3 and d_4 are related to the basic numerical invariants d_1 and p by the formulas

$$
\begin{aligned}
(1.1)\qquad 2d_2 &= (d_1 - 2)(d_1 - 1) - 2p \\
6d_3 &= (d_1 - 4)[(d_1 - 3)(d_1 - 2) - 6p] \\
24d_4 &= (d_1 - 6)(d_1 - 5)[(d_1 - 4)(d_1 - 3) - 12p] + 12p(p + 9) .
\end{aligned}
$$

These formulas are our central topic.

The first two formulas may seem familiar already. Indeed, they are identical to corresponding formulas for ruled surfaces S in $\mathbb{P}^3$. In fact, the formulas for Z and the ones for S may be derived from each other as follows. Given Z, cut it with a general hyperplane and apply the corresponding formulas to the section S; conversely, given S, erect a cone Z. The formulas for surfaces are derived in [Baker, 1933; Ex. 7, p. 165]. Apparently they go back to Salmon's bare-handed work on reciprocal surfaces done sometime before 1848. Of course,

[1]Partially supported by the NSF under MCS 79-06895.

Salmon did not use the concept of genus. It was not introduced until 1857 by Riemann, and it was not used to classify algebraic curves until 1864 when Clebsch gave a formula for plane curves identical to the one above for d_2. Clebsch's formula and the one above may be derived from each other by a further application of the method of section and projection

The third formula was apparently not discovered until about 1930 by Roth. This great a delay is perhaps surprising because, roughly between 1880 and 1900, Schubert, Segre and others obtained other enumerative formulas for plane-forms (and for higher-dimensional ruled varieties), Schubert investigated other multiple-coincidence formulas, and Severi, when working on the double-point formula for arbitrary varities, called attention to the loci of higher-order multiple points, see [Segre, 1912]. Roth [1931, Formula (9), p. 121] gave the third formula in a slightly different form (along with other formulas). Earlier James [1928, Formula (27), p. 521] gave a dual form of the formula (it enumerates the 4-fold tangent hyperplanes to a ruled surfac in $\mathbb{P}^4$). Both Roth and James used the correspondence principle and Cayley's functional method.

The formulas must sometimes be taken with a grain of salt, for they can fail outright or they can be technically valid yet perhaps mean less than expected. For example, consider the plane pencil; that is, Z is swept out by the linear pencil of planes through a line. Here $d_1 = 1$ and $p = 0$, so the formulas yield $d_2 = 0$ and $d_3 = -1$ and $d_4 = 5$. One the face of it, only the value -1 of d_3 is absurd. However, the negativity does have some significance; it indicates the presence of some pathology in the locus of triple points and puts the value 5 of d_4 in suspicion. The value 0 of d_2 suggests that there are no singularities and so d_3 and d_4 ought to be 0. In fact, of course, Z is a 3-plane, so Z is smooth and d_2, d_3 and d_4 are all 0.

For a second example, consider the cone Z whose vertex is a line F and whose base is a plane curve $\tilde{Z}$ of degree d and geometric genus p; the rulings are the joins of F with the points of $\tilde{Z}$. As was stated above (and will be proved in (3.)), the formula for d_2 is valid and coincides with the Clebsch formula for $\tilde{Z}$. Hence $d_2 = 0$ if and only if $\tilde{Z}$ is smooth, if and only if Z is normal. However, if $\tilde{Z}$ is a smooth conic, then the double locus of Z is F, which is nonempty, although $d_2 = 0$. Moreover, the triple and quadruple loci are empty, yet the formulas yield $d_3 = 0$ and $d_4 = 1$. If $\tilde{Z}$ is an (irreducible) cubic, then the triple locus of Z is F and the quadruple locus is empty, yet the formulas yield $d_2 = 1$, $d_3 = 0$ and $d_4 = 0$ if $p = 0$ and $d_2 = 0$, $d_3 = 1$ and $d_4 = 2$ if $p = 1$; the values $d_3 = 0$ and $d_4 = 1$ are obviously wrong.

In previous treatments, little or no attention was paid to the conditions guaranteeing the validity of the formulas or to precisely what they mean. (Do, however, see [Roth, 1931, 1.3].) What is more, the first derivations were not rigorous. In the present article, the formulas and some generalizations of them will be given a thorough treatment by applying modern multiple-point theory of maps. Moreover, this article can serve as an introduction to multiple-point theory.

In Section 2, the formulas will be derived formally from the corresponding general multiple-point formulas. The analogous work for surfaces was done in [Kleiman, 1977] on p. 367 for the first formula and on p. 389 for the second. The possibility of deriving the third formula similarly was mentioned in the footnote on p. 389 (which was added to the printer's proofs). Roberts [1980, (6.5)] derived from a general multiple-point formula (6.3) of his own another one of Roth's formulas; this one, Roth's Formula (12), enumerates the "stationary" triple points of Z .

Section 3 is concerned with the meaning and validity of the formulas, and it has the most technical mathematics. First, some multiple-point theory will be reviewed, including sufficient conditions for the validity of the general formulas and some special properties of maps with no $\overline{S}_2$-singularities. Then the conditions will be checked for the plane-forms in a fairly large class. Thus there are indeed many plane-forms for which the formulas do hold! Moreover, it will be shown that for most of these plane forms the cycle of i-fold points is the reduced cycle on the closure of the ordinary i-fold points.

Section 4 presents some observations and examples, which help round out the story. First come some observations about the condition that a general point of Z lie in just one ruling, then some concerning the intersection F of all the rulings. In particular, it will be seen that the obstruction to the validity of the formulas in examples like those above always lies in the size of F as well as in the sizes of the multiple-point loci. Finally, there is a discussion of some examples of ruled varieties, which are characterized by certain simple properties. This work involves among other things a more refined form of the formula for d_2, which deals with cycles modulo rational equiva- lence. In particular, it will be proved that the plane pencil is the only example of a smooth plane form and that the quadric surface is the only ruled surface in 3-space with no point common to all the rulings.

The point of the present work is not only to provide a fuller and more rigorous treatment of the three formulas, it is not only to refine and generalize them, but it is also to do these things as an applicatio of general multiple-point theory. Thus the conceptual framework, the technical validity and the practical usefulness of the general theory are put to the test. Moreover, examples of plane-forms like those above illustrate "pathological" behavior within the general theory. When a general theory yields old ad hoc formulas as special cases and more knowledge about their meaning and validity, then we have gained a deeper understanding and more power and, in developing the theory, we have made a significant advance.

2. Derivation.

It is no more difficult to treat r-folds Z in $Y = \mathbb{P}^{r+n}$ swept out by an irreducible 1-parameter family of (r-1)-planes or "rulings" $(r \geqslant 1,\ n \geqslant 0)$. So the greater generality will be pursued from now on.

(2.1) Setup. The technical setup which will be used throughout the remainder of the article is indicated in the following diagram and elaborated below it.

$$\begin{array}{ccc} X = \mathbb{P}(E) & \xrightarrow{\quad f \quad} & Y = \mathbb{P}(V) \\ \Big\downarrow g & \searrow^{f'} & \Big\uparrow \\ C & & Z = f(X) \end{array}$$

Here C is the parameter curve of the family; assume that it is smooth, complete and irreducible, and let p denote its genus. Here Y is the ambient $\mathbb{P}^{r+n}$, given via a vector space V of dimension $(r + n + 1)$. Here X is the total space of the family, given via a locally free sheaf E on C of rank r. Here g and f are the structure maps. So they induce an embedding

$$(f,g) : X \hookrightarrow Y \times C ,$$

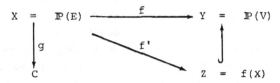

which corresponds to a surjection,

$$u: V_C \longrightarrow E .$$

(It is u that singles out E from among its twists by invertible sheaves on C.) So far, the setup is that of virtually any flat, irreducible 1-parameter family of (r-1)-planes in Y. The setup will now be refined with additional hypotheses, and some simple but basic observations will be made rounding out the picture.

The triple X, g, f corresponds (bijectively) to a map from C into the Grassmannian $\mathrm{Grass}_{r-1}(Y)$ of (r-1)-planes in Y. Consider the normalized image C_1. It parametrizes another flat family, but one that induces the original family via the natural map $\nu: C \to C_1$. Both families sweep out the same variety Z, and Z alone comes into play in the formulas (1.1). However, if ν is not an isomorphism, then a general point of Z will not lie in exactly one member of the original family. So assume that ν is an isomorphism; that is, assume that C is mapped birationally into the Grassmannian.

Assume that Z spans Y or, in other words, that the induced map $V \to H^0(E)$ is injective.

Assume that Z is of dimension r. Then the map $f': X \to Z$ is of finite degree, because it is surjective and X and Z are irreducible of the same dimension. It need not be birational, but birationality is not a stringent requirement, see (4.1). Moreover, birationality does hold whenever f factors into an embedding in an auxiliary projective space followed by a general projection; as is the case for most of the Z studied in Section 3. Birationality just means that a general point of Z lies in exactly one ruling or "generatrix." Birationality is necessary for the application of multiple-point theory to f. So assume it.

Consider a section D of Z by a general (n + 1)-plane. It is a curve, which is irreducible by Bertini's theorem. Moreover, D is a "directrix;" that is, D meets each ruling exactly once. Indeed, otherwise D would have a linear space as a component; alternatively, note that the variety of (n + 1)-planes that meet an (r-1)-plane in a line or more has codimension at least 2.

The geometric genus of the directrix D is p. Indeed, consider the inverse image $D' = f^{-1}D$. Each component of D' has dimension at least 1, because D is a local complete intersection. No component of D' can lie entirely in a fiber $f^{-1}z$, because otherwise D would meet some (in fact, every) ruling at z and at some other point. Since $f'\colon X \to Z$ is birational, therefore D' is irreducible and maps birationally onto D. Hence D' and D are of the same geometric genus. On the other hand, g carries D' birationally onto C (so isomorphically since c is smooth), because D meets each ruling $g^{-1}y$ $(y \in C)$ scheme-theoretically once. Therefore D, D' and C are of the same geometric genus, namely p.

Let h denote the pullback under $f\colon X \to Y$ of the hyperplane class, say, modulo rational equivalence. Then

$$h = f^*c_1O_Y(1) = c_1O_X(1)$$

because obviously $f^*O_Y(1) = O_X(1)$. Denote the degree of Z by d_1. Then

$$d_1 = \int_X h^r$$

because the degree of the map $f'\colon X \to Z$, which appears absent on the right, is equal to 1.

(2.2) Multiple-point formulas. Let $f\colon X \to Y$ be a proper map between smooth and irreducible quasi-projective varieties. Denote the codimens[ion] of f by n, so

$$n = \mathrm{cod}(f) = \dim(Y) - \dim(X).$$

Some multiple-point theory for f will now be reviewed; for more information, see (3.1), (3.2) and [Kleiman, 1980, 1981, 1982].

A point x of X is called an *i-fold point* of f if, roughly, there exist $i-1$ other points with the same image under f as x. If in fact all i points are distinct in X, then x is called a *strict i-fold point*. In general, some of the i points may lie infinitely near others, and this ramification must be taken into account. For double-points $(i=2)$, the matter is completely understood (thanks principally to Laksov).

For treating higher singularities $(i \geq 3)$, two methods are known. Both methods work when the map f has no $\overline{S}^2$-*singularities* (that is,

$\Omega_f^1(x)$ has dimension at most 1 for all x) and also in several cases when f has some. For greater generality, a new method must be found. Now, the map f of (2.1) has no $\overline{S}^2$-singularities, because (f,g) embeds each $f^{-1}z$ in the curve $C\times\{g\}$. So in principle, both methods could be used to treat ruled varieties. However, the method of the Hilbert scheme does not yet yield a quadruple-point formula, while the method of iteration does. (For more information about the advantages and disadvantages of each method, see (2.7) and (2.8).) Therefore the method of iteration will be used here.

The locus M_i of i-fold points supports a natural positive cycle, whose class (again, say, modulo rational equivalence) will be denoted by m_i . In particular, m_1 denotes the fundamental class of X . Each m_i is given by a multiple-point formula in terms of the preceding m's and the Chern classes of the map f . These classes $c_j = c_j(f)$ are defined as the classes of the (virtual) normal bundle of f . The normal bundle is defined for any map that is a local complete intersection, but in the present case X and Y are smooth and the total Chern class $c_* = \Sigma c_j$ is given in terms of the total Chern classes (of the tangent bundles) of X and Y by

$$c_* = f^*c_*(Y)/c_*(X) .$$

Now, the first few multiple-point formulas are

(2.2.1)
$$m_2 = f^*f_*m_1 - c_n m_1$$
$$m_3 = (f^*f_*m_2 - 2c_n m_2) + (\sum_{i=1}^{n} 2^i c_{n-i}) m_1$$
$$m_4 = (f^*f_*m_3 - 3c_1 m_3) + 6c_2 m_2 - 6(c_1c_2 + 2c_3) m_1 .$$

Here the double-point formula and the triple-point formula are given for arbitrary n , but the quadruple-point formula only for $n = 1$. The cases $n = 2,3$ have also been worked out, but no pattern in n is apparent and the case $n = 1$ is of primary interest in the present work.

(2.3) Lemma. Return to the setup of (2.1) and put

$$e = g^*c_1E , \quad t = g^*c_1C .$$

Then these Chern classes satisfy the relations,

(2.3.1) $$e^2 = 0 , \quad et = 0 , \quad t^2 = 0 \quad \text{and} \quad eh^{r-1} = h^r .$$

Moreover, the total Chern class c_* of f is given by the following expression (which is independent of r):

$$(2.3.2)\qquad c_* = 1 + \Sigma_{i\geq 1}\left[\binom{n+1}{i}h - \binom{n+1}{i-1}t + \binom{n}{i-1}e\right]h^{i-1}$$

Proof. The relations (2.3.1) hold because C is a curve and E has rank

There is a subsheaf F of E that is free, locally a summand, and of corank 1. For example, such an F is generated by the linear functionals defining the general (n+1)-plane section D of Z. Thus there is an exact sequence

$$0 \to F \to E \to L \to 0$$

in which L is invertible, and obviously $L = \det(E)$. Dualizing this sequence, pulling it up to X, tensoring with $O_X(1)$ and taking Chern classes yields

$$c_*((g^*E^*)(1)) = (1 - e + h)(1 + h)^{r-1}.$$

Therefore, the standard formulas,

$$c_*(X) = g^*c_*(C)\ c_*((g^*E^*)(1)),$$
$$c_*(Y) = (1 + c_1(O_Y(1)))^{r+n+2},$$

and the definition $c_* = f^*c_*(Y)/c_*(X)$ yield

$$c_* = (1 + h)^{n+2}/(1 + t)(1 + h - e).$$

The relations $t^2 = 0$ and $e^2 = 0$ and the Taylor expansion of $(1 + h - e)^{-1}$ in e yield

$$c_* = (1 + h)^{n+2}(1 - t)[(1 + h)^{-1} + (1 + h)^{-2}e].$$

It is now a straightforward matter to obtain the asserted expression, (2.3.2).

(2.4) Lemma. Retain the setup of (2.1) and the additional notation e, t of (2.3).

(i) Then, $\int_X h^r = d_1$, $\int_X eh^{r-1} = d_1$, and $\int_X th^{r-1} = 2 - 2p$

(ii) Let x be the class of a q-cycle on X. Then

$$f^*f_*x = (\int_X xh^q)h^{r+n-q}.$$

Proof. (i) The first formula is treated in (2.1), and the second follows immediately from it and the relation $eh^{r-1} = h^r$ of (2.3). Now, in (2.1) it is proved that the inverse image D' of a general (r-1)-plane in Y maps birationally onto C. It follows that

$$\int_X th^{r-1} = \int_C c_1(C) .$$

Alternatively, this formula may be derived directly by additivity from the formula $\int_X g^*[y]h^{r-1} = 1$ where $y \in C$.

(ii) By Bezout's theorem, f_*x is equal to its degree times the class of a linear space of codimension $r + n - q$. By the projection formula,

$$\deg(f_*x) = \int_X xh^q .$$

(2.5) Theorem. In the setup of (2.1) and additional notation e,t of (2.3), the multiple-point formulas (2.2.1) become

$$m_2 = (d_1 - n - 1)(h^n) + (\tfrac{1}{2})(n+1)n(th^{n-1}) - n(eh^{n-1}) \quad \text{for } n \leq r ,$$

$$m_3 = [d_1^2 - (4n+3)d_1 - n(n+1)p + 2(n+1)(2n+1)]h^{2n}$$
$$+ n(n+1)(d_1 - 4n - 2)h^{2n-1}t + 2n(-d_1 + 3n + 2)h^{2n-1}e \quad \text{for } 2n \leq r ,$$

$$m_4 = (d_1^3 - 15d_1^2 + 74d_1 - 6d_1p + 36p - 120)h^3 + (3d_1^2 - 39d_1 - 6p + 132)h^2e$$
$$+ (3d_1^2 - 45d_1 - 6p + 180)h^2t \quad \text{for } n = 1 .$$

Proof. The assertion is a formal consequence of Lemmas (2.3) and (2.4); the computations are straightforward but tedious.

(2.6) Theorem. The formulas (1.1) of the introduction are formal corollaries of the general multiple-point formulas (2.2.1), with

$$(i!)d_i = \int_X m_i h^{r-i+1}$$

Proof. The assertion is a simple formal consequence of (2.5) with n=1 and (2.4.i).

(2.7) Remark. The factor of i! in (2.6) reflects an undesirable feature of the method of iteration, although this matter is of no consequence for numerical formulas like (1.1). The factor is present essentially because the cycles m_i enumerate not the i-fold points but the ordered i-tuples of points with the same image. The elimination of the factor through a more refined enumeration is one advantage of the method of the Hilbert scheme. Other advantages include less stringent hypotheses and substantially simpler computations. However, at present the method of the Hilbert scheme yields no ith-order formulas for $i \geq 4$ for ramifie maps, although in principle it may do so yet and do so more simply.

(2.8) Remark. Formulas for d_2 and d_3 for arbitrary n and r similar to (1.1) can be obtained just as simply from (2.5) and (2.4); these formulas are

$$2d_2 = (d_1 - n)[d_1 - (n+1)] - n(n+1)p \quad \text{for } n \leq r ,$$

$$6d_3 = [d_1 - 2n][d_1 - (2n+1)][d_1 - (2n+2)] - n(n+1)[3d_1 - 4(2n+1)] \quad \text{for } 2n \leq r$$

With more effort but in the same way, formulas for d_4 for $n = 2,3$ and for d_5 and d_6 for $n = 1$ can be obtained from the corresponding general multiple-point formulas, which are given in [Kleiman, 1980, 1981] In fact, in principle, it is possible to work out formulas for m and then d_i for arbitrary i, n and r; the method of iteration provide an effective mechanical procedure for doing so. Moreover, the work in Section 3 provides a large class of r-folds in $\mathbb{P}^{r+n}$ for which all these formulas hold. However, there seems to be little point is explicitly working out any of these formulas before they are needed.

3. Meaning and Validity.

In this section, it will be shown that Formulas (1.1) and their generalizations in (2.5) and (2.8) hold for many ruled varieties and mean they should. First, however, some more general multiple-point theory will be recalled and some new general results proved.

(3.1) The formulas and genericity. (See [Kleiman, 1980, 1981]) Let $f: X \to Y$ be a proper map between irreducible quasi-projective varietie (Less stringent but more involved hypotheses in fact suffice.) Basic to the method of iteration are the "derived" maps $f_i: X_{i+1} \to X_i$ of f They are defined inductively as follows: let f be f, let f_i be the composition $f_i = p_2 p$ in the following diagram

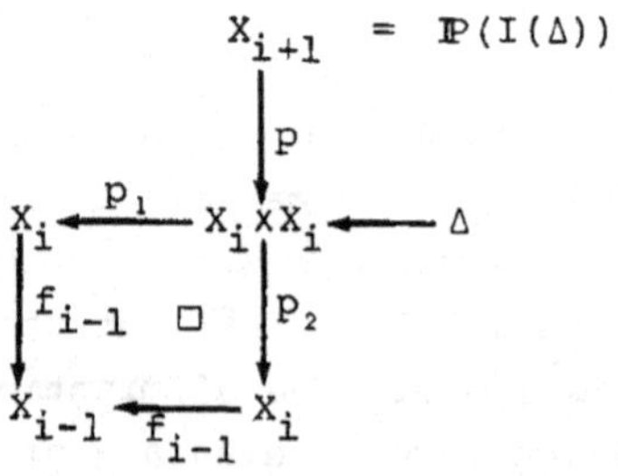

$$\boxed{f_i = p_2 p}$$

Here the square is cartesian, p_1 and p_2 are the projections, Δ is the diagonal, $I(\Delta)$ is its ideal, and p is the structure map. Note that the formation of the f_i commutes with every base-change $Y_1 \to Y$.

The i-fold point cycle m_i is defined as the pushout to X of the fundamental cycle of X_i,

$$(3.1.1) \qquad m_i = (f_1 \ldots f_{i-1})_*[X_i] .$$

Note that m_i has support in the locus M_i of i-fold points of f, which is equal to the set-theoretic image of X_i.

The corresponding class (again say modulo rational equivalence) is also denoted m_i and is called the i-fold point class. Each class m_i is given by an expression, called the i-fold-point formula, in terms of the preceding m's and the Chern classes of f (see (2.2)).

The double-point formula (i=2) is a special case of another general formula, the residual-intersection formula. The higher-order formulas ($i \geq 3$) are derived mechanically by recursion; pushing out to X the (i-1)-fold-point formula for f_1 and identifying the terms yields the i-fold-point formula for f.

The i-fold-point formula for f ($i \geq 2$) is valid when the i-maps $f_0 (=f)$, $f_1, \ldots, f_{i-1}$ are all local complete intersections of the same codimension. When this condition obtains, f is called i-generic. The condition to be i-generic is, however, overly stringent. So consider the condition that f be a local complete intersection, say of codimension n, and that there exist a closed set F of Y such that $f^{-1}F$ has codimension at least $(i-1)n+1$ in X and such that the restriction of f over $Y - F$ is i-generic. An f satisfying this condition is called practically i-generic and, for such an f too, the i-fold-point formula is valid.

Suppose that X is regularly embedded as a Y-scheme in a smooth, quasi-projective Y-scheme P. For example, if X and Y are smooth, then the graph map $X \to X \times Y$ is an appropriate embedding. Then f is a local complete intersection of codimension,

$$n = \text{cod}(f) = \dim(Y) - \dim(X) .$$

Moreover, each irreducible component X_i^1 of X_j is such that

$$\dim(X) - \dim(X'_j) \leqslant (j-1)n .$$

Equality holds here for $j = 2,\ldots,i$ if f is i-generic. Conversely, if equality holds for each component and for each j and if X is Cohen-Macaulay, then f is i-generic and each X_j is Cohen-Macaulay to Furthermore, if f is i-generic, then each X_j is regularly embedded i a smooth quasi-projective X_{j-1}-scheme.

(3.2) Curvilinear subschemes (See [Le Barz, 1980, §1].) A 0-dimensiona k-scheme A is called curvilinear if it is embeddable in a smooth 1-dimensional k-scheme. Clearly, A is curvilinear if and only if, at each point x of A, the embedding dimension, or $\dim(M_x/M_x{}^2)$, is at most 1. In other words, A is curvilinear if and only if the sheaf of differentials $\Omega^1_{A/k}$ has rank at most 1 at each point. Obviously, if A is curvilinear, then it is a local complete intersection.

Let X be a smooth k-scheme of pure dimension r. It is well known that the length-i subschemes of X which are local complete intersections form an open subset U of the Hilbert scheme $H = \mathrm{Hilb}^i_{X/}$ and that is smooth of pure dimension ir. Let W denote the universal family. Then the points of H over which the sheaf $\Omega^1_{X/H}$ has rank at most 1 form another open set U_1. Obviously, U_1 parametrizes the curvilinear length-i subschemes. So $U_1 \subset U$. Therefore U_1 is smooth and of pure dimension ir. Moreover, it is not hard to prove that the subschemes of X consisting of i distinct points form a dense open subset S of U_1.

(3.3) The case of no $\overline{S}_2$-singularities (see [Kleiman, 1982]). Let $f\colon X \to Y$ be a proper map with no $\overline{S}_2$-singularities between irreduci quasi-projective varieties, such as the map f of (2.1), see (2.2). Fix $i \geqslant 1$. Then there is a natural closed embedding of X_i into th i-fold self-product $X[i]$,

$$X_i \hookrightarrow X[i] = X\times_Y \cdots \times_Y X ,$$

the embedding is an isomorphism off the union of the diagonals, the derived map $f_{i-1}\colon X_i \to X_{i-1}$ is induced by the projection of $X[i]$ onto the last $i-1$ factors, and each permutation of the factors of $X[i]$ induces an automorphism of X_i over k.

Let $(x_1,\ldots,x_i)$ be a k-point of $X[i]$. Denote the common image of the x_i in Y by y, and consider the subscheme

$$A(x_1,\ldots,x_i) \;=\; \mathrm{Spec}\,(O_{f^{-1}y}/M_{x_1}\cdots M_{x_i}) \;\subset\; f^{-1}y$$

where the M's are the ideals of the x's. It can be proved that $(x_1,\ldots,x_i)$ lies in X_i if and only if $A(x_1,\ldots,x_i)$ has length-i. In fact, the $A(x_1,\ldots,x_i)$ of length-i are the fibers of a natural flat family A of length-i subschemes of X/Y parametrized by X_i. In addition, the structure sheaf O_A possesses an i-step filtration whose jth successive quotient is an invertible sheaf supported on the graph of $p_j|X_i$, where p_j is the jth projection of X[i], and A is universal for such families of filtered length-i subschemes of X/Y.

Assume that f is finite. Then, obviously, all the f_j are finite, too. Assume in addition that X and Y are smooth. Then in view of (3.1) it is clear that the following statements are true: (i) for each $j \geq 2$, each component M_j' of the locus M_j of j-fold points has codimension in X at most $(j-1)n$ where $n = \dim(Y) - \dim(X)$; (ii) for $j = 2,\ldots,i$, equality obtains in (i) for each M_j' if and only if f is i-generic; (iii) if f is i-generic, then for $j = 2,\ldots,i$ the support of the cycle m_j is all of M_j and the j-fold-point formula holds.

(3.4) <u>Proposition</u>. In the setup of (2.1) assume that f is finite. Then for $j \geq 2$ each component of the locus M_j of j-fold points of f is of dimension at least $r-(j-1)n$. If each component is just of this dimension for $j = 2,\ldots,i$, then f is i-generic, the support of the i-fold-point cycle m_i is all of M_i and the formulas for m_i and d_i in (2.5), (1.1) and (2.8) are valid.

<u>Proof</u>. The assertion is immediate from (3.3), (2.5), (2.6), and (2.8). Note, however, that the definition of d_i involved here is more refined than the one used in Section 1.

(3.5) <u>Proposition</u>. In the setup of (2.1) assume that Z is a cone with vertex F over a compatibly ruled variety $\tilde{Z}$; that is, the rulings of Z are the joins of F with the rulings of $\tilde{Z}$. Then for $j \geq 2$ the j-fold-point locus M_j and cycle m_j are the "rulingwise" joins of F with their counterparts $\tilde{M}_j$ and $\tilde{m}_j$, except that $M_j = F$ and $m_j = 0$ if $\tilde{M}_j$ is empty. If each component of M_j (or $\tilde{M}_j$) has the right codimension, $(j-1)n$, for $j = 2,\ldots,i$, if $\tilde{f}$ the counterpart of f is finite and if the dimension of F is at most $r-(i-1)n-2$, then f is practically i-generic, the support of m_i is all of M_i

and the formulas for m_i and d_i in (2.5), (1.1), and (2.8) are valid; moreover, the formula for d_i coincides with its counterpart.

Proof. Note that no component of m_j can lie in $f^{-1}F$; indeed, since $f^{-1}F = C x_k F$, by base change $X_j x_Y F = C[j] x_k F$. Note that $Y - F$ is a locally trivial bundle over $\tilde{Y}$, the counterpart of Y , and that the restriction of f over $Y - F$ is isomorphic to $\tilde{f} x_{\tilde{Y}} (Y - F)$. Note that, if $\dim(F)$ is at most $r - (i-1)n - 2$, then $\operatorname{cod}(f^{-1}F, X)$ is at least $(i-1)n + 1$. Finally note that p and d_i are obviously equal to their counterparts. The assertions are now easily obtained from (3.1), (3.3), (2.5), (2.6), and (2.8).

(3.6) Lemma. Let X be a smooth, r-dimensional, closed subvariety of a projective space P over k and assume that X lies in no hyperplane. Fix $i \geq 2$. Let $f: X \to Y$ with $Y = \mathbb{P}^{r+n}$ be a projection from a center L , and consider this condition on it: f has no $\overline{S}_2$-singularity and for $j = 2, \ldots, i$ each length-j subscheme of each fiber of f generates a $(j-1)$-plane M in P . Assume that the condition is satisfied if L belongs to a certain open subset of the Grassmannian of codimension-$(r+n+1)$ planes of P .

Then there is a smaller open set such that, if L belongs it, then the following statements hold: (i) f is i-generic and the j-fold-point formula holds for $j \leq i$; (ii) the j-tuples $(x_1, \ldots, x_j)$ of distinct points of X with the same image under f form a dense open subset D_j of X_j for $j \leq i$,; (iii) the j-tuples $(x_1, \ldots, x_j)$ of D_j such that f is not ramified at any of the x's form a dense open subset R_j for $j \leq i - 1$; (iv) in characteristic zero, X_j is reduced for $j \leq i$ and (v) in any characteristic, X_j is reduced and, for every j-tuple $(x_1, \ldots, x_j)$ of k-points of X in a certain dense open set of x_j , the embedded tangent spaces to X in P at the x's are embedded by f in Y as r-planes in general position through a point for $j \leq i$ provided that the embedded tangent spaces to X at some j-tuple of distinct points are in general position and that R_j is dense (by (iii) it is dense for $j < i$).

Proof. Fix j and consider a flat family $w: W \to H$ of length-j subschemes of X/k . Say $P = \mathbb{P}(B)$. Take the canonical map $B_P \to O_P(1)$, pull it back to W , and form the adjoint on H , getting

$$u: B_H \to w_* O_W(1) .$$

Denote the open set where u is surjective by U. The restriction $u|U$ defines a map into the Grassmannian,

$$G : U \to \mathrm{Grass}_{j-1}(P) .$$

Note that the k-points μ of U are just the k-points of H that represent length-j subschemes A of X that generate $(j-1)$-planes M of P; moreover, $G(\mu)$ represents M.

For each codimension-$(r+n+1)$ plane L, let $\sigma(L)$ denote the Schubert variety of $(j-1)$-planes M such that

$$\dim (M \cap L) \geqslant j - 2 .$$

Then $\sigma(L)$ has codimension $(r+n)(j-1)$. Moreover, if U is of pure-dimension, then each irreducible component of $G^{-1}\sigma(L)$ will have codimension $(r+n)(j-1)$ and it will meet any given dense open subset of U, whenever L belongs to an appropriate dense open subset of the Grassmannian; indeed, these statements hold because a Grassmannian admits a transitive group action (see [Kleiman, 1974]). By the same token, in characteristic zero, $G^{-1}\sigma(L)$ will be smooth if U is.

Apply the above observations as follows. Take $2 \leqslant j \leqslant i$ and take for H the open subscheme of $\mathrm{Hilb}^j_{X/k}$ parametrizing the curvilinear subschemes, see (3.2). Since X is smooth, irreducible, and r-dimensional, H is smooth, irreducible, and jr-dimensional. Let L be a general plane of codimension $(r+n+1)$. Then, by the above, each irreducible component of $G^{-1}\sigma(L)$ has dimension $r-n(j-1)$ and meets any given dense open subset of H; moreover, in characteristic zero, $G^{-1}\sigma(L)$ is smooth.

Consider X_j. Since, by hypothesis, f has no $\overline{S}_2$-singularities, the facts recalled in (3.3) imply that the j-tuples of distinct points of X form an open subset D_j of X_j and that there is a natural map $d : X_j \to U$, whose (set-theoretic) image is just $G^{-1}\sigma(L)$ and whose fibers are finite. Since X_j is complete, d is finite. Consider the open set S of H representing subschemes of distinct points. Then, by the above, $S \cap G^{-1}\sigma(L)$ is dense in $G^{-1}\sigma(L)$. Therefore $d^{-1}(S \cap G^{-1}\sigma(L))$ is dense in X_j and equal to D_j. Thus (ii) holds.

As recalled in (3.1), each irreducible component of X_j has dimension at least $r-n(j-1)$. Now $d : X_j \to U$ is finite. Hence, since $d(X_j)$ is of pure dimension $r-n(j-1)$, so is X_j. This is true for $j \leq i$. Therefore f is i-generic and the j-fold-point formula holds (see (3.1)); that is, (i) holds. Since X is smooth, so Cohen-Macaulay, and since f is j-generic, each X_j is Cohen-Macaulay.

Let $(x_1,\ldots,x_j)$ be a k-point of D_j. It is easy to see in view of (3.3) that f is ramified at one of the x's, say x_1, if and only if the point $(x_1,x_1,\ldots,x_j)$ of $X[j+1]$ lies in X_{j+1}. Suppose $j \leq i-1$. Then X_j is of pure dimension $r-n(j-1)$ and x_{j+1} is of pure dimension $r-n_j$. Therefore (iii) holds.

It is easy to see in view of (3.3) that the symmetric group acts freely on D_j and that the quotient is an open subset of $G^{-1}\sigma(L)$, even scheme-theoretically. Suppose the characteristic is zero. Then $G^{-1}\sigma(L)$ is smooth. Hence D_j is smooth. Since D_j is dense in X_j and since X_j is Cohen-Macaulay, X_j is reduced. Thus (iv) holds.

Assume the hypotheses of (v). Observe that then the unordered j-tuples of distinct points of X whose tangent spaces are in general position form a dense open subset of H, which meets $d(R_j)$ in a dense open subset T of it. Next, observe that the (rather intelligent) proof of Lemma 7 in [Roberts, 1971] yields more than is stated; in our context, it yields that for every j-tuple $(x_1,\ldots,x_j)$ of $d^{-1}T$, the embedded tangent spaces of the x's are embedded by f in Y as j r-planes in general position through a point. It is now a straightforw. matter to prove that $X[j]$ is smooth along $d^{-1}T$ (see for example [Roberts, 1975, Prop. 5.4] or [Kleiman, 1981, proof of Prop. 4.6]). Since $d^{-1}T$ is a dense open subset of both R_j and X_j and since X_j is Cohen-Macaulay, therefore X_j is reduced. Thus (iv) holds.

(3.7) Proposition. In the setup of (2.1), assume (1) f factors into an embedding of X into a projective space P followed by a general central projection and (2) for $j = 2,\ldots,i$ a length-j subscheme of X spans an (j-1)-plane of P whenever the restriction of g is an embedding of the subscheme into C. Then f is i-generic and the formulas for m_j and d_j in (2.5), (1.1) and (2.8) are valid for $j \leq i$.

Assume in addition either (3) $j \leq i$ and the characteristic is zer or (4) $j \leq i-1$ and the embedded tangent spaces to X in P at some j-tuple of distinct points are in general position. Then m_j is represented by $(j-1)!$ times the reduced cycle on M_j, which is equal to the closure of the strict j-fold points of f, and d_j is equal to the degree of the closure Z_j of the set of points of Z at which j distinct branches meet. Moreover, if (3') $j \leq i-1$ and the characteristic is zero (resp. if (4)), then Z_j is the closure of the set of points at which exactly j distinct smooth branches meet (resp. meet transversally).

Proof. Clearly g restricts to an embedding on each fiber $f^{-1}y$. Hence the hypotheses of (3.6) are satisfied. The conclusions of (3.6), together with (2.5), (2.6) and (2.8) yield the assertion.

(3.8) The class. Each member is given as in (2.1), but the general setup is qualified as follows. The parameter curve C is equipped with a distinguished embedding in a projective space $\mathbb{P}(V')$. The sheaf E on C is such that $E(-1)$ is generated by its global sections, and a particular space V'' of sections that generate is fixed. Finally, V is a general $(r+n+1)$-dimensional subspace of $V''\otimes V'$, and $u: V_C \to E$ is the induced map. Hence f factors into a sequence of embeddings,

$$X \hookrightarrow \mathbb{P}(V'') \times \mathbb{P}(V') \hookrightarrow \mathbb{P}(V'' \otimes V') = P$$

followed by a general projection into $Y = \mathbb{P}(V)$. Because C is embedded in $\mathbb{P}(V')$, the induced map $C \to \mathrm{Grass}_{r-1}(Y)$ is also an embedding.

Call X i-twisted in P if the embedding of C in $\mathbb{P}(V')$ is $(i-1)$-twisted in the sense of [van zur Gathen, 1980, 1.3], that is, if any length-i subscheme of C spans an $(i-1)$-plane. Call X strongly i-twisted in P if the embedding of C in $\mathbb{P}(V')$ is $(i+1)$-twisted and if the tangent lines to C at some i-tuple of distinct points are in general position. Note that X is strongly i-twisted if it is 2i-twisted.

The embedding of C in $\mathbb{P}(V')$ is i-twisted if it is defined by a complete linear system of degree at least $2p+i$, or if it is defined by a general complete linear system of degree at least $p+2i$. The first assertion is a simple consequence of the Riemann-Roch theorem; it is a straightforward generalization of the standard case, $i = 1$. The second assertion results from a short argument using the Jacobian. Both assertions are proved in [von zur Gathen, 1980, 1.7].

Condition (2) of (3.7) is satisfied when X is i-twisted, and Condition (4) is satisfied but with $i+1$ for i when X is strongly i-twisted. Indeed, it is a simple basic property of the Segre embeddin that the product of a $(j-1)$-plane and an $(r-1)$ plane spans a $(jr-1)$-plane.

Note that if X is i-twisted or strongly i-twisted in P then it is j-twisted or strongly j-twisted in P for $j \leq i$.

(3.9) Theorem. Consider a ruled r-fold Z in $Y = \mathbb{P}^{r+n}$ belonging to the class described in (3.8). If the normalization X of Z is i-twisted in its given ambient space P, then $f: X \to Y$ is i-generic and the formulas for m_i and d_i in (2.5), (1.1) and (2.8) are valid.

If X is i-twisted and the characteristic is zero or if X is strongly i-twisted, then m_i is represented by $(i-1)!$ times the redu cycle on M_i, which is equal to the closure of the strict i-fold points of f, and d_i is equal to the degree of the closure Z_i of of the set of points of Z at which i distinct branches meet. Moreover, if X is $(i+1)$-twisted and the characteristic is zero (resp. if X is strongly i-twisted), then Z_i is the closure of the set of points at which exactly i distinct smooth branches meet (resp. meet transversally.

Proof. The assertion follows immediately from (3.8) and (3.7).

(3.10) Remark. It is not hard to see that there is a number d depending only on r, p and i such that, if the degree d_1 of X is at least d, then Condition (2) of (3.7) automatically holds (G. Sacchiero, private communication, 25/7/1981). The idea of the proof is to construct a unisecant which is linearly normal and nonspecial and which contains the given lengh-j subscheme by cutting X with a suitab linear space (which contains as many rulings as possible). This method of satisfying Condition (2) has the rather desirable advantage of dealing directly with any ruled r-fold given as a central projection of its normalization, which must only be linearly normal and nonspecial; there is no need for the kind of reembedding of X implicit in (3.9). Howev it seems that (3.9) yields the existence of ruled r-folds of smaller degree for which the formulas hold.

Hilbert, in this 15th problem [Hilbert, 1900], asked us "to establish rigorously and with an exact determination of the limits of their validity those geometrical numbers which," like the d_i, were "determined" by the early enumerative geometers. Certainly we must prove that these numbers are valid in some, if not most, cases to give them enumerative significance, and in the more cases, the better; however, an exact determination of the limits of their validity seems to be rather difficult.

4. Complements.

Throughout this section, the notation and hypotheses will be that of (2.1), except that the hypothesis that $f': X \to Z$ be birational will be dropped in (4.1) (and nowhere else).

(4.1) Observations concerning the birationality of f'.

(i) Assume that f' is not birational. Assume that the characteristic is zero or that the separable degree of f' is greater than 1. Then Z is an r-plane and equal to Y.

(ii) Drop the hypothesis that f' be birational. Let T be an r-plane in Y, and let T^* denote the dual space viewed as a subvariety of of the Grassmannian $\mathrm{Grass}_{r-1}(Y)$. Then Z is equal to T if and only if the image C^* of C lies in T^*. Moreover, when these equivalent conditions obtain, then Z, T and Y coincide, f' is separable, and $\deg(f') = \deg(C^*)$.

Proof. (i)(a) Let A and B be distinct (r-1)-planes contained in Z. Suppose that $A \cap B$ contains a smooth point z of Z. Then both A and B lie in the embedded tangent space T to Z at z, which is an r-plane. Hence A and B span T and $A \cap B$ is an (r-2)-plane.

(b) The hypotheses imply that there is a nonempty open subset U of Z such that, for each $z \in U$, the fiber $f'^{-1}g$ has at least two distinct points. By the hypotheses of (2.1), there is a finite subset S of C such that $C - S$ is mapped isomorphically into $\mathrm{Grass}_{r-1}(Y)$. Let $U' = U - fg^{-1}(S)$. Then U' is a nonempty open subset of Z, and each $z \in U'$ lies in at least two distinct rulings. Fix a ruling A that meets U'. The points of C representing rulings that meet A form a closed set, namely, $g(f^{-1}A)$. If this set were not all of C, it would be finite and, as B ranges over the rulings distinct from A, the intersections $A \cap B$ would not sweep out $A \cap U'$ for reasons of dimension, a contradiction. Hence every ruling B meets A and the intersections $A \cap B$ for $B \neq A$ sweep out A.

(c) Fix two distinct rulings A and B containing a common smooth point z of the open set U' of (b). Then A and B span an r-plane T by (a). Let D be a general ruling. Then $A \cap D$ and $B \cap D$ are distinct (r-2)-planes by virtue of (a) and (b). Since these (r-2)-planes lie in T, so does their span D. Hence every plane of the family lies in T. Thus $Z = T$. Finally, since Z spans Y by a hypothesis of (2.1), therefore $Z = Y$.

(ii) Obviously, Z is equal to T if and only if each ruling lies in T, if and only if the image C^* of C lies in T^*. Assume that these equivalent conditions obtain. Then Z and T coincide with Y because Z spans Y by a hypothesis of (2.1). Now, let z be a general point of T and let H denote the corresponding hyperplane in T^*. It is not hard to see that $g : X$ carries $f^{-1}z$ isomorphically onto the inverse image of $H \cap C^*$. Since H is general, $H \cap C^*$ is smooth and consists of $\deg(C^*)$ distinct points and it is isomorhpic to its inverse image in C. Therefore f' is separable and $\deg(f') = \deg(C^*)$.

(4.2) <u>Observations concerning the finiteness of</u> f.

(i) f is not finite if and only if Z is a cone. In fact, a fiber $f^{-1}z$ is infinite in just these two cases: (a) Z is a cone but not a r-plane, and z lies on the vertex; (b) Z is an r-plane, the family ruling Z is a (linear) pencil, and z lies on the axis. In both cases, z lies on every ruling.

<u>Proof.</u> Assume that $f^{-1}z$ is infinite. Then $g(f^{-1}z)$ is infinite and closed in C, so it is all of C. Thus z lies on every ruling. Now, let y be any point of Z other than z. Then y is in some ruling, and it contains z; hence the line joining y and z lies in Z. Therefore Z is a cone and z lies on the vertex. If Z is an r-plane, then, since f' is birational, (4.1,ii), implies that the family is a pencil and so, since z lies on every member, it lies on the axis.

Conversely, assume that Z is a cone and let z be a point of the vertex. If every ruling passes through z, then $f^{-1}z$ is infinite. Suppose that some ruling R does not pass through z. Then the join of R and z is an r-plane contained in Z, so equal to Z. Hence (4.1,ii) implies that the family is a pencil and so, since z does not lie on R, it does not lie on the axis.

(ii) Let F denote the fundamental locus of f; by definition, F consists of the points z of Z such that $f^{-1}z$ is infinite. By (i), F is a linear space; in fact, F is just the intersection of all the rulings.

Assume that F is nonempty and that Z is not an r-plane. Then it follows from (i) that Z is a cone with vertex F, that the base $\tilde{Z}$ is ruled by the projections of the rulings of Z and that $\tilde{f}$, the counterpart of f, is finite. Hence (3.5) is applicable.

(iii) f is practically 2-generic if and only if the singular locus S of Z has codimension at least n. Indeed, use (i), (ii) and apply (3.5) or (3.4), noting that $f' : X \to Z$ is an isomorphism off $S \cup F$ by Zariski's main theorem.

Assume Z is a "form" or "primal;" that is, $n = 1$. Then, in particular, f is always practically 2-generic. However, even if f is finite, it need not be practically 3-generic. Indeed, an example may be constructed simply by modifying the construction in (3.6) by replacing the embedding of C in $\mathbb{P}(V')$ by any birational map with a triple-point.

(4.3) <u>Example</u>: <u>The Segre variety</u>. This is the variety $Z = \mathbb{P}^{r-1} \times \mathbb{P}^1$ embedded in $\mathbb{P}^{2r-1}$ (so $n = r-1$) with degree $d_1 = r$; it is naturally ruled with $f' = 1_Z$ and $g = p_2$. Trivially f is i-generic for all i, and it is easy to check directly that all the formulas in (2.5) and (2.8) hold and yield 0.

The Segre variety is the only r-fold Z ruled by (r-1)-planes in $\mathbb{P}^{r+n}$ such that $n \leq r-1$ and f is an embedding; in fact, it is the only Z such that $n \leq r-1$, the singular locus of Z has codimension n at least, $d_2 = 0$ and f is finite. Indeed, since f is finite, $f^*O(1)$ is ample; hence $d_2 = 0$ implies $m_2 = 0$. By (4.2,iii) f is practically

2-generic, so by (3.1) the formula for m_2 in (2.5) holds. Multiplyi both sides of this formula by eh^{r-1-n} and using (2.3.1) and (2.4,i) yields $d_1 = n + 1$. Then the formula for d_2 in (2.8) yields $p = 0$ Hence E splits into a direct sum of r invertible sheaves $O(e_i)$. Since E is generated by its global sections, $e_i \geq 0$ for all i. If $e_i = 0$ for some i, then Z would be a cone and f would not be finite; hence $e_i \geq 1$ for all i and so $\Sigma e_i \geq r$. On the other hand, $\Sigma e_i = d_1$ by (2.4.1). However $d_1 = n + 1 \leq r$. Hence $e_i = 1$ for all i and $n = r - 1$. Therefore Z is the Segre variety.

If $n = r$, if the singular locus of Z is finite (or empty) and i f is finite (for example, the latter conditions obtain in the setup of (3.7)), then f is an embedding if and only if

$$(d_1 - n)(d_1 - n - 1) = n(n + 1)p .$$

Indeed, the formula for d_2 in (2.8) holds by (4.2,iii) and (3.4); hence $d_2 = 0$. On the other hand, it is easy to see in view of (3.1) and (3.3) that f is an embedding if and only if X_2 is empty, and that, when $n = r$ and f is finite, then X_2 is empty if and only if $d_2 = 0$.

Finally, assume that Z is normal and that f is finite. Then f is an embedding by Zariski's main theorem. Hence, if $n \leq r-1$, the Z is the Segre variety.

(4.4) <u>Example</u>: <u>The hyperplane pencil</u>. This is the linear family of $(r-1)$-planes through an $(r-2)$-plane F in an r-plane Y. Here Z is equal to Y, so $n = 0$ and Z is smooth. However, f is not smooth; in fact, the locus M_i of i-fold points of f is obviously equal to $f^{-1}F$ for all $i \geq 2$. (By 3.5), f is practically i-generi for all i and the formulas for m_2 and m_3 in (2.5) and those for d_2 and d_3 in (2.8) are valid. Since $d_1 = 1$ and $n = 0$, these formula yield 0's, as they should. Note that when the plane pencil was consid ered in the introduction, it was viewed as a plane form $(n=1)$; this poi of view is wrong for multiple-point theory and indeed the formulas for d_3 and d_4 gave incorrect results.

The hyperplane pencil and the Segre variety are the only two cases in which $n \leq r-1$ and Z is smooth; the latter case obtains if f is finite, the former if not. Indeed, if f is finite, then is the Segre variety by (4.3). If f is not finite, then we have hyperplane pencil by (4.2,i) because a smooth cone of dimension r s an r-plane.

4.5) Example: The Segre cone. This is the cone Z of dimension ≥ 3 whose base is the Segre variety $\mathbb{P}^n \times \mathbb{P}^1$ with $n \geq 1$. Its rulings are the joins of the vertex F with the rulings of the base. Obviously $d_1 = n+1$. Obviously, the locus M_i of i-fold points of f is equal to $f^{-1}F$ for all $i \geq 2$. Since cod $(F,Z) = n + 2$, by (3.5) then f is practically 2-generic but not practically i-generic for $i \geq 3$. Again by (3.5) the formula for m_2 in (2.5) and that for d_2 in (2.8) are valid, and it is easy to check using (2.4,i) that these formulas yield $m_2 = 0$ and $d_2 = 0$. On the other hand, it is clear directly that $m_2 = 0$, whence $d_2 = 0$.

Thus f is not an embedding although it is practically 2-generic (in fact, it is 2-generic) and $m_2 = 0$. Here $1 \leq n \leq r-1$. However, the case $n = 0$ is covered by the hyperplane pencil discussed in (4.4).

The Segre cone, the hyperplane pencil and the Segre variety are the only cases such that $n \leq r - 1$, such that the intersection F of all the rulings has codimension $n + 2$ at least, and such that either (a) the singular locus has codimension n at least and $d_2 = 0$; or (b) Z is normal off F. Notice that after all F has codimension just $n + 2$ (empty if $n = r - 1$) and that the three cases are distinguished by the value of n.

Indeed, if F is empty, then f is finite by (4.2.i) and Z is the Segre variety by (4.3). If Z is an r-plane, then we have a hyperplane pencil by (4.2,i). If F is nonempty and if Z is not an r-plane, then (4.2,ii) asserts that Z is a cone with vertex F over a compatibly ruled base $\tilde{Z}$ and that (3.5) holds. Now, obviously the codimension of $\tilde{Z}$ in its ambient space is also n, and obviously,

$$\dim(\tilde{Z}) = \operatorname{cod}(F,Z) \geq n + 1 .$$

Therefore (3.5) and (4.3) imply that $\tilde{Z}$ is the Segre variety. Thus Z is the Segre cone.

In particular, the only ruled form Z $(n = 1)$ such that $d_2 = 0$ and such that the intersection F of all the rulings has codimension in Z at least 3 is the smooth quadric surface in $\mathbb{P}^3$ if F is empty or the cone over this surface with vertex F if F is nonempty (whence F has codimension just 3). Of course, there is no ruled form such that cod (F) = 1 , and the only one such that cod (F) = 2 is, by (4.2,ii), a plane curve $\tilde{Z}$ if F is empty or the cone over a plane curve $\tilde{Z}$ with vertex F if F is nonempty; here $d_1 \geq 2$, and d_2 = if and only if $\tilde{Z}$ is smooth, if and only if Z is normal.

References

Baker, Henry F. [1933]: Principles of Geometry. Volume VI, Introducti to the theory of algebraic surfaces and higher loci, Cambridge University Press (1933), republished by Frederick Unger Publishing Co., New York (1960).

Hilbert, David [1900]: "Mathematical problems," translated by Dr. Mary Winston Newson, Bull. Am. Math. Soc. 50(1902), 437-479.

James, C.G.F. [1928]: "On the multiple tangents and multisecants of scrolls in higher space," Proc. Lond. Math. Soc. (2) 27(1927-28), 513-540.

Kleiman, Steven L. [1974]: "The transversality of a general translate, Compositio Math. 38(1974), 287-297.

_____ [1977]: The enumerative theory of singularities. Real and compl singularities, Oslo 1976, P. Holm, ed., pp. 297-396. Sijthoff & Noordhoof (1977).

_____ [1980]: "Multiple-point formulas for maps," originally prepared i connection with the celebration of the 10th anniversary of the founding of the Univ. Simon Bolivar, Caracas, held Jan. 28-31, 198 appeared in Vol. 3, Geometria algebrica y teoria de numeros, A. B. Altman coordinator, of the Simposia series of the university; reproduced with minor changes for a conference report on enumerati geometry and classical algebraic geometry, Nice, June (1981), Birkhäuser Boston, in press.

_____ [1981]: "Multiple-point formulas I: iteration," Acta Math. 147 (1981), 13-49.

_____ [1982]: "Multiple-point formulas II: the Hilbert scheme," in preparation.

Roberts, Joel [1971]: "Generic projections of algebraic varieties," Am. J. Math. 93(1971), 191-215.

_____ [1975]: "Singularity subschemes and generic projections," Trans. A Math. Soc. 212 (1975), 229-268.

_____ [1980]: "Some properties of double point schemes," Compositio Math. 41(1980), 61-94.

Roth, Leonard [1931]: "On plane-forms in four dimension," Proc. London Math. Soc. (2) 33 (1931-2), 115-144.

Segre, Corrado [1912]: "Mehrdimensionale Räume," Encylopädie der Mathematischen Wissenschaften mit einschluss ihrer anwendungen, III, 2, 2, C7 (Abgeschlossen Ende 1912), B. G. Teubner, Leipzig (1921-1934), 669-972.

von zur Gathen, Joachim [1980]: Sekantenräume von kurven. Inaugural-Dissertation Univ. Zurich (1980).

FONDAZIONE C.I.M.E.
CENTRO INTERNAZIONALE MATEMATICO ESTIVO
INTERNATIONAL MATHEMATICAL SUMMER CENTER

"Theory of Invariants"

is the subject of the First 1982 C.I.M.E. Session.

The Session, sponsored by the Consiglio Nazionale delle Ricerche, will take place under the scientific direction of Prof. FRANCESCO GHERARDELLI (Università di Firenze, Italy), at the Villa «La Querceta», Montecatini Terme, Italy, *from June 10 to June 18, 1982.*

Courses

a) ***Classical Theory of Invariants and Compactifications of Algebraic Symmetric Spaces.*** (8 lectures in English).
Prof. Corrado DE CONCINI (Università di Pisa, Italy).

b) ***Geometric Invariant Theory and Applications to Moduli Problems.*** (8 lectures in English).
Prof. David GIESEKER (IAS, Princeton, USA).

1. Introduction to geometric invariant theory.
2. Moduli of stable bundles on a smooth curve.
3. Moduli of stable curves.
4. Degeneration techniques in the study of the moduli space of stable vector bundles on a smooth curve.

References:

1. MUMFORD, D., Stability of Projective Varieties. ***L'Enseignement Mathématique XXIII*** fasc. 1-2, 1977.
2. NEWSTEAD, P.E., Lectures on Introduction to Moduli Problems and Orbit Spaces. New Delhi: Narosa Pub. House, 1978 (Tata Lecture Notes).

c) ***Infinite Root Systems, Representations of Quivers and Invariant Theory.*** (8 lectures in English).
Prof. Victor KAC (MIT, Cambridge, USA).

Contents:

Infinite root system and Weyl group.
Rosenlicht theorem, Vinberg's lemma and Weil conjectures.
Quivers and their representations, reflection functors, description of dimensions of indecomposable representations.
Schur representations and the problem of classification of prehomogeneous vector spaces.

References:

1. GABRIEL P., Unzerlegbare Darstellungen. ***Manuscripta Math. 6,*** 71-103 (1972)
2. KAC V.G., Infinite root systems, representations of graphs and invariant theory, ***Inventiones Math. 56,*** 57-92 (1980).
3. RINGEL C.M., Representations of K-species and bimodules. ***J. of Algebra. 41,*** 269-302 (1976).

Seminars

A number of seminars and special lectures will be offered during the Session.

FONDAZIONE C.I.M.E.
CENTRO INTERNAZIONALE MATEMATICO ESTIVO
INTERNATIONAL MATHEMATICAL SUMMER CENTER

"Thermodynamics and Constitutive Equations"

is the subject of the Second 1982 C.I.M.E. Session.

The Session, sponsored by the Consiglio Nazionale delle Ricerche, will take place under the scientific direction of Prof. GIUSEPPE GRIOLI (Università di Padova, Italy) at Noto, Italy, ***from June 23 to July 2, 1982.***

Courses

a) ***Thermodynamcs and Constitutive Relations*** (8 lectures in English).
Prof. Bernard D. COLEMAN (Carnegie-Mellon University, USA).

Lect. 1. Elementary applications of the Clausius-Duhem inequality: theories of elastic and viscous materials, various types of heat conductors, and materials with internal variables.

References:

1. B.D. COLEMAN & W. NOLL, ***Arch. Rational Mech. Anal. 13,*** 167-178 (1963).
2. B.D. COLEMAN & V.J. MIZEL, ***Arch. Rational Mech. Anal. 13,*** 245-261 (1963); ***J. Chem. Phys. 40,*** 1116-1125 (1964).
3. B.D. COLEMAN, ***Proprietà di Media e Teoremi di Confronto in Fisica Matematica,*** C.I.M.E. Session at Bressanone, 1963.
4. B.D. COLEMAN & M.E. GURTIN, ***J. Chem. Phys. 47,*** 597-613 (1967).
5. C. TRUESDELL, ***Rational Thermodynamics,*** McGraw-Hill, New York, (1969)

Lect. 2. Viscoelastic materials and theories of fading memory.

References:

1. B.D. COLEMAN & W. NOLL, ***Arch. Rational Mech. Anal. 6,*** 355-370 (1960); ***Rev. Modern Phys. 33,*** 239-249 (1961).
2. B.D. COLEMAN & V.J. MIZEL, ***Arch. Rational Mech. Anal. 23,*** 87-123 (1966); ***ibid. 29,*** 18-31 (1968); ***ibid. 30,*** 172-196 (1968).
3. B.D. COLEMAN & D.R. OWEN, ***Arch. Rational Mech. Anal. 55,*** 275-299 (1974).

Lect. 3. Thermodynamics of materials with memory.

References:

1. B.D. COLEMAN, ***Arch. Rational Mech. Anal. 17,*** (1964); ***ibid. 17,*** 230-254 (1964).
2. B.D. COLEMAN & V.J. MIZEL, ***Arch. Rational Mech. Anal. 27,*** 255-274 (1967).
3. W.A. DAY, ***The Thermodynamics of Simple Materials with Fading Memory,*** Springer Tracts in Natural Philosophy, Vol. 22, Springer-Verlag, Berlin, (1962).

Lect. 4. Thermodynamics of electromagnetic fields in dissipative media: magnetically active and electrically polarizable materials exhibiting dispersion and absorption, and various non-linear generalizations of them.

References:

1. B.D. COLEMAN & E.H. DILL, ***Arch. Rational Mech. Anal. 51,*** 1-53 (1973); ***Z.A.M.P. 22,*** 691-702 (1971).

Lect. 5. A theory of thermodynamics in which the existence and regularity of entropy and free energy as functions of state are proved rather than assumed, and conditions necessary and sufficient for the uniqueness of these functions can be given.

References:

1. B.D. COLEMAN & D.R. OWEN, ***Arch. Rational Mech. Anal. 54,*** 1-104 (1972).

Lect. 6 & 7. Examples of materials for which the regularity of entropy and free energy functions is not ***a priori*** evident and requires verification.

References:

1. B.D. COLEMAN & D.R. OWEN, ***Arch. Rational Mech. Anal. 59,*** 25-51 (1975); ***ibid. 70,*** 339-354 (1979); ***Rendiconti dell'Accademia Nazionale dei Lincei*** (VIII) ***61,*** 77-81 (1976); ***Annali di Matematica pura ed applicata*** (IV) ***108,*** 189-196 (1976).
2. B.D. COLEMAN, M. FABRIZIO, & D.R. OWEN, On the Thermodynamics of Second Sound in Dielectric Crystals, ***Arch. Rational Mech. Anal.,*** in press.

Lect. 8. Approaches to thermodynamics in which «absolute temperature» is a derived concept.
References:

1. C. TRUESDELL, ***The Tragicomical History of Thermodynamics,*** 1822-1854, New York, Springer-Verlag, (1980)
2. C. TRUESDELL & S. BHARATHA, The Concepts and Logic of Classical Thermodynamics as a Theory of Heat Engines, New York, Springer-Verlag, (1977).
3. J. SERRIN, ***Arch. Rational Mech. Anal. 70,*** 355-371 (1979); ***Lectures on Thermodynamics,*** University of Naples.
4. B.D. COLEMAN, D.R. OWEN, & J. SERRIN, ***Arch. Rational Mech. Anal. 77,*** 103-142 (1982).

b) Title to be communicated.
Prof. C.W. DAFERMOS (Brown University, USA).

c) ***Rational Thermodynamcs of Mixtures*** (8 lectures in English).
Prof. I. MULLER (TU Berlin).

1. Basic Concepts
1.1 Thermodynamic Fields; 1.2 Equations of Balance; 1.3 Constitutive Equations.

2. Thermodynamics
2.1 Entropy Principle; 2.2 Chemical Potentials; 2.3 Diffusion and Thermal Diffusion.

3. Sound Propagation in Mixtures
3.1 The Speed of Diffusion; 3.2 First and Second Sound.

4. Application to Liquid Helium
4.1 Status of Landau's Theory; 4.2 Helium in Rotation.

5. Kinetic Theory
5.1 Boltzmann Equation and Equations of Transfer; 5.2 Maxwellian Iteration; 5.3 Speed of Heat Conduction, Entropy Flux and Material Objectivity.

6. Outlook.

Literature:

On Equation of Balance:
TRUESDELL, C. & TOUPIN, The Chemical Field Theories, Handbuch der Physik III/1, Springer 1960
TRUESDELL, C., Rational Thermodynamics, McGraw-Hill (1969).

On the Entropy Inequality:
MULLER, I., Thermodynamik, Grundlagen der Materialtheorie, Bertelsmann Universitatsverlag, Dusseldorf (1973).
LIU, I-SHIH, Method of Lagrange Multipliers for Exploitation of the Entropy Principle. ***Arch. Rat. Mech. Anal. 40,*** (1972).
MULLER, I., Thermodynamics and Statistical Mechanics of Fluids and Mixtures of Fluids. Lecture Notes of CNR Summer School in Bevi and Canana.

On Liquid Helium:
LANDAU, L.D., The Theory of Superfluidity of Helium II, ***J. Phys.*** (V.S.S.R.) ***5,*** (1941).
LANE, C.T., Superfluid Physics, McGraw-Hill (1962).

On the Kinetic Theory:
CHAPMAN, S. & COWLING, T.G., Mathematical Theory of non-uniform Gases. Cambridge University Press (1961).
MULLER, I., On the Frame Dependence of Stress and Heat Flux, ***Arch. Rat. Mech. Anal. 45,*** (1972).

On the Speed of Heat Conduction:
CATTANEO, C., Sulla conduzione del calore, ***Atti del seminario mat. e fis. Univ. Modena. 3,*** (1948).
MULLER, I., Zur Ausbreitungsgeschwindigkeit von Storungen in kontinuierlichen Medien. Aachener Dissertation (1966).

Seminars

A number of seminars and special lectures will be offered during the Session.

FONDAZIONE C.I.M.E.
CENTRO INTERNAZIONALE MATEMATICO ESTIVO
INTERNATIONAL MATHEMATICAL SUMMER CENTER

"Fluid Dynamics"

is the subject of the Third 1982 C.I.M.E. Session.

The Session, sponsored by the Consiglio Nazionale delle Ricerche, will take place under the scientific direction of Prof. HUGO BEIRAO DA VEIGA (Università di Trento, Italy) at the «Villa Monastero», Varenna, Italy, ***from August 22 to September 1, 1982.***

Courses

a) ***Construction of weak solutions for Hyperbolic Problems*** (6 lectures in English).
Prof. C. BARDOS (Université Paris-Nord, France)

1. Study of a single conservation law and of the Hamilton-Jacobi equation by the viscosity method.
2. Hyperbolic systems. The Rankine Hugoniot condition and the different forms of the Entropy condition.
3. The construction of Di Perna of the solution of a two by two conservation Law in one space variable.
4. Numerical method. The Courant-Friedrichs-Lewy condition and the order of numerical method. Examples of numerical schemas.

Basic references

1. WHITHAM, Linear and non linear waves.
2. COURANT AND HILBERT, Methods of Mathematical Physics.
3. RITCHINGER AND MORTON, Difference Methods of Initial Value Problems.
4. P.L. LIONS, Generalized solutions of the Hamilton Jacobi equation, Pitman.

On the other hand many articles can be used as

1. R.J. DI PERNA, Convergence of approximate solution to conservation law (preprint).
2. P.D. LAX, Shock waves and entropy. In ***«Contribution to non linear functional analysis»*** (ZARANTONELLO), Academic Press, 603-634 (1971).
3. A. MAJDA and S. OSHER, Numerical viscosity and the entropy condition, ***Comm. Pure Appl. Math. 32,*** 797-838 (1979).
4. S.N. KRUCKOV, First order quasi linear equation with several independent variables, ***Math. U.S.S.R. Sbornik. 10,*** 217-249 (1970).
5. C. DAFERMOS, The entropy rate admissibility criterion for solutions of hyperbolic conservation law, ***J. Diff. Eq.,*** n. 2 (1973).
6. A.Y. LEROUX, A numerical conception of entropy for quasi linear equations.
7. A. HARTER, J.M. HYMAN and P.D. LAX, On finite difference approximation and entropy condition for shocks, ***Comm. Pure Appl. Math. 29,*** 297-322 (1976).

b) ***Compressible Fluids*** (8 lectures in English).
Prof. Andrew MAJDA (University of California, Berkeley, USA)

1. The local existence of smooth solutions to the compressible equations
2. Compressible and incompressible fluids
3. Formation of shock waves in smooth solutions
4. Existence and stability of multidimensional shock fronts.

References

for 1)-2) 1. T. KATO, ***Arch. Rat. Mech. Anal. 58,*** (1975), 181-205
2. S. KLAINERMAN-A. MAIDA, ***C.P.A.M.. 24*** (1981), 481-524
for 3) 3. P.D. LAX, S.I.A.M. Regional Conf. Series, 13
4. S. KLAINERMAN-A. MAJDA, ***C.P.A.M.. 23*** (1980), 241-263
for 4) 5. A. MAJDA, ***Bull. A.M.S.. 4*** (1981), 342-344
3. A. MAJDA, ***Memoirs A.M.S..*** (to appear).

c) ***The concepts of Continuum Thermomechanis*** (8 lectures in English).
Prof. J. SERRIN (University of Minnesota, USA)

1. Thermodynamical structure
2. Laws of thermodynamics

3. Accumulation theorem
4. Continuum thermomechanics and invariance
5. Balance laws of continuum thermomechanics
6. The Clausius-Duhem inequality
7. Example: The Navier-Stokes equations
8. Example: The shock layer in gas dynamics.

The basic literature reference for the subject would be:

1. Mathematical theory of Fluid Dynamics, Handbuch der Physik, Vol. 8/1
2. Notes on Thermodynamics, University of Minnesota, 1981.

Seminars

A number of seminars and special lectures will be offered during the Session.